# Automatic Speech and Speaker Recognition

# Automatic Speech and Speaker Recognition

## Large Margin and Kernel Methods

**Joseph Keshet**

*IDIAP Research Institute, Martigny, Switzerland*

**Samy Bengio**

*Google Inc., Mountain View, CA, USA*

A John Wiley and Sons, Ltd, Publication

*Registered office*
John Wiley & Sons Ltd, The Atrium, Southern Gate, Chichester, West Sussex, PO19 8SQ,
United Kingdom

For details of our global editorial offices, for customer services and for information about how to apply
for permission to reuse the copyright material in this book please see our website at www.wiley.com.

*Library of Congress Cataloging-in-Publication Data*

    Automatic speech and speaker recognition : large margin and kernel methods /
edited by Joseph Keshet, Samy Bengio.
      p. cm.
    Includes bibliographical references and index.
    ISBN 978-0-470-69683-5 (cloth)
1. Automatic speech recognition. I. Keshet, Joseph. II. Bengio, Samy.
    TK7895.S65A983 2009
    006.4'54–dc22
                                                     2008038551

A catalogue record for this book is available from the British Library.

ISBN 9780470696835 (H/B)

Set in 10/12pt Times by Sunrise Setting Ltd, Torquay, UK.
Printed in Great Britain by CPI Antony Rowe, Chippenham, Wiltshire

# Contents

## 9    Large Margin Methods for Part-of-Speech Tagging                    139
*Yasemin Altun*

## 10  A Proposal for a Kernel Based Algorithm for Large Vocabulary Continuous
   **Speech Recognition**                                                 159
*Joseph Keshet*

## IV  Applications                                                       173

## 11  Discriminative Keyword Spotting                                    175
*David Grangier, Joseph Keshet and Samy Bengio*

## 12  Kernel-based Text-independent Speaker Verification                 195
*Johnny Mariéthoz, Samy Bengio and Yves Grandvalet*

# List of Contributors

**Yasemin Altun**
Department Schölkopf
Max Planck Institute for Biological Cybernetics
Tübingen, Germany
yasemin.altun@tuebingen.mpg.de

**Francis R. Bach**
INRIA – Willow project
Département d'Informatique
Ecole Normale Supérieure
Paris, France
francis.bach@mines.org

**Samy Bengio**
Google Research
Google Inc.
Mountain View, CA, USA
bengio@google.com

**Dan Chazan**
Department of Electrical Engineering
The Technion Institute of Technology
Haifa, Israel
dan_chazan@yahoo.com

**Koby Crammer**
Department of Computer and Information Science
University of Pennsylvania
Philadelphia, PA, USA
crammer@cis.upenn.edu

**Mark J. F. Gales**
Department of Engineering
University of Cambridge
Cambridge, UK
mjfg@eng.cam.ac.uk

**Yves Grandvalet**
Heudiasyc, Unité Mixte 6599
CNRS & Université de Technologie de Compiègne
Compiègne, France
yves.grandvalet@utc.fr

**David Grangier**
Department of Machine Learning
NEC Laboratories America, Inc.
Princeton, NJ, USA
dgrangier@nec-labs.com

**Michael I. Jordan**
Computer Science Division and Department of Statistics
University of California
Berkeley, CA, USA
jordan@eecs.berkeley.edu

**Joseph Keshet**
IDIAP Research Institute
Martigny, Switzerland
jkeshet@idiap.ch

**Johnny Mariéthoz**
IDIAP Research Institute
Martigny, Switzerland
marietho@idiap.ch

**Brian Roark**
Center for Spoken Language Understanding
Division of Biomedical Computer Science
Oregon Health & Science University
Portland, Oregon, USA
roark@cslu.ogi.edu

**Lawrence K. Saul**
Department of Computer Science and Engineering
University of California
San Diego, CA, USA
saul@cs.ucsd.edu

**Fei Sha**
Computer Science Department
University of Southern California
Los Angeles, CA, USA
feisha@usc.edu

**Shai Shalev-Shwartz**
Toyota Technological Institute at Chicago
Chicago, USA
shai@tti-c.org

**Yoram Singer**
Google Research
Google Inc.
Mountain View, CA, USA
singer@google.com

**Nathan Srebo**
Toyota Technological Institute at Chicago
Chicago, USA
nati@uchicago.edu

# Preface

This is the first book dedicated to uniting research related to speech and speaker recognition based on the recent advances in large margin and kernel methods. The first part of the book presents theoretical and practical foundations of large margin and kernel methods, from Support Vector Machines to large margin methods for structured learning. The second part of the book is dedicated to acoustic modeling of continuous speech recognizers, where the grounds for practical large margin sequence learning are set. The third part introduces large margin methods for discriminative language modeling. The last part of the book is dedicated to the application of keyword spotting, speaker verification and spectral clustering.

The book is an important reference to researchers and practitioners in the field of modern speech and speaker recognition. The purpose of the book is twofold: first, to set the theoretical foundation of large margin and kernel methods relevant to the speech recognition domain; second, to propose a practical guide on implementation of these methods to the speech recognition domain. The reader is presumed to have basic knowledge of large margin and kernel methods and of basic algorithms in speech and speaker recognition.

Joseph Keshet
*Martigny, Switzerland*
Samy Bengio
*Mountain View, CA, USA*

# Part I

# Foundations

# 1

# Introduction

## Samy Bengio and Joseph Keshet

One of the most natural communication tools used by humans is their voice. It is hence natural that a lot of research has been devoted to analyzing and understanding human uttered speech for various applications. The most obvious one is *automatic speech recognition*, where the goal is to transcribe a recorded speech utterance into its corresponding sequence of words. Other applications include *speaker recognition*, where the goal is to determine either the claimed identity of the speaker (verification) or who is speaking (identification), and speaker segmentation or diarization, where the goal is to segment an acoustic sequence in terms of the underlying speakers (such as during a dialog).

Although an enormous amount of research has been devoted to speech processing, there appears to be some form of local optimum in terms of the fundamental tools used to approach these problems. The aim of this book is to introduce the speech researcher community to radically different approaches based on more recent kernel based machine learning methods. In this introduction, we first briefly review the predominant speech processing approach, based on hidden Markov models, as well as its known problems; we then introduce the most well known kernel based approach, the Support Vector Machine (SVM), and finally outline the various contributions of this book.

## 1.1 The Traditional Approach to Speech Processing

Most speech processing problems, including speech recognition, speaker verification, speaker segmentation, etc., proceed with basically the same general approach, which is described here in the context of speech recognition, as this is the field that has attracted most of the research in the last 40 years. The approach is based on the following statistical framework.

A sequence of acoustic feature vectors is extracted from a spoken utterance by a front-end signal processor. We denote the sequence of acoustic feature vectors by $\bar{\mathbf{x}} = (\mathbf{x}_1, \mathbf{x}_2, \ldots, \mathbf{x}_T)$,

*Automatic Speech and Speaker Recognition: Large Margin and Kernel Methods*   Joseph Keshet and Samy Bengio
© 2009 John Wiley & Sons, Ltd

where $\mathbf{x}_t \in \mathcal{X}$ and $\mathcal{X} \subset \mathbb{R}^d$ is the domain of the acoustic vectors. Each vector is a compact representation of the short-time spectrum. Typically, each vector covers a period of 10 ms and there are approximately $T = 300$ acoustic vectors in a 10 word utterance. The spoken utterance consists of a sequence of words $\bar{v} = (v_1, \ldots, v_N)$. Each of the words belongs to a fixed and known vocabulary $\mathcal{V}$, that is, $v_i \in \mathcal{V}$. The task of the speech recognizer is to predict the most probable word sequence $\bar{v}'$ given the acoustic signal $\bar{\mathbf{x}}$. Speech recognition is formulated as a *Maximum a Posteriori* (MAP) decoding problem as follows:

$$\bar{v}' = \arg \max_{\bar{v}} P(\bar{v}|\bar{\mathbf{x}}) = \arg \max_{\bar{v}} \frac{p(\bar{\mathbf{x}}|\bar{v}) P(\bar{v})}{p(\bar{\mathbf{x}})}, \tag{1.1}$$

where we used Bayes' rule to decompose the posterior probability in Equation (1.1). The term $p(\bar{\mathbf{x}}|\bar{v})$ is the probability of observing the acoustic vector sequence $\bar{\mathbf{x}}$ given a specified word sequence $\bar{v}$ and it is known as *the acoustic model*. The term $P(\bar{v})$ is the probability of observing a word sequence $\bar{v}$ and it is known as *the language model*. The term $p(\bar{\mathbf{x}})$ can be disregarded, since it is constant under the max operation.

The acoustic model is usually estimated by a Hidden Markov Model (HMM) (Rabiner and Juang 1993), a kind of graphical model (Jordan 1999) that represents the joint probability of an observed variable and a hidden (or latent) variable. In order to understand the acoustic model, we now describe the basic HMM decoding process. By decoding we mean the calculation of the $\arg \max_{\bar{v}}$ in Equation (1.1). The process starts with an assumed word sequence $\bar{v}$. Each word in this sequence is converted into a sequence of basic spoken units called *phones*[1] using a pronunciation dictionary. Each phone is represented by a single HMM, where the HMM is a probabilistic state machine typically composed of three states (which are the hidden or latent variables) in a left-to-right topology. Assume that $\mathcal{Q}$ is the set of all states, and let $\bar{q}$ be a sequence of states, that is $\bar{q} = (q_1, q_2, \ldots, q_T)$, where it is assumed there exists some latent random variable $q_t \in \mathcal{Q}$ for each frame $\mathbf{x}_t$ of $\bar{\mathbf{x}}$. Wrapping up, the sequence of words $\bar{v}$ is converted into a sequence of phones $\bar{p}$ using a pronunciation dictionary, and the sequence of phones is converted to a sequence of states, with in general at least three states per phone. The goal now is to find the most probable sequence of states.

Formally, the HMM is defined as a pair of random processes $\bar{q}$ and $\bar{\mathbf{x}}$, where the following first order Markov assumptions are made:

1. $P(q_t|q_1, q_2, \ldots, q_{t-1}) = P(q_t|q_{t-1})$;

2. $p(\mathbf{x}_t|\mathbf{x}_1, \ldots, \mathbf{x}_{t-1}, \mathbf{x}_{t+1}, \ldots, \mathbf{x}_T, q_1, \ldots, q_T) = p(\mathbf{x}_t|q_t)$.

The HMM is a *generative model* and can be thought of as a generator of acoustic vector sequences. During each time unit (frame), the model can change a state with probability $P(q_t|q_{t-1})$, also known as the *transition probability*. Then, at every time step, an acoustic vector is emitted with probability $p(\mathbf{x}_t|q_t)$, sometimes referred to as the *emission probability*. In practice the sequence of states is not observable; hence the model is called hidden. The probability of the state sequence $\bar{q}$ given the observation sequence $\bar{\mathbf{x}}$ can be found using Bayes' rule as follows:

$$P(\bar{q}|\bar{\mathbf{x}}) = \frac{p(\bar{\mathbf{x}}, \bar{q})}{p(\bar{\mathbf{x}})},$$

---

[1] A *phone* is a consonant or vowel speech sound. A *phoneme* is any equivalent set of phones which leaves a word meaning invariant (Allen 2005).

where the joint probability of a vector sequence $\bar{\mathbf{x}}$ and a state sequence $\bar{q}$ is calculated simply as a product of the transition probabilities and the output probabilities:

$$p(\bar{\mathbf{x}}, \bar{q}) = P(q_0) \prod_{t=1}^{T} P(q_t|q_{t-1}) \, p(\mathbf{x}_t|q_t), \qquad (1.2)$$

where we assumed that $q_0$ is constrained to be a non-emitting initial state. The emission density distributions $p(\mathbf{x}_t|q_t)$ are often estimated using diagonal covariance Gaussian Mixture Models (GMMs) for each state $q_t$, which model the density of a $d$-dimensional vector $\mathbf{x}$ as follows:

$$p(\mathbf{x}) = \sum_i w_i \mathcal{N}(\mathbf{x}; \boldsymbol{\mu}_i, \boldsymbol{\sigma}_i), \qquad (1.3)$$

where $w_i \in \mathbb{R}$ is positive with $\sum_i w_i = 1$, and $\mathcal{N}(\cdot; \boldsymbol{\mu}, \boldsymbol{\sigma})$ is a Gaussian with mean $\boldsymbol{\mu}_i \in \mathbb{R}^d$ and standard deviation $\boldsymbol{\sigma}_i \in \mathbb{R}^d$. Given the HMM parameters in the form of the transition probability and emission probability (as GMMs), the problem of finding the most probable state sequence is solved by maximizing $p(\bar{\mathbf{x}}, \bar{q})$ over all possible state sequences using the *Viterbi algorithm* (Rabiner and Juang 1993).

In the training phase, the model parameters are estimated. Assume one has access to a training set of $m$ examples $\mathcal{T}_{\text{train}} = \{(\bar{\mathbf{x}}_i, \bar{v}_i)\}_{i=1}^m$. Training of the acoustic model and the language model can be done in two separate steps. The acoustic model parameters include the transition probabilities and the emission probabilities, and they are estimated by a procedure known as the *Baum–Welch algorithm* (Baum *et al.* 1970), which is a special case of the Expectation-Maximization (EM) algorithm, when applied to HMMs. This algorithm provides a very efficient procedure to estimate these probabilities iteratively. The parameters of the HMMs are chosen to maximize the probability of the acoustic vector sequence $p(\bar{\mathbf{x}})$ given a virtual HMM composed as the concatenation of the phone HMMs that correspond to the underlying sequence of words $\bar{v}$. The Baum–Welch algorithm monotonically converges in polynomial time (with respect to the number of states and the length of the acoustic sequences) to local stationary points of the likelihood function.

Language models are used to estimate the probability of a given sequence of words, $P(\bar{v})$. The language model is often estimated by $n$-grams (Manning and Schutze 1999), where the probability of a sequence of $N$ words ($\bar{v}_1, \bar{v}_2, \ldots, \bar{v}_N$) is estimated as follows:

$$p(\bar{v}) \approx \prod_t p(v_t|v_{t-1}, v_{t-2}, \ldots, v_{t-N}), \qquad (1.4)$$

where each term can be estimated on a large corpus of written documents by simply counting the occurrences of each $n$-gram. Various smoothing and back-off strategies have been developed in the case of large $n$ where most $n$-grams would be poorly estimated even using very large text corpora.

## 1.2 Potential Problems of the Probabilistic Approach

Although most state-of-the-art approaches to speech recognition are based on the use of HMMs and GMMs, also called Continuous Density HMMs (CD-HMMs), they have several drawbacks, some of which we discuss hereafter.

- Consider the logarithmic form of Equation (1.2),

$$\log p(\bar{\mathbf{x}}, \bar{q}) = \log P(q_0) + \sum_{t=1}^{T} \log P(q_t|q_{t-1}) + \sum_{t=1}^{T} \log p(\mathbf{x}_t|q_t). \qquad (1.5)$$

There is a known structural problem when mixing densities $p(\mathbf{x}_t|q_t)$ and probabilities $P(q_t|q_{t-1})$: the global likelihood is mostly influenced by the emission distributions and almost not at all by the transition probabilities, hence temporal aspects are poorly taken into account (Bourlard *et al.* 1996; Young 1996). This happens mainly because the variance of densities of the emission distribution depends on $d$, the actual dimension of the acoustic features: the higher $d$, the higher the expected variance of $p(\bar{\mathbf{x}}|\bar{q})$, while the variance of the transition distributions mainly depend on the number of states of the HMM. In practice, one can observe a ratio of about 100 between these variances; hence when selecting the best sequence of words for a given acoustic sequence, only the emission distributions are taken into account. Although the latter may well be very well estimated using GMMs, they do not take into account most temporal dependencies between them (which are supposed to be modeled by transitions).

- While the EM algorithm is very well known and efficiently implemented for HMMs, it can only converge to local optima, and hence optimization may greatly vary according to initial parameter settings. For CD-HMMs the Gaussian means and variances are often initialized using K-Means, which is itself also known to be very sensitive to initialization.

- Not only is EM known to be prone to local optimal, it is basically used to maximize the likelihood of the observed acoustic sequence, in the context of the expected sequence of words. Note however that the performance of most speech recognizers is estimated using other measures than the likelihood. In general, one is interested in minimizing the number of errors in the generated word sequence. This is often done by computing the Levenshtein distance between the expected and the obtained word sequences, and is often known as the *word error rate*. There might be a significant difference between the best HMM models according to the maximum likelihood criterion and the word error rate criterion.

Hence, throughout the years, various alternatives have been proposed. One line of research has been centered around proposing more discriminative training algorithms for HMMs. That includes Maximum Mutual Information Estimation (MMIE) (Bahl *et al.* 1986), Minimum Classification Error (MCE) (Juang and Katagiri 1992), Minimum Phone Error (MPE) and Minimum Word Error (MWE) (Povey and Woodland 2002). All these approaches, although proposing better training criteria, still suffer from most of the drawbacks described earlier (local minima, useless transitions).

The last 15 years of research in the machine learning community has welcomed the introduction of so-called large margin and kernel approaches, of which the SVM is its best known example. An important role of this book is to show how these recent efforts from the machine learning community can be used to improve research in the speech processing domain. Hence, the next section is devoted to a brief introduction to SVMs.

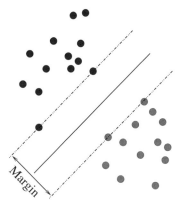

Figure 1.1 Illustration of the notion of margin.

## 1.3 Support Vector Machines for Binary Classification

The most well known kernel based machine learning approach is the SVM (Vapnik 1998). While it was not developed in particular for speech processing, most of the chapters in this book propose kernel methods that are in one way or another inspired by the SVM.

Let us assume we are given a training set of $m$ examples $\mathcal{T}_{\text{train}} = \{(\mathbf{x}_i, y_i)\}_{i=1}^m$ where $\mathbf{x}_i \in \mathbb{R}^d$ is a $d$-dimensional input vector and $y_i \in \{-1, 1\}$ is the target class. The simplest binary classifier one can think of is the linear classifier, where we are looking for parameters ($\mathbf{w} \in \mathbb{R}^d$, $b \in \mathbb{R}$) such that

$$\hat{y}(\mathbf{x}) = \text{sign}(\mathbf{w} \cdot \mathbf{x} + b). \tag{1.6}$$

When the training set is said to be linearly separable, there is potentially an infinite number of solutions ($\mathbf{w} \in \mathbb{R}^d$, $b \in \mathbb{R}$) that satisfy (1.6). Hence, the SVM approach looks for the one that maximizes the *margin* between the two classes, where the margin can be defined as the sum of the smallest distances between the separating hyper-plane and points of each class. This concept is illustrated in Figure 1.1.

This can be expressed by the following optimization problem:

$$\min_{\mathbf{w}, b} \quad \frac{1}{2} \|\mathbf{w}\|^2 \tag{1.7}$$

$$\text{subject to } \forall i \;\; y_i(\mathbf{w} \cdot \mathbf{x}_i + b) \geq 1.$$

While this is difficult to solve, its following dual formulation is computationally more efficient:

$$\max_{\boldsymbol{\alpha}} \quad \sum_{i=1}^m \alpha_i - \frac{1}{2} \sum_{i=1}^m \sum_{j=1}^n y_i y_j \alpha_i \alpha_j \mathbf{x}_i \cdot \mathbf{x}_j \tag{1.8}$$

$$\text{subject to } \begin{cases} \forall i \;\; \alpha_i \geq 0 \\ \sum_{i=1}^m \alpha_i y_i = 0. \end{cases}$$

One problem with this formulation is that if the problem is not linearly separable, there might be no solution to it. Hence one can relax the constraints by allowing errors with an additional hyper-parameter $C$ that controls the trade-off between maximizing the margin and minimizing the number of training errors (Cortes and Vapnik 1995), as follows:

$$\min_{\mathbf{w},b} \quad \frac{1}{2}\|\mathbf{w}\|^2 + C \sum_i \xi_i \tag{1.9}$$

$$\text{subject to} \quad \begin{cases} \forall i \quad y_i(\mathbf{w} \cdot \mathbf{x}_i + b) \geq 1 - \xi_i \\ \forall i \quad \xi_i \geq 0 \end{cases}$$

which has dual formulation

$$\max_{\alpha} \quad \sum_{i=1}^{m} \alpha_i - \frac{1}{2} \sum_{i=1}^{m} \sum_{j=1}^{n} y_i y_j \alpha_i \alpha_j \mathbf{x}_i \cdot \mathbf{x}_j \tag{1.10}$$

$$\text{subject to} \quad \begin{cases} \forall i \quad 0 \leq \alpha_i \leq C \\ \sum_{i=1}^{m} \alpha_i y_i = 0. \end{cases}$$

In order to look for nonlinear solutions, one can easily replace $\mathbf{x}$ by some nonlinear function $\boldsymbol{\phi}(\mathbf{x})$. It is interesting to note that $\mathbf{x}$ only appears in dot products in (1.10). It has thus been proposed to replace all occurrences of $\boldsymbol{\phi}(\mathbf{x}_i) \cdot \boldsymbol{\phi}(\mathbf{x}_j)$ by some kernel function $k(\mathbf{x}_i, \mathbf{x}_j)$. As long as $k(\cdot, \cdot)$ lives in a reproducing kernel Hilbert space (RKHS), one can guarantee that there exists some function $\boldsymbol{\phi}(\cdot)$ such that

$$k(\mathbf{x}_i, \mathbf{x}_j) = \boldsymbol{\phi}(\mathbf{x}_i) \cdot \boldsymbol{\phi}(\mathbf{x}_j).$$

Thus, even if $\boldsymbol{\phi}(\mathbf{x})$ projects $\mathbf{x}$ in a very high (possibly infinite) dimensional space, $k(\mathbf{x}_i, \mathbf{x}_j)$ can still be efficiently computed.

Problem (1.10) can be solved using off-the-shelf quadratic optimization tools. Note however that the underlying computational complexity is at least quadratic in the number of training examples, which can often be a serious limit for most speech processing applications.

After solving (1.10), the resulting SVM solution takes the form of

$$\hat{y}(\mathbf{x}) = \text{sign}\left( \sum_{i=1}^{m} y_i \alpha_i k(\mathbf{x}_i, \mathbf{x}) + b \right) \tag{1.11}$$

where most $\alpha_i$ are zero except those corresponding to examples in the margin or misclassified, often called *support vectors* (hence the name of SVMs).

## 1.4   Outline

The book has four parts. The first part, *Foundations*, covers important aspects of extending the binary SVM to speech and speaker recognition applications. Chapter 2 provides a detailed review of efficient and practical solutions to large scale convex optimization problems one encounters when using large margin and kernel methods with the enormous datasets used in speech applications. Chapter 3 presents an extension of the binary SVM to multiclass, hierarchical and categorical classification. Specifically, the chapter presents a more complex setting in which the possible labels or categories are many and organized.

The second part, *Acoustic Modeling*, deals with large margin and kernel method algorithms for sequence prediction required for acoustic modeling. Chapter 4 presents a large margin algorithm for forced alignment of a phoneme sequence to a corresponding speech signal, that is, proper positioning of a sequence of phonemes in relation to a corresponding continuous speech signal. Chapter 5 describes a kernel wrapper for the task of phoneme recognition, which is based on the Gaussian kernel. This chapter also presents a kernel based iterative algorithm aimed at minimizing the Levenshtein distance between the predicted phoneme sequence and the true one. Chapter 6 reviews the use of dynamic kernels for acoustic models and especially describes the Augmented Statistical Models, resulting from the generative kernel, a generalization of the Fisher kernel. Chapter 7 investigates a framework for large margin parameter estimation for continuous-density HMMs.

The third part of the book is devoted to *Language Modeling*. Chapter 8 reviews past and present work on discriminative training of language models, and focuses on three key issues: training data, learning algorithms, features. Chapter 9 describes different large margin algorithms for the application of part-of-speech tagging. Chapter 10 presents a proposal for large vocabulary continuous speech recognition, which is solely based on large margin and kernel methods, incorporating the acoustic models described in Part II and the discriminative language models.

The last part is dedicated to *Applications*. Chapter 11 covers a discriminative keyword spotting algorithm, based on a large margin approach, which aims at maximizing the area under the receiver operating characteristic (ROC) curve, the most common measure to evaluate keyword spotters. Chapter 12 surveys recent work on the use of kernel approaches to text-independent speaker verification. Finally, Chapter 13 introduces the main concepts and algorithms together with recent advances in learning a similarity matrix from data. The techniques in the chapter are illustrated on the blind one-microphone speech separation problem, by casting the problem as one of segmentation of the spectrogram.

# References

Allen JB 2005 *Articulation and Intelligibility*. Morgan & Claypool.

Bahl LR, Brown PF, de Souza PV and Mercer RL 1986 Maximum mutual information of hidden Markov model parameters for speech recognition. *Proceedings of the IEEE International Conference on Acoustics, Speech and Signal Processing, ICASSP*, pp. 49–53.

Baum LE, Petrie T, Soules G and Weiss N 1970 A maximization technique occurring in the statistical analysis of probabilistic functions of Markov chains. *The Annals of Mathematical Statistics* **41**(1), 164–171.

Bourlard H, Hermansky H and Morgan N 1996 Towards increasing speech recognition error rates. *Speech Communication* **18**, 205–231.

Cortes C and Vapnik V 1995 Support-vector networks. *Machine Learning* **20**(3), 273–297.

Jordan MI (ed.) 1999 *Learning in Graphical Models*. MIT Press.

Juang BH and Katagiri S 1992 Discriminative learning for minimum error classification. *IEEE Transactions on Signal Processing* **40**(12), 3043–3054.

Manning CD and Schutze H 1999 *Foundations of Statistical Natural Language Processing*. MIT Press.

Povey D and Woodland PC 2002 Minimum phone error and I-smoothing for improved discriminative training. *Proceedings of the IEEE International Conference on Acoustics, Speech and Signal Processing, ICASSP.*

Rabiner L and Juang BH 1993 *Fundamentals of speech recognition*, 1st edn. Prentice Hall.

Vapnik VN 1998 *Statistical Learning Theory*. John Wiley & Sons.

Young S 1996 A review of large-vocabulary continuous speech recognition. *IEEE Signal Processing Mag.* pp. 45–57.

# 2

# Theory and Practice of Support Vector Machines Optimization

## Shai Shalev-Shwartz and Nathan Srebo

Support Vector Machines (SVMs), and more generally $L_2$-regularized linear predictors, are effective and popular discriminative learning tools. The task of learning an SVM is cast as an optimization problem of jointly minimizing an empirical loss term and an $L_2$-norm regularization term. Solving the optimization problem is a challenging task particularly when the number of training examples is large. In this chapter we describe two stochastic approaches for solving the optimization problem posed by $L_2$ linear predictors, focusing on cases in which the number of training examples is very large. In particular, we review a stochastic gradient descent approach and a stochastic coordinate dual ascent method. Our goal is to provide optimization procedures that are both efficient and easy to implement.

## 2.1 Introduction

In this chapter, we consider the computational problem of training SVMs and related linear prediction models. We focus on computational methods appropriate for training using very large data sets. To set up the stage for understanding the relative merits of different computational approaches, we begin in Section 2.2 with a brief introduction to SVMs, and more generally, to $L_2$-regularized linear prediction. In theory, the optimization problem posed by SVM is a convex optimization problem that can be solved using off-the-shelf optimizers which find an extremely accurate solution. In practice, such solvers are only appropriate for training using fairly small data sets. In Section 2.3, we discuss the optimization problem from a machine learning perspective, emphasizing that our true goal is to find a predictor with low generalization error. Based on this observation, we argue that we do not need to find an extremely accurate solution to the optimization problem. We present methods that

*Automatic Speech and Speaker Recognition: Large Margin and Kernel Methods*   Joseph Keshet and Samy Bengio
© 2009 John Wiley & Sons, Ltd

find approximate solutions while their runtime increases only moderately with the size of the data set. The runtime dependence on the optimization accuracy is worse than that of traditional methods, but, as discussed above, from a machine learning perspective this should not pose a major problem. In particular, in Section 2.4 we describe a stochastic gradient descent approach and in Section 2.5 we present a stochastic coordinate dual ascent method. We chose these two approaches as they are both efficient and easy to implement.

**Notation**   We denote scalars with lower case letters (e.g. $w$ and $\alpha$), and vectors with bold face letters (e.g. $\mathbf{w}$ and $\boldsymbol{\alpha}$). The transpose of a column vector $\mathbf{w}$ is denoted by $\mathbf{w}^\mathsf{T}$. The $i$th element of a vector $\mathbf{w}$ is denoted by $w_i$. The inner product between two $d$-dimensional vectors $\mathbf{x}$ and $\mathbf{w}$ is denoted by $\langle \mathbf{x}, \mathbf{w} \rangle := \sum_{i=1}^{d} x_i w_i$. The set of real numbers is denoted by $\mathbb{R}$. For any $k \geq 1$, the set of integers $\{1, \ldots, k\}$ is denoted by $[k]$. The $L_2$ norm of a vector $\mathbf{x}$ is denoted by $\|\mathbf{x}\| := \sqrt{\langle \mathbf{x}, \mathbf{x} \rangle}$. The expected value of a random variable $Z$ is denoted by $\mathbb{E}Z$. When considering the expectation $\mathbb{E}h(z)$ of a function $h$ applied to the random variable $Z$, we sometimes omit $Z$ from the notation and use the shorthand $\mathbb{E}h$.

## 2.2   SVM and $L_2$-regularized Linear Prediction

The goal of SVM training is to learn a *linear predictor* $\mathbf{w}$ which can be used to predict the label $y \in \mathcal{Y}$ of a given instance $\mathbf{x} \in \mathcal{X}$. We say that $\mathbf{w}$ is a linear predictor if the prediction $h_{\mathbf{w}}(\mathbf{x}) \in \mathcal{Y}$ made based on the instance $\mathbf{x}$ depends on $\mathbf{w}$ linearly.

### 2.2.1   Binary Classification and the Traditional SVM

In the simplest case of binary classification, the prediction is based on a fixed *feature map* $\boldsymbol{\phi}$, mapping each instance to a vector $\boldsymbol{\phi}(\mathbf{x})$. A label $\mathbf{y} \in \{\pm 1\}$ is then predicted according to the sign of $\langle \mathbf{w}, \boldsymbol{\phi}(\mathbf{x}) \rangle$, that is, $h_{\mathbf{w}}(\mathbf{x}) = \text{sign}\langle \mathbf{w}, \boldsymbol{\phi}(\mathbf{x}) \rangle \in \{\pm 1\}$. Typically, a bias term $b \in \mathbb{R}$ is also allowed, with $h_{\mathbf{w},b}(\mathbf{x}) = \text{sign}(\langle \mathbf{w}, \boldsymbol{\phi}(\mathbf{x}) \rangle + b)$. For the time being we do not allow a bias, and defer its introduction to Section 2.2.5.

In order to learn $\mathbf{w}$, we use a *training set* $(\mathbf{x}_1, y_1), \ldots, (\mathbf{x}_m, y_m)$ of $m$ labeled examples. We then seek the predictor $\mathbf{w}$ minimizing the $L_2$-regularized empirical loss:

$$f(\mathbf{w}) = \frac{\lambda}{2} \|\mathbf{w}\|^2 + \frac{1}{m} \sum_{i=1}^{m} \ell(\langle \mathbf{w}, \boldsymbol{\phi}(\mathbf{x}_i) \rangle, y_i),$$

where $\lambda$ is a regularization trade-off parameter and $\ell(z, y)$ is a loss function. In the standard SVM formulation for binary classification, this is the hinge loss $\ell(z, y) = \max\{0, 1 - yz\}$. Note that we did not directly refer to the prediction rule $h_{\mathbf{w}}$ in the optimization objective and that minimizing the loss $\ell(\langle \mathbf{w}, \boldsymbol{\phi}(\mathbf{x}_i) \rangle, y)$ differs from simply minimizing the number of prediction errors made by $h_{\mathbf{w}}(\mathbf{x})$. Indeed in training time, which is the focus of the current chapter, we are concerned mostly with this loss function rather than with the prediction rule $h_{\mathbf{w}}$.

Our hope in using the SVM learning rule, $\arg \min_{\mathbf{w}} f(\mathbf{w})$, is that we can find a predictor $\mathbf{w}$ that both has low norm, and attains low empirical loss on the training samples. Minimizing the empirical loss without any further constraints can easily lead to *overfitting* – that is, finding a (complicated) predictor that has high overall error, but happens to have low error on the training data. Overfitting can be avoided by limiting the class of predictors allowed, in our case to the class of low norm predictors. We actually do this indirectly, by including also a norm term in the objective and thus preferring a low-norm predictor over other predictors with lower training error but higher norm. Regularizing the norm of $\mathbf{w}$ ensures that if $\mathbf{w}$ attains low average loss on a random training sample, it will also generalize and attain low expected loss, and thus also low prediction error, over unseen examples (Schölkopf and Smola 2002).

### 2.2.2 More General Loss Functions

More generally, and especially when more complex labels are involved, we allow the prediction to depend on multiple inner products with $\mathbf{w}$. Formally, with $\mathcal{H}$ being a Hilbert space in which $\mathbf{w}$ resides, we allow a fixed feature map $\mathbf{\Phi} : \mathbf{x} \to \mathcal{H}^k$ which maps each instance into $k$ vectors in $\mathcal{H}$ (when $\mathcal{H}$ is finite dimensional, we can think of $\mathbf{\Phi}(\mathbf{x})$ as a matrix). Often we have $k = |\mathcal{Y}|$, with one feature vector for each candidate label, which we denote $\mathbf{\Phi}(\mathbf{x}, y)$. In any case, prediction is based on the $k$ inner products $\langle \mathbf{w}, \boldsymbol{\phi}(\mathbf{x})_i \rangle$. We use $\langle \mathbf{w}, \mathbf{\Phi}(\mathbf{x}) \rangle \in \mathbb{R}^k$ to denote the vector of these inner products. If we think of $\mathbf{\Phi}(\mathbf{x})$ as a matrix, we have $\langle \mathbf{w}, \mathbf{\Phi}(\mathbf{x}) \rangle = \mathbf{w}^\mathsf{T} \mathbf{\Phi}(\mathbf{x})$. We again have $h_{\mathbf{w}}(\mathbf{x}) = h(\langle \mathbf{w}, \mathbf{\Phi}(\mathbf{x}) \rangle)$, where we now allow a more general rule $h : \mathbb{R}^k \to \mathcal{Y}$. The optimization objective also has the same form as before:

$$f(\mathbf{w}) = \frac{\lambda}{2} \|\mathbf{w}\|^2 + \frac{1}{m} \sum_{i=1}^{m} \ell(\langle \mathbf{w}, \mathbf{\Phi}(\mathbf{x}_i) \rangle, y_i), \tag{2.1}$$

except now the loss function $\ell : \mathbb{R}^k \times \mathcal{Y} \to \mathbb{R}$ takes as input a vector of prediction values. The exact form of the loss function can vary. However, we will require that the loss function be convex in its first argument, $\langle \mathbf{w}, \mathbf{\Phi}(\mathbf{x}_i) \rangle$, so that the optimization objective $f(\mathbf{w})$ is convex with respect to $\mathbf{w}$.

### 2.2.3 Examples

**Example 2.2.1 (Binary classification with the hinge loss)** *In the binary classification problem mentioned before, $\mathcal{Y} = \{+1, -1\}$ and $\mathbf{\Phi}(\mathbf{x})$ is a single vector (i.e. $k = 1$). The prediction rule is $h(z) = \mathrm{sign}(z)$ and the loss is given by $\ell(z, y) = \max\{0, 1 - yz\}$. This is the 'standard' soft-margin SVM formulation (except perhaps for the lack of a bias term).*

**Example 2.2.2 (Binary classification with the log-loss)** *Instead of the hinge loss, other loss functions might be used with binary labels $\mathcal{Y} = \{+1, -1\}$. A popular choice is the log-loss given by: $\ell(z, y) = \log(1 + \exp(-yz))$. The prediction rule is as before.*

**Example 2.2.3 (Regression with the $\epsilon$-insensitive loss)** *We now turn to regression problems over the reals, $\mathcal{Y} = \mathbb{R}$. The standard Support Vector Regression formulation still uses a mapping $\mathbf{\Phi}$ to a single vector $\boldsymbol{\phi}(\mathbf{x})$ (i.e. $k = 1$), with a loss function given by $\ell(z, y) = \max\{0, |y - z| - \epsilon\}$. The prediction rule simply follows $h(z) = z$.*

**Example 2.2.4 (Cost-sensitive multiclass categorization)** *In multi-class categorization problems, $\mathcal{Y}$ is a finite discrete set of classes. This time, we use a mapping to $k = |\mathcal{Y}|$ feature vectors, which we index using the possible labels: $\mathbf{\Phi}(\mathbf{x}) = (\boldsymbol{\phi}(\mathbf{x}, y))_{y \in \mathcal{Y}}$. This results in a vector $\mathbf{z} = \langle \mathbf{w}, \mathbf{\Phi}(\mathbf{x}) \rangle$ of prediction values indexed by $\mathcal{Y}$. The label corresponding to the* *largest inner product is chosen:*

$$h(\mathbf{z}) = \arg \max_{y \in \mathcal{Y}} z_y. \tag{2.2}$$

*The cost-sensitive loss function is given by*

$$\ell(\mathbf{z}, y) = \max_{y' \in \mathcal{Y}} \delta(y', y) - z_y + z_{y'}, \tag{2.3}$$

*where $\delta(y', y)$ is the cost of predicting $y'$ instead of $y$.*

**Example 2.2.5 (Sequence prediction)** *Sequence prediction is similar to cost-sensitive multi-class categorization, but the set of targets, $\mathcal{Y}$, can be very large. For example, in phoneme recognition tasks, $\mathcal{X}$ is the set of all speech utterances and $\mathcal{Y}$ is the set of all phoneme sequences. Therefore, $|\mathcal{Y}|$ is exponential in the length of the sequence. Nevertheless, as we show later, if the functions $\Phi$ and $\delta$ adhere to a specific structure then we can still solve the resulting optimization problem efficiently. More specifically, if the optimization problem given in the definition of the loss function (see Equation (2.3)) can be solved efficiently, then the optimization problem required for training can be solved efficiently as well.*

### 2.2.4   Kernels

One of the main benefits of SVMs is the ability to describe the feature mapping $\Phi$ implicitly, by providing only a *kernel function K* specifying the inner product between feature vectors. This can be done by considering the representer theorem (Kimeldorf and Wahba 1971), which in our case can be stated as follows:

**Theorem 2.2.6** *Let $\mathbf{w}^\star$ be the minimizer of $f(\mathbf{w})$ defined in Equation (2.1), then $\mathbf{w}^\star$ can be written as $\sum_{i=1}^{m} \sum_{j=1}^{k} \alpha_{i,j} \boldsymbol{\phi}(\mathbf{x}_i)_j$ for some set of coefficients $\alpha_{i,j} \in \mathbb{R}$.*

*Proof.* Let $C$ denote the span of the vectors $\{\boldsymbol{\phi}(\mathbf{x}_i)_j : 1 \leq i \leq m, 1 \leq j \leq k\}$. We can write any vector, and in particular $\mathbf{w}^\star$, as a sum $\mathbf{w}_1 + \mathbf{w}_2$ where $\mathbf{w}_1 \in C$ while $\mathbf{w}_2$ is orthogonal to $C$; Therefore, for all $i$, $j$ we have $\langle \boldsymbol{\phi}(\mathbf{x}_i)_j, \mathbf{w}_2 \rangle = 0$ and thus $\ell(\langle \mathbf{w}^\star, \boldsymbol{\phi}(\mathbf{x}_i) \rangle, y_i) = \ell(\langle \mathbf{w}_1, \boldsymbol{\phi}(\mathbf{x}_i) \rangle, y_i)$. In addition, $\|\mathbf{w}^\star\|^2 = \|\mathbf{w}_1\|^2 + \|\mathbf{w}_2\|^2$. Therefore, $f(\mathbf{w}^\star) - f(\mathbf{w}_1) = (\lambda/2)\|\mathbf{w}_2\|^2$. The optimality of $\mathbf{w}^\star$ now implies that $\mathbf{w}_2$ must be the zero vector, which concludes our proof.                                                                                 □

A consequence of the representer theorem is that we can limit our attention only to weight vectors of the form $w = \sum_{i=1}^{m} \sum_{j=1}^{k} \alpha_{i,j} \boldsymbol{\phi}(\mathbf{x}_i)_j$, parameterized by the coefficients $\boldsymbol{\alpha} = \{\alpha_{i,j}\}$. For such vectors, we have

$$\|\mathbf{w}\|^2 = \sum_{i,j} \sum_{i',j'} \alpha_{i,j} \alpha_{i',j'} k((x_i, j), (x_{i'}, j'))$$

where $k((x_i, j), (x_{i'}, j')) = \langle \boldsymbol{\phi}(x_i)_j, \boldsymbol{\phi}(x_{i'})_{j'} \rangle$ and

$$\langle \Phi(\mathbf{x}), \mathbf{w} \rangle = \sum_{i',j'} \alpha_{i',j'} k(x_i, (x_{i'}, j')),$$

where we denote $k(x_i, (x_{i'}, j')) = (k((x_i, 1), (x_{i'}, j')), \ldots, k((x_i, k), (x_{i'}, j')))$. We can therefore write our optimization problem as a minimization problem over the coefficients $\{\alpha_{i,j}\}$:

$$f(\boldsymbol{\alpha}) = \frac{1}{2} \sum_{i,j} \sum_{i',j'} \alpha_{i,j} \alpha_{i',j'} k((x_i, j), (x_{i'}, j'))$$

$$+ \frac{1}{m} \sum_i \ell \left( \sum_{i',j'} \alpha_{i',j'} k(x_i, (x_{i'}, j')), y_i \right). \qquad (2.4)$$

This expression can be written much more compactly for the simpler case of $k = 1$ (i.e. a single feature vector for each example). Let $K$ be the $m \times m$ *Gram matrix* with $K_{i,j} = k(x_i, x_j)$, then, for $k = 1$:

$$f(\boldsymbol{\alpha}) = \frac{\lambda}{2} \boldsymbol{\alpha}^\mathsf{T} K \boldsymbol{\alpha} + \frac{1}{m} \ell((K\boldsymbol{\alpha})_i, y_i). \qquad (2.5)$$

The important advantage of writing the objective in terms of $\boldsymbol{\alpha}$, and optimizing over this parameterization, is that the objective Equation (2.4) depends on the feature mapping only through the inner products $\langle \boldsymbol{\phi}(x_i)_j, \boldsymbol{\phi}(x_{i'})_{j'} \rangle$. It is often easier to describe the inner products by providing the function $k((x_i, j), (x_{i'}, j'))$, and this is enough for solving the optimization problem, and indeed also for using the learned coefficients $\boldsymbol{\alpha}$ to make predictions. Furthermore, the optimization objective specified in Equation (2.4) does not depend on the dimensionality of the feature mapping. We can therefore use a very high dimensional, or even infinite dimensional, feature space $\mathcal{H}$.

It should be emphasized that the optimization objective $f(\boldsymbol{\alpha})$ specified in Equation (2.4) and Equation (2.5) is *not* the dual of our initial $f(\mathbf{w})$. Rather this is a different way of parameterizing our original problem $\arg \min_\mathbf{w} f(\mathbf{w})$. (i.e. a different parameterization of the primal). We discuss the dual problem (which is a different optimization problem) in Section 2.5.1.

## 2.2.5   Incorporating a Bias Term

In many applications, the weight vector $\mathbf{w}$ is augmented with a bias term which is a scalar, typically denoted as $b$. For binary classification, the prediction becomes $\mathrm{sign}(\langle \mathbf{w}, \mathbf{x} \rangle + b)$ and the loss is accordingly defined as

$$\ell((\mathbf{w}, b), (\mathbf{x}, y)) = \max\{0, 1 - y_i(\langle \mathbf{w}, \mathbf{x}_i \rangle + b)\}. \qquad (2.6)$$

The bias term often plays a crucial role in binary classification problems when the distribution of the labels is uneven as is typically the case in text processing applications where the negative examples vastly outnumber the positive ones. The algorithms we present below assume that there is no learning of the bias term (namely, the bias term is known in advance). We now briefly describe two different approaches to learning the bias term and underscore the advantages and disadvantages of each approach.

The first approach is rather well known and its roots go back to early work on pattern recognition (Duda and Hart 1973). This approach simply amounts to adding one more feature to each feature vector $\boldsymbol{\phi}(\mathbf{x}, y)$ thus increasing the dimension by 1. The artificially added feature always take the same value. We assume without loss of generality that the value of the constant feature is 1. Once the constant feature is added the rest of the algorithm remains intact, thus the bias term is not explicitly introduced. Note however that by equating the last component of $\mathbf{w}$ with $b$, the norm-penalty counterpart of $f$ becomes $\|\mathbf{w}\|^2 + b^2$. The disadvantage of this approach is thus that we solve a slightly different optimization problem. On the other hand, an obvious advantage of this approach is that it requires no modifications to the algorithm itself rather than a modest increase in the dimension.

In the second approach, we rewrite the optimization problem as

$$\min_{b \in \mathbb{R}} \left( \min_{\mathbf{w}} \frac{\lambda}{2} \|\mathbf{w}\|^2 + \frac{1}{m} \sum_{i=1}^{m} \ell((\mathbf{w}, b), (\mathbf{x}_i, y_i)) \right).$$

Denote the expression inside the outer-most parentheses by $g(b)$. Then, we can rewrite the optimization problem as $\min_b g(b)$. Since the function $\ell$ is jointly convex with respect to $\mathbf{w}$ and $\mathbf{b}$, the function $g(b)$ is a convex function (see Boyd and Vandenberghe 2004, page 87). Therefore, a simple binary search for the value of $b$ is guaranteed to find an $\epsilon$ close solution in no more than $O(\log(1/\epsilon))$ evaluations of the function $g(b)$. Each evaluation of the function $g(b)$ requires solving an SVM problem with a fixed bias term. Therefore, incorporating the bias term increases the runtime by a multiplicative factor of $O(\log(1/\epsilon))$.

## 2.3   Optimization Accuracy From a Machine Learning Perspective

In the previous section, we phrased training as an optimization problem of finding a minimizer of $f(\mathbf{w})$. As long as the loss function $\ell(z, y)$ is convex in $z$, the optimization problems corresponding to training are convex optimization problems and can be solved using standard approaches. Even if only kernel information is available rather than an explicit feature mapping, we can still directly minimize the SVM objective using the formulation given in Equation (2.4).

If the loss is differentiable then the problem is an unconstrained differentiable optimization problem and can be solved using standard gradient-based descent methods (Nocedal and Wright 2006). We might therefore sometimes choose to replace a non-differentiable loss such as the hinge loss with a 'smoothed' approximation of the loss (e.g. rounding the non-differentiable 'corners'). Gradient-based methods can then be used (Chapelle 2007). This yields a very simple and flexible approach for a wide variety of loss functions.

Alternatively, if the loss is piecewise linear, as is the case for the hinge loss used in the standard SVM formulation, the optimization problem can be written as a quadratic program (QP) (Schölkopf and Smola 2002) and a standard QP solver can be used (Boyd and Vandenberghe 2004; Nocedal and Wright 2006). Although off-the-shelf QP solvers are only appropriate for training using fairly small data sets, much effort has been directed toward developing QP solvers specifically optimized for the standard SVM optimization problem. Notable examples are SMO (Platt 1998), SVM-Light (Joachims 1998), and LibSVM (Fan *et al.* 2005).

Viewing SVM training as a standard optimization problem and applying the methods discussed above can be appropriate for mid-sized problems, involving thousands and possibly tens of thousands of examples. However, since the entire data set is processed at each iteration of the above methods, the runtime of these methods increases, sometimes sharply, with the size of the data set.

On the other hand, the high degree of optimization accuracy achieved by standard optimization approaches might not be necessary in SVM training. The focus of traditional optimization approaches is obtaining fast convergence to the optimum $f(\mathbf{w}^*) = \min_{\mathbf{w}} f(\mathbf{w})$. Typically linear, or even quadratic, convergence is guaranteed. That is, a solution with $f(\tilde{\mathbf{w}}) \leq f(\mathbf{w}^*) + \epsilon$ is obtained after $O(\log(1/\epsilon))$ or even $O(\log\log(1/\epsilon))$ iterations. These methods are therefore especially well suited in situations in which very accurate solutions are necessary. As mentioned above, this might come at the expense of a very sharp runtime dependence on the number of training examples or other problem parameters.

To understand how important optimization accuracy is from a machine learning perspective, it is important to remember that our true objective is not actually $f(\mathbf{w})$. Our true objective is the generalization error – the expected error of $\mathbf{w}$ on future, yet unseen, instances. Of course, we cannot actually observe or calculate this objective at training time, and so we use the average loss on the training examples as an estimator for this expected error. To control the variance of the training loss as an estimator we also regularize the norm of $\mathbf{w}$ by adding the term $\frac{1}{2}\lambda\|\mathbf{w}\|^2$ to our objective, arriving at our optimization objective $f(\mathbf{w})$.

Although we cannot minimize the generalization error directly, it is still insightful to understand runtime in terms of the resulting guarantee on this true objective. That is, using some optimization method, what is the required runtime to get low generalization error? To do so, it is useful to consider a rough decomposition of the generalization error into three terms (this decomposition can be made more precise – see Shalev-Shwartz and Srebro (2008) for further details):

- **The approximation error** This is the best expected error possible with a low-norm linear predictor (using the feature mapping or kernel we committed to). That is, this is the best we could hope for, even if we could magically minimize the expected error directly, but still restricted ourselves to low-norm linear predictors.

- **The estimation error** This is the additional error introduced by the fact that we are using only a finite data set. That is, we are using the average training loss, which is only an empirical estimate of the expected loss.

- **Optimization error** This is the additional error due to the inaccuracy of the optimization. That is, this is the difference in *generalization* error between the true optimum $\mathbf{w}^* = \arg\min_{\mathbf{w}} f(\mathbf{w})$ and the solution $\tilde{\mathbf{w}}$ found by the optimization algorithm, which might have an objective value which is $\epsilon$ worse than $\mathbf{w}^*$.

Based on this error decomposition of the generalization error, the only component that can be controlled by the optimization algorithm is the optimization error, which is of the same order as the optimization accuracy, $\epsilon$. The estimation error, on the other hand, is controlled by the data set size – as the number $m$ of training examples increases, our estimate is more accurate, and the estimation error decreases as $O(1/\sqrt{m})$. The combined contribution to our true objective, the generalization error, is then $O(\epsilon + 1/\sqrt{m})$. It is therefore reasonable to expect to keep the optimization accuracy $\epsilon$ on roughly the same order as $1/\sqrt{m}$. In particular, having a runtime dependence on $1/\epsilon$, as opposed to at most $\log(1/\epsilon)$ in the methods discussed above, can be preferable to even a linear runtime dependence on the data set size.

Therefore, as an alternative to the traditional optimization methods, in this chapter we consider simple methods that take advantage of the optimization objective being an average over many examples and the fact that even rough accuracy can be enough for us. We present methods where the runtime of each iteration does not necessarily increase with the size of the data set. The overall runtime of the method either does not at all increase with the size of the data set, or increases only moderately (in particular, when formulation Equation (2.4) is used, and the number of optimization variables depends on the number of training examples). The runtime dependence on the optimization accuracy will indeed be much worse than the logarithmic dependence of traditional approaches, but, as discussed above, this dependence is generally preferable to dependence on the data set size, especially for very large data sets. Indeed, the methods we present are well-suited for problems with even millions of training examples, for which traditional methods are not applicable.

## 2.4 Stochastic Gradient Descent

In this section we present a stochastic gradient descent (SGD) approach for training SVMs. SGD is an extremely simple optimization procedure. But despite its simplicity, SGD can be an excellent choice for SVM training. It is in fact state-of-the-art for training large scale SVM when the feature vectors are given explicitly, especially when the feature vectors are sparse (Bottou 2008; Shalev-Shwartz *et al.* 2007).

Each SGD step is extremely cheap computationally, with a computational cost proportional to the size of the (sparse) representation of a single training feature vector. The flip side is that the solutions converges rather slowly to the true optimal solution, and, as we show below, the number of SGD iterations required to get within $\epsilon$ of optimal scales as $O(1/\epsilon)$. However, as discussed in the previous section, such a 'slow' convergence rate is tolerable, and is far out-weighed by the low computational cost per iteration.

Like many other methods, SGD is an iterative method where in each iteration we calculate an updated weight vector $\mathbf{w}_t$. At iteration $t$, we randomly choose a training example $(\mathbf{x}_i, y_i)$, with $i = i(t) \in [m]$ uniformly and independently selected at each iteration. We then consider the estimate of the optimization objective $f(\mathbf{w})$ based on the single training example $(\mathbf{x}_i, y_i)$:

$$f_t(\mathbf{w}) = \frac{\lambda}{2}\|\mathbf{w}\|^2 + \ell(\langle \mathbf{w}, \boldsymbol{\Phi}(\mathbf{x}_i)\rangle, y_i). \tag{2.7}$$

Since $i$ is chosen uniformly at random from $[m]$, we have that $\mathbb{E} f_t(\cdot) = f(\cdot)$. We then update $\mathbf{w}$ using the sub-gradient of $f_t(\mathbf{w})$. More specifically, we set $\mathbf{w}_{t+1} = \mathbf{w}_t - \eta_t \nabla_t$, where $\nabla_t$ is a sub-gradient of $f_t$ at $\mathbf{w}_t$ and $\eta_t$ is a scalar (the learning rate). A sub-gradient is a generalization of the more familiar gradient that is also applicable to non-differentiable functions. In Section 2.4.1 we formally define this notion and show how to easily calculate a sub-gradient for several loss functions. Following Shalev-Shwartz *et al.* (2007), we set $\eta_t = (1/\lambda (t + 1))$. Additionally, since the gradient of the function $(\lambda/2)\|\mathbf{w}\|^2$ at $\mathbf{w}_t$ is $\lambda\mathbf{w}_t$, we can rewrite $\nabla_t = \lambda\mathbf{w}_t + \mathbf{v}_t$, where $\mathbf{v}_t$ is a sub-gradient of $\ell(\langle \mathbf{w}, \boldsymbol{\Phi}(\mathbf{x}_i)\rangle, y_i)$ at $\mathbf{w}_t$. Therefore, the update of the SGD procedure can be written as

$$\mathbf{w}_{t+1} = \left(1 - \frac{1}{t}\right)\mathbf{w}_t - \frac{1}{(t + 1)\lambda}\mathbf{v}_t. \tag{2.8}$$

Unraveling the recursive definition of $\mathbf{w}_{t+1}$ we can also rewrite $\mathbf{w}_{t+1}$ as

$$\mathbf{w}_{t+1} = -\frac{1}{\lambda} \sum_{t'=1}^{t} \frac{1}{t'+1} \prod_{j=t'+2}^{t+1} \left(1 - \frac{1}{j}\right) \mathbf{v}'_t = -\frac{1}{\lambda} \sum_{t'=1}^{t} \frac{1}{t'+1} \prod_{j=t'+2}^{t+1} \frac{j-1}{j} \mathbf{v}_{t'}$$

$$= -\frac{1}{\lambda(t+1)} \sum_{t'=1}^{t} \mathbf{v}_{t'}. \tag{2.9}$$

In fact, in the pseudo-code for the SGD procedure in Figure 2.1, we use the representation $\theta_t = \sum_{t'=1}^{t-1} \mathbf{v}_{t'}$, from which we can calculate $\mathbf{w}_t = (1/\lambda t)\theta_t$. This is done to emphasize that the current weight vector is an accumulation of sub-gradients. More importantly, as explained below, this also enables a more efficient implementation when the features $\Phi(\mathbf{x})$ are sparse.

---

**Input:** A sequence of loss functions $\ell(\langle \mathbf{w}, \Phi(\mathbf{x}_1) \rangle, y_1), \ldots, \ell(\langle \mathbf{w}, \Phi(\mathbf{x}_m) \rangle, y_m)$

Regularization parameter $\lambda$

**Initialize:** $\theta_1 = \mathbf{0}$

**Do:**

Choose $i \in [m]$ uniformly at random

Calculate sub-gradient $\mathbf{v}_t$ of $\ell(\langle \mathbf{w}, \Phi(\mathbf{x}_i) \rangle, y_i)$ at $\mathbf{w}_t = \frac{1}{\lambda t}\theta_t$ (see Section 2.4.1)

Update: $\theta_{t+1} = \theta_t - \mathbf{v}_t$

UNTIL stopping condition is met (see Section 2.4.2)

---

Figure 2.1  A stochastic gradient descent procedure for solving SVM.

**Using kernels**  So far we have described the algorithm assuming we have a direct access to the features $\Phi(x)$ and it is practical to represent the weight vector $\mathbf{w}$ explicitly. We now turn to the case in which the features are described implicitly through a kernel function. In this case we cannot represent the weight vector $\mathbf{w}$ explicitly. The important observation here is that sub-gradients $\mathbf{v}_t$, and so also $\mathbf{w}_t$, can always be expressed as a linear combination of feature vectors.

We begin with considering the simpler case in which each example is represented with a single feature vector $\phi(x)$ as in the standard SVM. In this case, it is easy to verify that any sub-gradient of $\ell(\langle \mathbf{w}, \phi(\mathbf{x}_i) \rangle, y_i)$ with respect to $\mathbf{w}$ can be written as $\mathbf{v}_t = g_t \phi(\mathbf{x}_i)$, where $g_t \in \mathbb{R}$ is a sub-gradient of $\ell(z, y_i)$ with respect to $z$, calculated at $z = \langle \mathbf{w}, \phi(\mathbf{x}_i) \rangle$. Following Equation (2.9), we see that our weight vector $\mathbf{w}_t$ is a linear combination of examples chosen in the iterations so far:

$$\mathbf{w}_t = -\frac{1}{\lambda t} \sum_{t'=1}^{t-1} g_{t'} \phi(\mathbf{x}_{i(t')}), \tag{2.10}$$

where $\mathbf{x}_{i(t)}$ is the example chosen at iteration $t$. We can therefore represent the weight vector by keeping track of the coefficients $g_t$ associated with each example. That is, we represent

$$\mathbf{w}_t = -\frac{1}{\lambda t} \sum_{i=1}^{m} \alpha_i \phi(\mathbf{x}_i), \tag{2.11}$$

where $\boldsymbol{\alpha}$ is a (sparse) vector with $\alpha_i = \sum_{t \mid i(t)=i} g_t$. Following the discussion in Section 2.2.4, we can now calculate the norm of $\mathbf{w}_t$ and inner products $\langle \mathbf{w}_t, \boldsymbol{\phi}(\mathbf{x}_i) \rangle$ using only kernel evaluations, and implement the SGD method without referring to the actual feature vectors $\boldsymbol{\phi}(\mathbf{x}_i)$.

Next, we consider the slightly more cumbersome case where $k > 1$, i.e. each example is represented by multiple feature vectors, perhaps one feature vector for each possible label. In this case a sub-gradient $\mathbf{v}_t$ of $\ell(\langle \mathbf{w}, \boldsymbol{\Phi}(\mathbf{x}_i) \rangle, y_i)$ can be written as

$$\mathbf{v}_t = \sum_{j=1}^{k} g_{t,j} \boldsymbol{\phi}(\mathbf{x}_i)_j, \qquad (2.12)$$

where $\mathbf{g}_t$ is a sub-gradient of $\ell(\mathbf{z}, y_i)$ with respect to $\mathbf{z}$ at $\mathbf{z} = \langle \mathbf{w}, \boldsymbol{\Phi}(\mathbf{x}_i) \rangle \in \mathbb{R}^k$. Again, each sub-gradient is spanned by feature vectors, and we have

$$\mathbf{w}_t = -\frac{1}{\lambda t} \sum_{i=1}^{m} \sum_{j=1}^{k} \alpha_{i,j} \boldsymbol{\phi}(\mathbf{x}_i)_j \qquad (2.13)$$

with $\alpha_{i,j} = \sum_{t \mid i(t)=i} g_{t,j}$. The required norm and inner products can, as before, be calculated using the kernel, and a kernalized implementation of the SGD method is possible. Note that at each iteration, only a single set of coefficients $\alpha_{i(t),j}$ is changed.

**Sequence prediction**  The SGD procedure is applicable for solving sequence prediction problems (see Example 2.2.5). To see this, we note that to efficiently calculate the sub-gradient $\mathbf{v}_t$ we need only find the label $\hat{y} = \arg \max_y \delta(y, y_i) - z_{y_i} + z_y$ (see the table in Section 2.4.1). Since this problem is equivalent to the problem of evaluating the loss function we conclude that SGD can be implemented as long as we can efficiently calculate the loss function.

**Sparse feature vectors**  The observation that the sub-gradients are spanned by feature vectors is also very beneficial when the feature vectors are given explicitly, and are very sparse. That is, each vector $\boldsymbol{\phi}(\mathbf{x}_i)$ has only a few non-zero entries. This would then mean that the sub-gradients $v_t$ are sparse. Examining the SGD pseudo-code in Figure 2.1, we see that the required operations are inner products of (the potentially dense) $\theta_t$ with the sparse vectors $\boldsymbol{\phi}(\mathbf{x}_i)_i$, and adding the sparse vector $\mathbf{v}_t$ into $\theta_t$. The total number of operations required for performing one iteration of SGD is therefore $O(d)$, where $d$ is the number of non-zero elements in $\boldsymbol{\Phi}(\mathbf{x}_i)$.

## 2.4.1  Sub-gradient Calculus

Given a convex function $g(\mathbf{w})$, a sub-gradient of $g$ at $\mathbf{w}_0$ is a vector $\mathbf{v}$ that satisfies

$$\forall \mathbf{w}, \; g(\mathbf{w}) - g(\mathbf{w}_0) \geq \langle \mathbf{v}, \mathbf{w} - \mathbf{w}_0 \rangle.$$

Two properties of sub-gradients are given below.

1. If $g(\mathbf{w})$ is differentiable at $\mathbf{w}_0$, then the gradient of $g$ at $\mathbf{w}_0$ is the unique sub-gradient of $g$ at $\mathbf{w}_0$.

2. If $g(\mathbf{w}) = \max_i g_i(\mathbf{w})$ for $r$ differentiable functions $g_1, \ldots, g_r$, and $j = \arg \max_i g_i(\mathbf{w}_0)$, then the gradient of $g_j$ at $\mathbf{w}_0$ is a sub-gradient of $g$ at $\mathbf{w}_0$.

Based on the above two properties, we now explicitly show how to calculate a sub-gradient for several loss functions. In the following table, we use the notation $z := \langle \mathbf{w}_t, \mathbf{\Phi}(\mathbf{x}_i) \rangle$.

| Loss function | Sub-gradient |
|---|---|
| $\ell(z, y_i) = \max\{0, 1 - y_i z\}$ | $\mathbf{v}_t = \begin{cases} -y_i \boldsymbol{\phi}(\mathbf{x}_i) & \text{if } y_i z < 1 \\ \mathbf{0} & \text{otherwise} \end{cases}$ |
| $\ell(z, y_i) = \log(1 + e^{-y_i z})$ | $\mathbf{v}_t = -\dfrac{y_i}{1 + e^{y_i z}} \boldsymbol{\phi}(\mathbf{x}_i)$ |
| $\ell(z, y_i) = \max\{0, |y_i - z| - \epsilon\}$ | $\mathbf{v}_t = \begin{cases} \boldsymbol{\phi}(\mathbf{x}_i) & \text{if } z - y_i > \epsilon \\ -\boldsymbol{\phi}(\mathbf{x}_i) & \text{if } y_i - z > \epsilon \\ \mathbf{0} & \text{otherwise} \end{cases}$ |
| $\ell(z, y_i) = \max_{y \in \mathcal{Y}} \delta(y, y_i) - z_{y_i} + z_y$ | $\mathbf{v}_t = \boldsymbol{\phi}(\mathbf{x}_i, \hat{y}) - \boldsymbol{\phi}(\mathbf{x}_i, y_i)$ where $\hat{y} = \arg \max_y \delta(y, y_i) - z_{y_i} + z_y$ |

## 2.4.2 Rate of Convergence and Stopping Criteria

We have so far discussed the iterations of the SGD approach, but did not discuss an appropriate stopping condition. In order to do so, we first present a guarantee on the number of iterations required in order to achieve a certain optimization accuracy. We then present a stopping condition based on this guarantee, as well as a more *ad hoc* stopping condition based on validation.

We say that an optimization method finds an $\epsilon$-accurate solution $\tilde{\mathbf{w}}$ if $f(\tilde{\mathbf{w}}) \leq \min_\mathbf{w} f(\mathbf{w}) + \epsilon$. The following theorem, adapted from Shalev-Shwartz *et al.* (2007), bounds the number of iterations required by the SGD method to obtain an $\epsilon$-accurate solution.

**Theorem 2.4.1** *Assume that the SGD algorithm runs $T$ iterations on a training set $\{(\mathbf{x}_i, y_i)\}_{i=1}^m$. Assume that the loss function $\ell(\langle \mathbf{w}, \mathbf{\Phi}(\mathbf{x}_i) \rangle, y_i)$ is convex with respect to $\mathbf{w}$ and that for all $t$, $\|\mathbf{v}_t\| \leq R$. Let $s$ be an integer s.t. $(T/s)$ is also in integer and let $r_1, \ldots, r_s$ be a sequence of indices where for each $i$, $r_i$ is randomly picked from $\{(T/s)(i-1)+1, \ldots, (T/s)\,i\}$. Then, with probability of at least $1 - e^{-s}$, exists $j \in [s]$ such that*

$$f(\mathbf{w}_{r_j}) \leq f(\mathbf{w}^\star) + \frac{2sR^2(1 + \log(T))}{\lambda T}.$$

The guarantee of Theorem 2.4.1 motivates an *a priori* stopping condition, namely stopping after a predetermined number of iterations which ensures the desired accuracy:

*A priori* **stopping condition**    Let $\epsilon > 0$ be a desired accuracy parameter and $\delta \in (0, 1)$ be a desired confidence parameter. We can set $T$ and $s$ in advance so that

$$\frac{2s R^2 (1 + \log(T))}{\lambda T} \leq \epsilon \quad \text{and} \quad e^{-s} \leq \delta. \tag{2.14}$$

Theorem 2.4.1 tells us that with probability of at least $1 - \delta$, the SGD procedure will find an $\epsilon$-accurate solution.

**Stopping condition using validation**    This *a priori* stopping condition is based on the guarantee of Theorem 2.4.1, which is a worst-case guarantee. Although when the stopping condition 2.14 is satisfied, we are guaranteed (with high probability) to have an $\epsilon$-accurate solution, it is certainly possible that we will reach this desired level of accuracy much sooner. Moreover, as discussed in Section 2.3, we should be less concerned about the optimization accuracy, and more concerned about the expected prediction error on future examples. This motivates using a prediction error on a held-out validation set as a stopping guideline. That is, set some training examples apart as a validation set and monitor the prediction error on the validation set once in a while. For example, if we set $10k$ examples as a validation set and calculate the loss function on the validation set after each $10k$ iterations, the overall runtime increases by a factor of 2. We can then terminate training when the loss on the validation set stop decreasing (see also Bottou 2008).

Neither of the above stopping conditions is very satisfying. The *a priori* approach might be too pessimistic, while the validation approach is rather time consuming and there is no rigorous way of deciding when the validation error stopped decreasing. Although the validation approach does often suffice, in the next section we present an alternative optimization approach with a better stopping criterion.

## 2.5   Dual Decomposition Methods

In the previous section we described a simple SGD approach for solving the SVM optimization problem. The main two advantages of the SGD procedure are its simplicity and its efficiency. However, as we discussed, its biggest drawback is the lack of a good stopping condition. Additionally, when using kernels, the computational cost of each SGD step does scale with the number of training examples, diminishing its attractiveness (although it is still competitive with other approaches).

In this section we describe a method for solving the SVM training optimization problem using the *dual* problem, which we will shortly introduce. Specifically, we consider a simple dual decomposition approach. Decomposition methods such as SMO (Platt 1998) and SVM-Light (Joachims 1998) switch to the dual representation of the SVM optimization problem, and follow a coordinate ascent optimization approach for the dual problem. Dual decomposition methods have two advantages. First, as we show below, the duality gap upper bounds the suboptimality of our intermediate solution. As a result, we have a clear stopping criterion. Second, in dual methods, we sometime remove vectors from the support set (that is, the set of examples for which $\alpha_i$ is non-zero, and so they participate in the resulting weight vector) and therefore we may find more sparse solutions. This property is desired when working with nonlinear kernels. The disadvantage of traditional decomposition methods, such as SMO and SVM-Light, is that their running time is sometimes much slower than SGD methods, especially when working with linear kernels. The procedure we describe in this section uses randomization for selecting a working example. Experimentally, the performance of this method is often comparable to that of SGD methods.

For simplicity, our focus in this section is on the problem of binary classification with the hinge loss, that is, the primal optimization problem is to minimize the following objective function:

$$f(\mathbf{w}) = \frac{\lambda}{2} \|\mathbf{w}\|^2 + \frac{1}{m} \sum_{i=1}^{m} \max\{0, 1 - y_i \langle \mathbf{w}, \boldsymbol{\phi}(\mathbf{x}_i) \rangle\}.$$

Dual decomposition approaches were devised for more complex problems. We refer the reader to Crammer and Singer (2001), Globerson *et al.* (2007), Shalev-Shwartz and Singer (2006), Taskar *et al.* (2003) and Tsochantaridis *et al.* (2004).

## 2.5.1 Duality

We have so far studied direct minimization of the objective $f(\mathbf{w})$. We refer to this problem, arg $\min_{\mathbf{w}} f(\mathbf{w})$, as the *primal* problem.[1] We can rewrite the primal problem as the following constrained optimization problem:

$$\min_{\mathbf{w}, \boldsymbol{\xi}} \frac{\lambda}{2} \|\mathbf{w}\|^2 + \frac{1}{m} \sum_{i=1}^{m} \xi_i \quad \text{s.t.} \quad \xi_i \geq 0, 1 - y_i \langle \mathbf{w}, \boldsymbol{\phi}(\mathbf{x}_i) \rangle \leq \xi_i.$$

Introducing Lagrange multipliers on the constraints (see, e.g., Boyd and Vandenberghe 2004; Schölkopf and Smola 2002), one can derive the *dual* problem:

$$\arg \max_{\boldsymbol{\alpha} \in [0,1]^m} \mathcal{D}(\boldsymbol{\alpha}) \tag{2.15}$$

where the dual objective is given by

$$\mathcal{D}(\boldsymbol{\alpha}) = \frac{1}{m} \left( \sum_{i=1}^{m} \alpha_i - \frac{1}{2\lambda m} \sum_{i=1}^{m} \sum_{j=1}^{m} \alpha_i \alpha_j y_i y_j \langle \boldsymbol{\phi}(\mathbf{x}_i), \boldsymbol{\phi}(\mathbf{x}_j) \rangle \right). \tag{2.16}$$

Duality theory tells us that for any weight vector $\mathbf{w}$ and a feasible dual vector $\boldsymbol{\alpha} \in [0, 1]^m$ we have $\mathcal{D}(\boldsymbol{\alpha}) \leq f(\mathbf{w})$ and equality holds if and only if $\mathbf{w}$ is the optimizer of the primal problem and $\boldsymbol{\alpha}$ is an optimizer of the dual problem. Furthermore, if $\boldsymbol{\alpha}$ is an optimizer of the dual problem then the following vector is an optimizer of the primal problem:

$$\mathbf{w} = \frac{1}{\lambda m} \sum_{i=1}^{m} \alpha_i y_i \boldsymbol{\phi}(\mathbf{x}_i). \tag{2.17}$$

The quantity $f(\mathbf{w}) - \mathcal{D}(\boldsymbol{\alpha})$ is called the *duality gap* and it upper bounds the suboptimality of $\mathbf{w}$ since

$$f(\mathbf{w}) - f(\mathbf{w}^\star) \leq f(\mathbf{w}) - \mathcal{D}(\boldsymbol{\alpha}).$$

---

[1] It should be emphasized that the problem arg $\min_{\boldsymbol{\alpha}} f(\boldsymbol{\alpha})$ given in Equation (2.4) is not a dual problem but rather a different parametrization of the primal problem.

One advantage of the dual problem is that it can be easily decomposed. That is, we can choose a subset of the dual variables and optimize $\mathcal{D}(\boldsymbol{\alpha})$ over these variables while keeping the rest of the dual variables intact. In the most simple approach, at each iteration we optimize $\mathcal{D}(\boldsymbol{\alpha})$ over a single element of $\boldsymbol{\alpha}$. The optimum of $\mathcal{D}(\boldsymbol{\alpha})$ over a single variable has a closed form solution and therefore the cost of each iteration is small. Similar approaches were suggested by numerous authors (see, e.g., Censor and Zenios 1997; Hildreth 1957; Mangasarian and Musicant 1999). We now describe the resulting algorithm.

Initially, we set $\boldsymbol{\alpha}^1 = \mathbf{0}$. At iteration $t$, we define

$$\mathbf{w}_t = \frac{1}{\lambda m} \sum_{j=1}^{m} \alpha_j^t \, y_j \, \boldsymbol{\phi}(\mathbf{x}_j).$$

Choose $i \in [m]$ uniformly at random and let

$$\mathbf{w}_t^{\backslash i} = \mathbf{w}_t - \frac{1}{\lambda m} \alpha_i^t y_i \boldsymbol{\phi}(\mathbf{x}_i) = \frac{1}{\lambda m} \sum_{j \neq i} \alpha_j^t \, y_j \, \boldsymbol{\phi}(\mathbf{x}_j).$$

We can rewrite the dual objective function as follows:

$$\mathcal{D}(\boldsymbol{\alpha}) = \frac{1}{m} \sum_{j=1}^{m} \alpha_j - \frac{\lambda}{2} \left\| \mathbf{w}_t^{\backslash i} + \frac{1}{\lambda m} \alpha_i y_i \boldsymbol{\phi}(\mathbf{x}_i) \right\|^2$$

$$= \frac{\alpha_i}{m} \left( 1 - y_i \langle \mathbf{w}_t^{\backslash i}, \boldsymbol{\phi}(\mathbf{x}_i) \rangle \right) - \frac{\|\boldsymbol{\phi}(\mathbf{x}_i)\|^2}{2\lambda m^2} \alpha_i^2 + C,$$

where $C$ does not depend on $\alpha_i$. Optimizing $\mathcal{D}(\boldsymbol{\alpha})$ with respect to $\alpha_i$ over the domain $[0, 1]$ gives the update rule

$$\alpha_i^{t+1} = \max \left\{ 0, \min \left\{ 1, \frac{\lambda m (1 - y_i \langle \mathbf{w}_t^{\backslash i}, \boldsymbol{\phi}(\mathbf{x}_i) \rangle)}{\|\boldsymbol{\phi}(\mathbf{x}_i)\|^2} \right\} \right\}.$$

A pseudo-code of the resulting algorithm is given in Figure 2.2. Note that we now have a clear stopping condition, $f(\mathbf{w}_t) - \mathcal{D}(\boldsymbol{\alpha}^t) \leq \epsilon$, which follows directly from the fact that the duality gap upper bounds the suboptimality,

$$f(\mathbf{w}_t) - f(\mathbf{w}^\star) \leq f(\mathbf{w}_t) - \mathcal{D}(\boldsymbol{\alpha}^t).$$

This stands in contrast to the SGD approach in which we did not have such a clear stopping condition.

**Nonlinear kernels** We can rewrite the update rule for $\alpha_i^{t+1}$ using the Gram matrix $K$ defined in Section 2.2.4 as follows. First, we note that

$$\lambda m \, y_i \langle \mathbf{w}_t^{\backslash i}, \boldsymbol{\phi}(\mathbf{x}_i) \rangle = \sum_{j \neq i} K_{i,j} \alpha_j^t = (K \, \boldsymbol{\alpha}^t)_i - K_{i,i} \, \alpha_i^t.$$

Additionally, we have $\|\boldsymbol{\phi}(\mathbf{x}_i)\|^2 = K_{i,i}$. Therefore,

$$\alpha_i^{t+1} = \max \left\{ 0, \min \left\{ 1, \frac{\lambda m - (K \, \boldsymbol{\alpha}^t)_i - K_{i,i} \, \alpha_i^t}{K_{i,i}} \right\} \right\}.$$

**Input:** A sequence of examples $(x_1, y_1), \ldots, (\mathbf{x}_m, y_m)$

Regularization parameter $\lambda$

Desired accuracy $\epsilon$

**Initialize:** $\boldsymbol{\alpha}^1 = (0, \ldots, 0)$

**Do:**

Set $\mathbf{w}_t = \dfrac{1}{\lambda m} \sum_{j=1}^{m} \alpha_j^t y_j \boldsymbol{\phi}(\mathbf{x}_j)$

Choose $i \in [m]$ uniformly at random

Set $\mathbf{w}_t^{\backslash i} = \mathbf{w}_t - \dfrac{1}{\lambda m} \alpha_i^t y_i \boldsymbol{\phi}(\mathbf{x}_i)$

Update: $\alpha_i^{t+1} = \max\left\{0, \min\left\{1, \dfrac{\lambda m (1 - y_i \langle \mathbf{w}_t^{\backslash i}, \boldsymbol{\phi}(\mathbf{x}_i) \rangle)}{\|\boldsymbol{\phi}(\mathbf{x}_i)\|^2}\right\}\right\}$

and for $j \neq i$, $\alpha_j^{t+1} = \alpha_j^t$

**Until:**

$f(\mathbf{w}_t) - \mathcal{D}(\boldsymbol{\alpha}^t) \leq \epsilon$

Figure 2.2  A dual coordinate ascend algorithm for solving SVM.

This update implies that $\alpha_i^{t+1}$ will become zero if $\mathbf{w}_t^{\backslash i}$ attains zero hinge loss on the $i$th example. That is, the update encourages sparse dual solutions. Compare this with the SGD approach, in which whenever we set $\alpha_i$ to a non-zero value it remains non-zero in all subsequent iterations.

**Convergence**    Several authors (e.g. Hsieh *et al.* 2008; Mangasarian and Musicant 1999) proved linear convergence rate for dual coordinate ascent. The basic technique is to adapt the linear convergence of coordinate ascent that was established by Luo and Tseng (1992). This convergence result tells us that after an unspecified number of iterations, the algorithm converges very fast to the optimal solution.

## 2.6  Summary

In this chapter we described two approaches for SVM optimization: stochastic gradient descent and stochastic dual coordinate ascent. These methods are efficient, easy to implement, and are adequate for SVM training in the presence of large data sets. The major advantage of the stochastic gradient descent approach is its simplicity – we show how it can be implemented for solving various $L_2$-regularized problems such as binary classification and sequence prediction. The advantages of the dual coordinate ascent approach are that it has a natural stopping criterion and it encourages sparse dual solutions.

# References

Bottou L 2008 Stochastic gradient descent examples. http://leon.bottou.org/projects/sgd (last accessed 4 December 2008).

Boyd S and Vandenberghe L 2004 *Convex Optimization*. Cambridge University Press.

Censor Y and Zenios S 1997 *Parallel Optimization: Theory, Algorithms, and Applications*. Oxford University Press, New York, NY.

Chapelle O 2007 Training a support vector machine in the primal. *Neural Computation* **19**(5), 1155–1178.

Crammer K and Singer Y 2001 On the algorithmic implementation of multiclass kernel-based vector machines. *Journal of Machine Learning Research* **2**, 265–292.

Duda RO and Hart PE 1973 *Pattern Classification and Scene Analysis*. John Wiley & Sons.

Fan RE, Chen PH and Lin CJ 2005 Working set selection using second order information for training SVM. *Journal of Machine Learning Research* **6**, 1889–1918.

Globerson A, Koo T, Carreras X and Collins M 2007 Exponentiated gradient algorithms for log-linear structured prediction. *Proceedings of the 24th International Conference on Machine Learning*, Oregon State University, OR, June 20–24, 2007.

Hildreth C 1957 A quadratic programming procedure. *Naval Research Logistics Quarterly* **4**, 79–85. Erratum, ibidem, p. 361.

Hsieh C, Chang K, Lin C, Keerthi S and Sundararajan S 2008 A dual coordinate descent method for large-scale linear SVM. *Proceedings of the 25th International Conference on Machine Learning*, Helsinki, Finland, July 5–9, 2008.

Joachims T 1998 Making large-scale support vector machine learning practical. *Advances in Kernel Methods—Support Vector Learning* (eds. B Schölkopf, C Burges and A Smola). MIT Press.

Kimeldorf G and Wahba G 1971 Some results on tchebycheffian spline functions. *Journal of Mathematical Analysis and Applications* **33**, 82–95.

Luo Z and Tseng P 1992 On the convergence of the coordinate descent method for convex differentiable minimization. *Journal of Optimization Theory and Applications* **72**, 7–35.

Mangasarian O and Musicant D 1999 Successive overrelaxation for support vector machines. *IEEE Transactions on Neural Networks* **10**.

Nocedal J and Wright S 2006 *Numerical Optimization*, 2nd edn. Springer.

Platt JC 1998 Fast training of Support Vector Machines using sequential minimal optimization. *Advances in Kernel Methods–Support Vector Learning* (eds. B Schölkopf, C Burges and A Smola). MIT Press.

Schölkopf B and Smola AJ 2002 *Learning with Kernels: Support Vector Machines, Regularization, Optimization and Beyond*. MIT Press.

Shalev-Shwartz S and Singer Y 2006 Efficient learning of label ranking by soft projections onto polyhedra. *Journal of Machine Learning Research* **7** (July), 1567–1599.

Shalev-Shwartz S, Singer Y and Srebro N 2007 Pegasos: Primal estimated sub-gradient solver for SVM. *Proceedings of the 24th International Conference on Machine Learning*, Oregon State University, OR, June 20–24, 2007.

Shalev-Shwartz S and Srebro N 2008 Svm optimization: Inverse dependence on training set size. *Proceedings of the 25th International Conference on Machine Learning*, Helsinki, Finland, July 5–9, 2008.

Taskar B, Guestrin C and Koller D 2003 Max-margin Markov networks. *Advances in Neural Information Processing Systems 17*.

Tsochantaridis I, Hofmann T, Joachims T and Altun Y 2004 Support vector machine learning for interdependent and structured output spaces. *Proceedings of the 21st International Conference on Machine Learning*, Banff, Alberta, Canada, July 4–8, 2004.

# 3

# From Binary Classification to Categorial Prediction

## Koby Crammer

The problem of binary classification is fundamental in many ways but it still does not describe the real-world well. We present a more complex setting in which the possible labels or categories are many and organized. This setup makes the learning task more difficult, not only qualitatively, but also quantitatively as well. This is because the learner has to take into consideration that some mistakes are more acceptable than others. In the next section we present a general framework for multicategorization problems and exemplify it with the task of automatic speech recognition. In Section 3.2 we present one linear model for the problem and in Section 3.3 we discuss loss functions for such problems. Next, in Section 3.5 and Section 3.6 we present two general algorithms for such problems, an extension for the Perceptron algorithm (Rosenblatt 1958) and one based on quadratic programming in the spirit of Passive–Aggressive (PA) algorithms (Crammer *et al.* 2006). We present a related batch formulation which extends support vectors machines (SVMs) (Cristianini and Shawe-Taylor 2000; Schölkopf and Smola 2002). We conclude with a discussion about related work and other approaches in Section 3.8.

## 3.1   Multi-category Problems

Let us start with a general description and definition of multi-category problems using automatic speed recognition (ASR) as an example. A common problem is to map acoustic signals of speech utterances to phonemes, which are often organized in a tree (see Figure 3.1). We like a way to define what category-predictions are preferable to others. Consider a fragment of a signal which should be mapped to the phoneme /ɔɪ/ (as in /*boy*/). One specification of a system is that it will predict correctly the correct phoneme, otherwise,

*Automatic Speech and Speaker Recognition: Large Margin and Kernel Methods*   Joseph Keshet and Samy Bengio
© 2009 John Wiley & Sons, Ltd

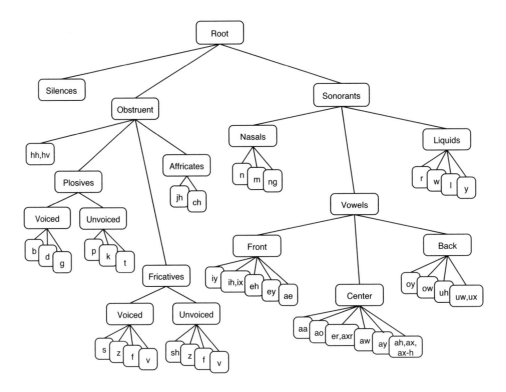

Figure 3.1 Hierarchical phonetic tree (reproduced by permission from Springer © Dekel, Keshet and Singer, *An Online Algorithm for Hierarchical Phoneme Classification*, Lecture Notes in Computer Science vol. 3361/2005, Machine Learning for Multimodal Interaction. Springer, Berlin/Heidelberg, 2005, pp. 146–158).

all the remaining categories are equally wrong. Another more involved specification is that in the case of a mistake we prefer that the system will output a similar phoneme such as /ow/ (as in /boat/) rather than an arbitrary phoneme. An even more complex preference groups phonemes into logical sets. In this case, if the system is not sure of the exact vowel we like the output to indicate that the phoneme is a vowel or even a center vowel. This makes sense when the predictor is a single component of a large system and a later component of the ASR system can correct some mistake or disambiguate between similar alternatives.

We denote by $\mathcal{C}$ the set of all possible categories (phonemes in our example) and describe and visualize preference between categories $c \in \mathcal{C}$ using directed acyclic graphs. The elements of the categories set $\mathcal{C}$ correspond to the nodes of the graph. Each ordered pair in the preference relation $(c, d) \in \mathbf{y} \subset \mathcal{C} \times \mathcal{C}$ translates into a directed edge $c \to d$ of the graph. No self loops in a graph represent reflexive relations and no loops of size two represent anti-symmetry relations. We do not enforce transitivity (if there is a pass from node $c$ to node $d$ then there is an edge between these two nodes). We refer to this type of preference as semi-orders, the formal definition of which is now given.

**Definition 3.1.1 (Semi-order)** *Let $C$ be a finite set. A semi-order $\pi$ is a subset of the product space $C \times C$ that its induced graph is a directed acyclic graph (DAG). Namely, for all $(a_1, b_1), (a_2, b_2), \ldots, (a_t, b_t) \in \pi$ such that $b_1 = a_2, b_2 = a_3, \ldots, b_{t-1} = a_t$ we have $b_t \neq a_1$.*

Semi-orders are less restricted than partial-orders. In both cases, we do require reflexivity and any-symmetry; however, partial-orders further assume transitivity, while semi-orders weaken transitivity and assume that no directed cycles exist. For example $\pi = \{(1, 2), (2, 3)\}$ is a semi-order but not a partial-order, since the axiom of transitivity implies that $(1, 3) \in \pi$.

The learning algorithm observes a set of examples $\{(\mathbf{x}_i, \mathbf{y}_i)\}_{i=1}^m$, each example $(\mathbf{x}, \mathbf{y})$ being constituted from a pair of an instance and a target, where a bold face symbol $\mathbf{y}$ designates a semi-order label rather than a simple label $y$. Each instance $\mathbf{x}$ is in an input space $\mathcal{X}$. To describe the target we define the following spaces. The set of all possible topics or categories was denoted above by $C$ and w.l.o.g. it is assumed to be $C = \{1, 2, \ldots, k\}$. We denote by $\mathcal{Y}$ the set of all semi-orders over $C$. Each instance is associated with a target label $\mathbf{y}$ taken from the set of semi-orders $\mathcal{Y}$ over the topics $C$. Formally:

$$\mathcal{Y} = \{\pi \ : \ \pi \in C \times C \text{ is a semi-order}\}.$$

The goal of the learning algorithm is to construct a mapping from the instance space to the prediction space.

We define the prediction space $\hat{\mathcal{Y}}$ to be a $k$ dimensional vector space $\mathbb{R}^k$. Each prediction $\hat{\mathbf{y}} \in \hat{\mathcal{Y}}$ naturally induces a total-order. If the $p$th coordinate of the vector $\hat{\mathbf{y}}$ is greater than the $q$th coordinate of $\hat{\mathbf{y}}$ we set the pair order $(p, q)$ to belong to the total-order. Formally:

$$\hat{\mathbf{y}} \in \hat{\mathcal{Y}} = \mathbb{R}^k \longmapsto \{(p, q) \ : \ \hat{\mathbf{y}}^p > \hat{\mathbf{y}}^q\}.$$

Ties are broken arbitrarily. We will abuse the notation and think of the prediction $\hat{\mathbf{y}}$ as the total-ordering it induces, rather than a vector of $k$ elements, and write $(p, q) \in \hat{\mathbf{y}}$, which is equivalent to $\hat{\mathbf{y}}^p > \hat{\mathbf{y}}^q$.

There is a many-to-many relation between the target space $\mathcal{Y}$ and the prediction space $\hat{\mathcal{Y}}$. A given prediction ($k$ dimensional vector) corresponds to some total-order, which may be consistent with many semi-orders (targets) by losing some of the order restrictions. *Vice versa*, each semi-order (target) is consistent with many total-orders by adding more order restrictions which can be represented with many predictions ($k$ dimensional vectors).

Consider for example a simple prediction problem with three possible classes $C = \{1, 2, 3\}$. First the prediction $\hat{\mathbf{y}} \in \mathbb{R}^3$ such that $\hat{\mathbf{y}} = (10, 8, 6)$ is consistent with the total-order $\{(1, 2), (1, 3), (2, 3)\}$ (since $\hat{\mathbf{y}}^1 > \hat{\mathbf{y}}^2$ and $\hat{\mathbf{y}}^2 > \hat{\mathbf{y}}^3$), which is consistent with the three semi-orders: $\mathbf{y} = \{(1, 2), (1, 3), \{(1, 2)\}$ and $\{(1, 3)\}$. The other way around, the semi-order $\mathbf{y} = \{(1, 2), (1, 3)\}$ does not specify which of categories 2 or 3 is preferable over the other. There are two possible total-orders consistent with the semi-order $\mathbf{y}$: either $\mathbf{y} \cup \{(2, 3)\}$ where 2 is preferred over 3, or $\mathbf{y} \cup \{(3, 2)\}$. This illustrates that more than a single total-order may be consistent with the same total-order. To illustrate the relation between total-orders and predictions, we choose the first case and consider the total-order $\{(1, 2), (1, 3), (2, 3)\}$. We note that every vector $\hat{\mathbf{y}} \in \mathbb{R}^3$ such that $\hat{\mathbf{y}}^1 > \hat{\mathbf{y}}^2$ and $\hat{\mathbf{y}}^2 > \hat{\mathbf{y}}^3$ is consistent with the chosen total-order; hence, a one-to-many relation between semi-orders and predictions.

As a consequence, given a prediction $\hat{\mathbf{y}}$ regarding an instance $\mathbf{x}$ or the corresponding total-order, there is no direct way to reconstruct the original target (semi-order) $\mathbf{y}$ associated with

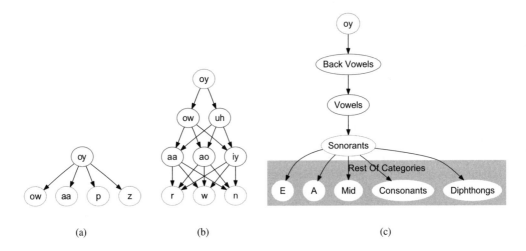

(a)                    (b)                    (c)

Figure 3.2 Three preference relations phonemes. (a) Organization of phonemes in two layers: correct phoneme and others. (b) Organization of phonemes in layers. (c) Organization of phonemes as a path from leaf to root, then all other phonemes.

the instance **x**. This gap is an inherent property of the model and, as we shall see, it is a byproduct of the class of prediction functions. Last, we assume that the empty semi-order, in which there are no order restrictions, is not a possible target, $y = \emptyset \notin \mathcal{Y}$. This is because any total-order is consistent with it and thus it does not challenge the learning algorithm.

A graphical illustration of the three possible preference relations is given in Figure 3.2. The left panel (Figure 3.2(a)) demonstrates the first specification in which the correct phoneme should be ranked above any other class. The induced graph consists of $k - 1$ pairs of the form $(r, s)$ for a fixed value of $r \in \mathcal{C}$ and all values of $s \neq r$. (For simplicity, only four edges are shown.) Any total-order in which the category $r \in \mathcal{C}$ is ranked above all other categories is consistent with this type of semi-order. The center panel (Figure 3.2(b)) demonstrates the second specification, in which phonemes are organized in layers. The top layer consists of the correct phoneme /oy/, the second layer consists of the most similar phonemes (all back-vowels, only two actually appear in the plot), the third layer consists of all other vowels (only three appear) and the fourth layer consists of all other phonemes (only three appear in the plot). There is an edge between any pair of phonemes from two consecutive layers. There are no order restrictions among categories from the same layer. In other words, the set of categories is divided into four disjoint subsets, All categories belonging to some subset are ranked above all the categories belonging to less preferred subsets.

Finally, the right panel (Figure 3.2(c)) illustrates the third preference specification as a tree. The phonemes are located only in the leaves of the tree. The internal nodes of the tree represent groups of categories. A father–child relation reflects specification. We add meta-categories, to reflect the internal node of the tree. A phoneme category (leaf) is viewed as a pass from the root to the corresponding node. This representation reflects the idea that some mistakes are more preferable than others: a meta-category which is an ancestor of the target is more acceptable than other categories or meta-categories.

We remind the reader that the above examples are specific cases designed to illustrate the power and flexibility of preference relations. Other choices may suit in other applications. General preference over categories can be any DAG. DAGs can be used to represent hierarchical ordering or specification of categories in a more complex structure. It should be used according to the exact problem in hand.

## 3.2 Hypothesis Class

The class of linear functions presented in Chapter 1 has many of the desired properties previewed there. To promote our hypothesis class we describe linear separators $\mathbf{w}$ using the notation of multicategorization. We denote the two possible categories by $\mathcal{C} = \{1, 2\}$, implement a linear separator $\mathbf{w}$ using two parameter vectors $\mathbf{w}_1 = \mathbf{w}$ and $\mathbf{w}_2 = -\mathbf{w}$ and define the prediction function $h : \mathcal{X} \to \mathbb{R}^2$ to be $h(\mathbf{x}) = ((\mathbf{w}_1 \cdot \mathbf{x}), (\mathbf{w}_2 \cdot \mathbf{x}))$. It is straightforward to verify that the prediction of $h$ is equivalent to the prediction of the linear separator $h(\mathbf{x}) = (\mathbf{w} \cdot \mathbf{x})$. Since the first element of $h(\mathbf{x})$ is larger than its second element, if and only if, $(\mathbf{w} \cdot \mathbf{x}) > 0$.

We use this interpretation and generalize the class of linear separators to multicategorization problems as follows. Given a problem with $k$ categories, we maintain a collection $\mathbf{W}$ of $k$ parameter vectors $\mathbf{w}_r$, one vector per category $r$. The weight vectors $\mathbf{w}_r$ are also called *prototypes*. Occasionally the notation is abused and the collection $\mathbf{W}$ is thought of as a concatenation of the $k$-vectors $\mathbf{w}_1, \ldots, \mathbf{w}_k$, that is $\mathbf{W} = (\mathbf{w}_1, \ldots, \mathbf{w}_k)$. In this case the norm of the collection $\mathbf{W}$ is defined as

$$\|\mathbf{W}\|^2 = \sum_r \|\mathbf{w}_r\|^2.$$

We define the hypothesis $h$ parameterized by $\mathbf{W}$ to be

$$h(\mathbf{x}) = ((\mathbf{w}_1 \cdot \mathbf{x}), \ldots, (\mathbf{w}_k \cdot \mathbf{x})) \in \mathbb{R}^k. \tag{3.1}$$

We denote by $h_r(\mathbf{x}) = (\mathbf{w}_r \cdot \mathbf{x})$. The model composed of the hypothesis class and the representation of input space is called *Single-Instance Multi-Prototype*. This is because the representation of an instance is fixed along categories whereas there is a vector prototype per category. There exist other models, such as the one presented in Chapter 9, which is popular in structured prediction. We refer the reader to *Concluding remarks* for more information.

These hypotheses divide the Euclidean space into cones. The angle of a cone depends both on the relative angle between pairs of two parameter vectors and on their norm. An illustration of the decision boundary in the plane appears in Figure 3.3. Both plots in the figure show the decision boundary induced by four prototype hypotheses. The colors code the four regions for which each of the categories is ranked at the top. A similar decomposition of space occurs when tracking categories ranked second, third and fourth. The left plot illustrates the decision boundary for a prediction rule in which the norms of all prototypes are equal to each other. The right-hand side plot shows the decision boundary for prototypes with different norms. The length of the arrows in the plot is proportional to the norm of the prototype.

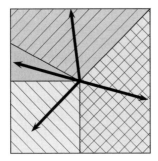 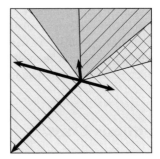

Figure 3.3 An illustration of a prediction function composed of four linear functions. The left plot shows the decision boundary for four linear functions with the same norm, whereas the right plot shows the decision boundary for linear functions that do not have the same norm.

## 3.3 Loss Functions

Let us consider a sample $\mathcal{T} = \left\{ (\mathbf{x}_i, y^i) \right\}_{i=1}^m$. Assume we are given two prediction functions taken from the set of all prediction functions $\mathcal{H} = \{(h_1(\mathbf{x}) \ldots h_k(\mathbf{x})) : h_r(\mathbf{x}) = (\mathbf{w}_r \cdot \mathbf{x})$ for $\mathbf{w}_r \in \mathbb{R}^n\}$ and we want to determine which is preferred. The simplest method to compare two hypotheses is to apply each of them to the sample $\mathcal{T}$ and compute the fraction of examples in which each of the hyperplanes was wrong. The hypothesis which makes fewer mistakes is the better. This may be well defined for binary classification problems, as a prediction is either correct or wrong. However, in our context of multi-category problems, it may not be enough as we need to quantize that some mistakes are more preferable than others. We thus introduce the notion of a loss function, which takes two arguments, an hypothesis $h$ and a labeled example $(\mathbf{x}, y)$, and measures the quality of prediction by a single non-negative number:

$$\ell : \mathcal{H} \times (\mathcal{X} \times \mathcal{Y}) \rightarrow \mathbb{R}_+. \tag{3.2}$$

A hypothesis which suffers a zero loss on an example is said to predict the label *perfectly*.

A proper loss function should be defined in order to evaluate prediction functions. In the problem of binary classification there is a single natural loss function, the zero-one loss which indicates whether an error has occurred or not. The loss functions for multicategorization we describe below reflect that some mistakes are more preferable than others by having a rich range of more than two possible values (as used in binary classification). In general, the loss functions described below are combinatorial in nature and count the number of subsets of the target semi-order that are not consistent with the predicted total-order.

Although these combinatorial loss functions are natural to use, they suffer from two drawbacks. First, they do not take into consideration the confidence information which is part of the prediction. More precisely, they consider only the relation between the elements of $\hat{\mathbf{y}}$ and their actual values. Thus if two hypotheses make an error when predicting a specific example, but for one of them the difference between the elements of $\hat{\mathbf{y}}$ is much greater than the other, their loss is equal. The other drawback of the combinatorial loss functions is their discontinuity in the parameter vectors $\mathbf{w}_r$. Thus, when we cast the learning algorithms as an optimization problems the outcome is often not-convex and hard to solve. Thus, later, in Section 3.4 we extend the notion of hinge loss (see e.g. Cristianini and Shawe-Taylor 2000; Schölkopf and Smola 2002) to our more complex setting.

## 3.3.1  Combinatorial Loss Functions

Let us illustrate the variety of possible loss functions using a multi-class multi-label problem. Targets are two-layer semi-orders, and predictions are total-ordering. Categories belonging to the top-layer are *relevant* to the input instance, while the other categories, belonging to the lower level, are *not-relevant*. In our example, assume that there are five categories ($\mathcal{Y} =$ {1, 2, 3, 4, 5}) and the first two categories are ranked above the other three categories ($\mathbf{y} =$ {(1, 3), (2, 3), (1, 4), (2, 4), (1, 5), (2, 5)}), and that some prediction $\hat{\mathbf{y}}$ induces the total-ordering : (1, 3), (3, 4), (4, 2), (2, 5) $\in \hat{\mathbf{y}}$. The loss functions we define below differ from each other by the resolution used to measure the performance of the prediction. The most high-level question we can ask is '*Is the prediction perfect?*' In our example the answer is negative since the pair (2, 3) $\in \mathbf{y}$ (2 is ranked above 3) is part of the semi-order whereas its opposite (3, 2) $\in \hat{\mathbf{y}}$ is contained in the prediction. A basic low-level question we ask is '*How many pairs of relevant–irrelevant categories are mismatched by the total-ordering?*' There are two such pairs (2, 3) and (2, 4). Unlike the previous question, which has only two possible answers, this question has seven possible answers, $|\mathbf{y}| + 1$. These loss functions partition the set of predictions (total-orderings) differently, in accordance with the possible values the loss function can attain. The first loss function favors perfect predictions over all other predictions, whereas the second loss function differentiates non-perfect predictions.

We can further ask two other natural questions: '*How many relevant categories are ranked below some irrelevant category?*' (answer: one, category 2) and '*How many irrelevant categories are ranked above some relevant category?*' (answer: two, categories 3, 4). These two loss functions are more detailed than the first loss function mentioned above, but less detailed than the second loss function. In both cases the possible number of answers is larger than 2 (number of possible answers for the first loss above) and less than 7 (number of possible answers for the second loss above).

Let us now formulate all these loss functions and other additional functions in a unified description. Let $\mathbf{y}$ be a semi-order over $\mathcal{Y}$, and $\hat{\mathbf{y}}$ be a prediction. A loss function is parameterized by a partition $\chi$ of the semi-order $\mathbf{y}$ into finite *disjointed* sets. Namely, $\bigcup_{\chi \in \chi} \chi = \mathbf{y}$ and $\chi_p \cap \chi_q = \emptyset$ for all $p \neq q$. The loss function is defined as

$$\mathbb{I}\left(\mathbf{y}; \hat{\mathbf{y}}, \chi\right) = \sum_{\chi \in \chi} [\![\{(r, s) \in \chi \; : \; (s, r) \in \hat{\mathbf{y}}\} \neq \emptyset]\!], \tag{3.3}$$

where $[\![A]\!]$ equals 1 is the predicate $A$ is true, and equals zero otherwise. Each summand is an indicator function of the event 'the set $\{(r, s) \in \chi \; : \; (s, r) \in \hat{\mathbf{y}}\}$ is not empty'. The set contains all the ordered pairs $(r, s)$ belonging to a specific cover $\chi \in \chi$ which are mismatched by the total-ordering $\hat{\mathbf{y}}$. In other words, the loss equals the number of cover elements $\chi \in \chi$ that contain at least a single ordered pair $(r, s)$ for which the total-ordering is not consistent. Since the family of cover loss functions is a sum of indicator functions of events we denote it by $\mathbb{I}$. Furthermore, if the cover $\chi$ is omitted it is assumed that a single cover $\chi = \{\mathbf{y}\}$ is used, and the loss indicates whether an error occurred or not, and in this case the loss is called *IsError*. On the other hand, if the cover contains only singleton elements, $\chi = \mathbf{y}$, the loss

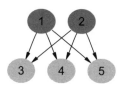

Figure 3.4  An illustration of the semi-order for a specific multi-class multi-label problem. There are 5 possible categories and the target is the first two categories.

counts the number of mismatched pairs, and is called *ErrorSetSize*. The four specific loss functions mentioned above are parameterized by different covers of the target semi-order $\mathbf{y}$. Specifically, for the first loss mentioned above we define a cover with a single element, $\chi_1 = \{\chi_1\}$, where $\chi_1 = \{(1, 3), (1, 4), (1, 5), (2, 3), (2, 4), (2, 5)\}$. For the second loss function we define a cover with six elements, each containing a single pair of the semi-order $y$, that is $\chi_2 = \{\chi_i\}_{i=1}^{6}$ where $\chi_1 = \{(1, 3)\}$, $\chi_2 = \{(1, 4)\}$, $\chi_3 = \{(1, 5)\}$, $\chi_4 = \{(2, 3)\}$, $\chi_5 = \{(2, 4)\}$ and $\chi_6 = \{(2, 5)\}$. For the third loss function, which counts the number of relevant categories ranked below an irrelevant category, we define a cover, with an element per relevant topic, and get, $\chi_3 = \{\chi_1, \chi_2\}$, $\chi_1 = \{(1, 3), (1, 4), (1, 5)\}$ and $\chi_2 = \{(2, 3), (2, 4), (2, 5)\}$. Similarly, for the fourth loss function, which counts the number of irrelevant categories ranked above a relevant category, we define a cover, with an element per irrelevant category, and get, $\chi_4 = \{\chi_1, \chi_2, \chi_3\}$, $\chi_1 = \{(1, 3), (2, 3)\}$, $\chi_2 = \{(1, 4), (2, 4)\}$ and $\chi_3 = \{(1, 5), (2, 5)\}$.

An illustration of the target appears in Figure 3.4. The two relevant categories 1 and 2 are marked with dark ellipses. The three remaining categories (3, 4, 5) are marked with light ellipses. A directed edge from node 1 to node 3 designates that category 1 is ranked above category 3 in the target. There is a one-to-one correspondence between relation pairs in the target semi-order and edges in the induced graph. The four covers corresponding to the four loss functions are depicted in Figure 3.5.

The effect of the cover $\chi$ used may be crucial: a smaller cover size leads to a loss function which is only sensitive to the existence of a mistake and is indifferent to its exact nature. In contrast, a large cover with small elements describes a loss that takes the 'small details' into account. The natural question that arises is what cover should be used? The answer is domain and task dependent. For example, in a problem of optical character recognition, if the correct digit is not ranked at the top of the list, it apparently does not matter how many categories are ranked above it. In other words, we would like to know whether an error occurred or not. However, with respect to the problem of document categorization, where each document is associated with a subset of relevant categories, it seems more natural to ask how many categories are mismatched by the total-order. To highlight this observation, we slightly modify our definitions and denote an example by a *triplet* $(\mathbf{x}, \mathbf{y}, \chi)$: an instance $\mathbf{x}$, a target semi-order $\mathbf{y}$ and a cover $\chi$ of $\mathbf{y}$. Thus the choice of loss is part of the problem description and is not a task of the learning algorithm. Since the loss functions are parameterized by a cover $\chi$ of the target $\mathbf{y}$ we call them *cover loss functions*. We write the loss function of Equation (3.3) as

$$\mathbb{I}\,(\mathbf{W}; (\mathbf{x}, \mathbf{y}, \chi)) = \sum_{\chi \in \chi} \llbracket\{(r, s) \in \chi \;:\; (\mathbf{w}_r \cdot \mathbf{x}) \leq (\mathbf{w}_s \cdot \mathbf{x})\} \neq \emptyset \rrbracket. \qquad (3.4)$$

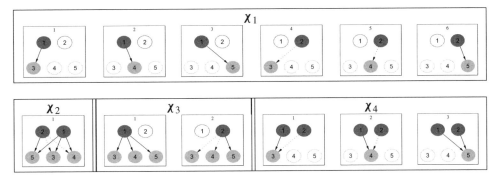

Figure 3.5 An illustration of covering loss computed for four different coverings. The target semi-order constitutes a set of six relation pairs, induced from a multi-class multi-label problem which ranks categories 1, 2 above categories 3, 4, 5. The prediction total-order is given by $\hat{\mathbf{y}} = \{(1, 3), (3, 4), (4, 2), (2, 5)\}$. The four covers $\chi_i$ for $i = 1, 2, 3, 4$ correspond to the four covers defined in the text, the index number of the title corresponds to the specific elements for the cover. Each plot corresponds to a specific element in a specific cover. A solid edge designates an order-pair in which the total-ordering $\hat{\mathbf{y}}$ is consistent, and a dotted edge designates a pair in which the total-order is not consistent with the semi-order. The value of the loss function is the number of plots which contain at least a single dotted edge. *Top:* The cover reflects the number of pairs-elements in the semi-order which the total-order is not consistent with (loss $= 2$). *Bottom Left:* The cover indicates whether the total-ordering is perfect or not (loss $= 1$, not perfect). *Bottom Center:* The cover reflects the number of relevant topics which are ranked below a relevant topic (loss $= 1$). *Bottom Right:* The cover reflects the number of non-relevant topics which are ranked above a relevant topic (loss $= 2$).

## 3.4 Hinge Loss Functions

The rich family of loss functions defined above suffers from similar drawbacks to the zero-one loss mainly used for binary classification. First, it does not take into consideration the absolute value of the inner product $(\mathbf{w}_r \cdot \mathbf{x})$ and only compares relations. Thus if two hypotheses make an error when predicting a specific example, but in one of them some quantities are larger than in the other, their loss is equal. The other drawback of the zero-one loss is its discontinuity in the parameters vector $\mathbf{W}$. Thus, when we cast the learning algorithms as optimization problems, the outcome is often hard to solve.

We thus combine this family of combinatorial loss functions together with the hinge loss (see e.g. Cristianini and Shawe-Taylor 2000; Schölkopf and Smola 2002). Focusing on the hypothesis class of linear prediction functions above, we assume $\hat{\mathbf{y}} = ((\mathbf{w}_1 \cdot \mathbf{x}), \ldots, (\mathbf{w}_k \cdot \mathbf{x}))$. A single summand of Equation (3.4) is bounded by

$$[\![\{(r, s) \in \chi \ : \ (s, r) \in \hat{\mathbf{y}}\} \neq \emptyset]\!] \leq \max_{(r,s) \in \chi} [1 - ((\mathbf{w}_r \cdot \mathbf{x}) - (\mathbf{w}_s \cdot \mathbf{x}))]_+.$$

Clearly, if $(r, s) \in \hat{\mathbf{y}}$ (which is equivalent to $(\mathbf{w}_r \cdot \mathbf{x}) > (\mathbf{w}_s \cdot \mathbf{x})$) then the left-hand side of the equation equals zero and the bound holds trivially. Otherwise, $(\mathbf{w}_r \cdot \mathbf{x}) \leq (\mathbf{w}_s \cdot \mathbf{x})$ and the left-hand side of the equation is 1. Furthermore, $((\mathbf{w}_r \cdot \mathbf{x}) - (\mathbf{w}_s \cdot \mathbf{x})) \leq 0$ and the right-hand side is at least 1. We thus bound the discrete loss of Equation (3.4) with the corresponding sum of hinge loss terms:

$$\mathbb{I}\,(\mathbf{W};\,(\mathbf{x},\,\mathbf{y},\,\chi)) \leq \mathbb{H}\,(\mathbf{W};\,(\mathbf{x},\,\mathbf{y},\,\chi)) = \sum_{\chi\in\chi}\ \max_{(r,s)\in\chi}\ [1 - ((\mathbf{w}_r \cdot \mathbf{x}) - (\mathbf{w}_s \cdot \mathbf{x}))]_+. \qquad (3.5)$$

Similar to cover loss functions we call the right-hand side of the above equation the *hinge cover loss* and it is denoted by $\mathbb{H}$. The two extreme cases need to be described in detail. On the one hand, if the cover $\chi$ contains a single element, then the sum in Equation (3.5) disappears and the loss becomes

$$\max_{(r,s)\in y}\ [1 - ((\mathbf{w}_r \cdot \mathbf{x}) - (\mathbf{w}_s \cdot \mathbf{x}))]_+,$$

which is larger than 1 if the prediction is not perfect. On the other hand, if the cover contains only singleton elements, then the max operator in Equation (3.5) vanishes and we obtain

$$\sum_{\chi\in y}[1 - ((\mathbf{w}_r \cdot \mathbf{x}) - (\mathbf{w}_s \cdot \mathbf{x}))]_+,$$

which bounds the number of mismatched pairs in the target semi-order. If a specific cover is not used, i.e. $\mathbb{H}\,(\mathbf{W};\,(\mathbf{x};\,\mathbf{y}))$ then it is assumed that the single cover $\chi = \{\mathbf{y}\}$ is used.

## 3.5 A Generalized Perceptron Algorithm

The Perceptron algorithm (Rosenblatt 1958) is a well known online algorithm for binary classification problems. The algorithm maintains a weight vector $\mathbf{w} \in \mathbb{R}^n$ used for prediction. Geometrically, the weight vector divides the space into two half spaces, and algebraically we compute the sign of the inner-product $(\mathbf{w} \cdot \mathbf{x})$. To motivate our algorithm let us now describe the Perceptron algorithm using the notation employed in this chapter.

To implement the Perceptron algorithm using a collection of two weight vectors $\mathbf{W}$ with one weight vector (prototype) per class, we set the first weight vector $\mathbf{w}_1$ to the hyperplane parameter $\mathbf{w}$ the Perceptron maintains and the weight vector $\mathbf{w}_2$ to $-\mathbf{w}$. We now modify $\mathbf{W} = (\mathbf{w}_1, \mathbf{w}_2)$ every time the algorithm misclassifies $\mathbf{x}$ as follows. If the correct label is 1 we replace $\mathbf{w}_1$ with $\mathbf{w}_1 + \mathbf{x}$ and $\mathbf{w}_2$ with $\mathbf{w}_2 - \mathbf{x}$. Similarly, we replace $\mathbf{w}_1$ with $\mathbf{w}_1 - \mathbf{x}$ and $\mathbf{w}_2$ with $\mathbf{w}_2 + \mathbf{x}$ when the correct label is 2 and $\mathbf{x}$ is misclassified. Thus, the weight vector $\mathbf{w}_y$ is moved toward the misclassified instance $\mathbf{x}$ while the other weight vector is moved away from $\mathbf{x}$. Note that this update implies that the total change to the two prototypes is zero. An illustration of this geometrical interpretation is given on the left-hand side of Figure 3.6. It is straightforward to verify that the algorithm is equivalent to the Perceptron algorithm.

The algorithm we now describe (and its corresponding analysis) employs a refined notion of mistake by examining all pairs of categories. Whenever $\mathbb{I}\,(\mathbf{W};\,(\mathbf{x},\,\mathbf{y},\,\chi)) > 0$ and the predicted ranking is not perfect, there must be at least one pair of categories $(r, s) \in \mathbf{y}$ whose ordering according to the predicted ranking disagrees with the feedback $\mathbf{y}$, i.e. category $r$ is ranked not-above category $s$ $((\mathbf{w}_r \cdot \mathbf{x}) \leq (\mathbf{w}_s \cdot \mathbf{x}))$ but $(r, s)$ belongs to the target label. We therefore define the *error-set* of $(\mathbf{W}, (\mathbf{x};\,\mathbf{y}))$ as the set of all pairs whose predicted ranking disagrees with the feedback. The error-set is formally defined as

$$E = \{(r, s) \in \mathbf{y} \ : \ (\mathbf{w}_r \cdot \mathbf{x}) \leq (\mathbf{w}_s \cdot \mathbf{x})\}. \qquad (3.6)$$

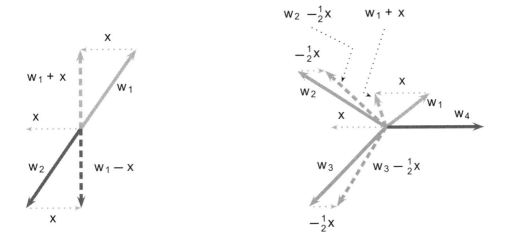

Figure 3.6 A geometrical illustration of the update for a binary problem (left) and a four-class multi-class single-label problem (right) using the Multicategorization Perceptron (MCP) algorithm.

Since our algorithm is a generalization of the Perceptron algorithm to multicategorization, we refer to it as the MCP algorithm, for Multicategorization Perceptron.

The error-set defined in Equation (3.6) plays a major role in updating the weight vectors. Generalizing the Perceptron's update which moves the (single) separating hyperplane toward the example $\mathbf{x}$, in our setting we would like to move each weight vector $\mathbf{w}_r$ toward $\mathbf{x}$ or away from it. However, there might be quite a few categories which are ranked correctly. These categories include all of the categories that are not involved with any inconsistency between the target semi-order and the predicted ranking. By definition, the indices of wrongly ordered pairs of categories constitute the error-set. For each pair $(r, s)$ of categories, we assign a weight denoted $\alpha_{r,s}$, constituting a $k \times k$ matrix. The value of $\alpha_{r,s}$ designates the effect of each categories-pair $(r, s)$ on the modification of the prediction function $\mathbf{W}$. We set the weights $\alpha_{r,s}$ to zero if neither the category-pair $(r, s)$ nor the category-pair $(s, r)$ are in $E$. For each element $\chi$ in the cover $\boldsymbol{\chi}$ which has at least one pair violated by the prediction (namely $\chi \cap E \neq \emptyset$), we set its value to be inversely proportional to the size of the intersection, that is:

$$
\alpha_{r,s} = \begin{cases} +\dfrac{1}{|\chi \cap E|} & (r, s) \in E \cap \chi \\[2mm] -\dfrac{1}{|\chi \cap E|} & (s, r) \in E \cap \chi \\[2mm] 0 & \text{otherwise.} \end{cases} \tag{3.7}
$$

By definition we have, $\sum_{(r,s) \in \mathbf{y}_i} \alpha_{r,s} = \mathbb{I}\,(\mathbf{W}; (\mathbf{x}; \mathbf{y}; \chi)) = |\{\chi \in \boldsymbol{\chi}_i \;:\; \chi \cap E \neq \emptyset\}|$.

We further define the amount by which the weight vectors are moved toward $\mathbf{x}$ or away from it. For each category indexed $r$ we define $\tau_r$ to be $\sum_s \alpha_{r,s}$ and add to $\mathbf{w}_r$ the input instance $\mathbf{x}$ scaled by $\tau_r$: $\mathbf{w}_r \leftarrow \mathbf{w}_r + \tau_r \mathbf{x}$. The sign of $\tau_r$ determines whether the $r$th weight vector is pushed towards $\mathbf{x}$ or away from it. Categories whose predicted rank is low are

**Initialize:** Set $\mathbf{W}_1 = 0$ : $\mathbf{w}_{1,1} = \mathbf{w}_{1,2} = \cdots = \mathbf{w}_{1,k} = 0$.

**Loop:** For $i = 1, 2, \ldots, m$

- Get a new instance: $\mathbf{x}_i \in \mathbb{R}^n$.
- Predict: $\hat{\mathbf{y}}_i = \left( \left( \mathbf{w}_{i,1} \cdot \mathbf{x}_i \right), \ldots, \left( \mathbf{w}_{i,k} \cdot \mathbf{x}_i \right) \right)$.
- Get a new target $\mathbf{y}_i$.
- Suffer loss: $\mathbb{L}_i = \mathbb{I} \left( \mathbf{W}_i; (\mathbf{x}_i; \mathbf{y}_i; \chi_i) \right)$.
- If $\mathbb{L}_i > 0$ :

     I. Set $E_i = \left\{ (r, s) \; : \; (r, s) \in \mathbf{y}_i, \; \left( \mathbf{w}_{i,r} \cdot \mathbf{x}_i \right) \le \left( \mathbf{w}_{i,s} \cdot \mathbf{x}_i \right) \right\}$.

     II. Form a parameter matrix $\alpha^i_{r,s} \; r, s = 1, \ldots, k$ as following:

$$
\alpha_{r,s} = \begin{cases} +\dfrac{1}{|\chi \cap E_i|} & (r, s) \in E_i \text{ and } (r, s) \in \chi \text{ for some } \chi \in \chi_i \\[2ex] -\dfrac{1}{|\chi \cap E_i|} & (s, r) \in E_i \text{ and } (s, r) \in \chi \text{ for some } \chi \in \chi_i \\[1ex] 0 & \text{otherwise.} \end{cases}
$$

     III. Set $\tau_{r,i} = \sum_{s=1}^{k} \alpha^i_{r,s}$ for $r = 1, \ldots, k$.

     IV. Update: $\mathbf{w}_{i+1,r} = \mathbf{w}_{i,r} + \tau_{r,i} \mathbf{x}_i$ for $r = 1, \ldots, k$.

**Output :** $h(\mathbf{x}) = \left( \left( \mathbf{w}_{m+1,r} \cdot \mathbf{x} \right), \ldots, \left( \mathbf{w}_{m+1,k} \cdot \mathbf{x} \right) \right)$.

Figure 3.7 Generalized Perceptron algorithm for multicategorization.

likely to be moved more aggressively toward $\mathbf{x}$, if they should be ranked above many other categories in accordance with the semi-order $\mathbf{y}$. Similarly, categories which should be ranked below many other categories in accordance with the semi-order $\mathbf{y}$ are pushed away from $\mathbf{x}$ in different proportions, depending on how high they are ranked. In other words, if $\alpha_{r,s} > 0$ then the inner product between $\mathbf{x}$ and $\mathbf{w}_r$ should be increased and the inner product between $\mathbf{x}$ and $\mathbf{w}_s$ should be decreased, and *vice versa* for the case $\alpha_{r,s} < 0$. The single coefficient $\tau_r$ balances these two opposite requirements for categories $r$, which should be ranked both above some categories $s$ and below some other categories $s$. The pseudo-code describing the family of category algorithms for bounded ranking-losses is given in Figure 3.7.

To further illustrate the update, let us look at a specific realization of the values $\alpha_{r,s}$. Concretely, let us assume that all the values are equal. We can therefore assume that $\alpha_{r,s} = 1$ for $(r, s) \in E$ and, at the end, scale all the values of $\tau_r$ by a constant. In this case, the amount by which we move each prototype directly depends on its location in the ranked list. The value $\tau_r$ of a category is equal to the difference between the number of categories which should be ranked above it, but they are not, and the number of categories which should be ranked below it, but they are not. Formally, the value of $\tau_r$ for this case is given by $|\{s \; : \; (r, s) \in E\}| - |\{s \; : \; (s, r) \in E\}|$.

A geometric illustration of this update for a multi-class single-labeled problem is given in the right-hand side of Figure 3.6. There are four classes in the example appearing in the figure. The target semi-order is $\mathbf{y} = \{(1, 2), (1, 3), (1, 4)\}$ and since $\mathbf{w}_1$ is not the most similar vector to $\mathbf{x}$, it is moved toward $\mathbf{x}$. The weight vectors $\mathbf{w}_2$ and $\mathbf{w}_3$ are also modified

**Parameter:**

- Clipping : $C$

**Initialize:** Set $\mathbf{W}_1 = 0$.
**Loop:** For $i = 1, 2, \ldots, m$

- Get a new instance: $\mathbf{x}_i \in \mathbb{R}^n$.
- Predict: $\hat{\mathbf{y}}_i = \left( \left( \mathbf{w}_{i,1} \cdot \mathbf{x}_i \right), \ldots, \left( \mathbf{w}_{i,k} \cdot \mathbf{x}_i \right) \right)$.
- Get a new target $\mathbf{y}_i$ and a new cover $\chi_i$ of it.
- Suffer loss: $\mathbb{L}_i = \mathbb{I}\left( \mathbf{W}_i; (\mathbf{x}_i; \mathbf{y}_i; \chi_i) \right)$.
- Set $\mathbf{W}_{i+1}$ to be the solution $\mathbf{W} = (\mathbf{w}_1, \ldots, \mathbf{w}_k)$ of the following optimization problem (Equation (3.9)) :
$$\min_{\mathbf{W}} \tfrac{1}{2} \|\mathbf{W} - \mathbf{W}_i\|_2^2 + C \mathbb{H}\left(\mathbf{W}; (\mathbf{x}_i; \mathbf{y}_i)\right).$$

**Output :** $h(\mathbf{x}) = \left( \left( \mathbf{w}_{m+1,r} \cdot \mathbf{x} \right), \ldots, \left( \mathbf{w}_{m+1,k} \cdot \mathbf{x} \right) \right)$.

**Figure 3.8** The Passive–Aggressive algorithm for multicategorization problems (primal formulation).

by subtracting $\mathbf{x}/2$ from each one. The last weight vector $\mathbf{w}_4$ is not in the error-set since $(\mathbf{w}_1 \cdot \mathbf{x}) > (\mathbf{w}_4 \cdot \mathbf{x})$ and therefore it is not modified.

The following theorem provides a bound of the cumulative ranking loss for the specific choice of loss functions and the parameter matrix $\alpha_{r,s}$ we described above.

**Theorem 3.5.1** *Let* $(\mathbf{x}_1, \mathbf{y}_1), \ldots, (\mathbf{x}_m, \mathbf{y}_m)$ *be an input sequence for the family of algorithms described in Figure 3.7, where* $\mathbf{x}_i \in \mathbb{R}^n$ *and* $\mathbf{y}_i \in \mathcal{Y} \times \mathcal{Y}$ *is a semi-order. Set* $R = \max_i \|\mathbf{x}_i\|$. *Let* $\mathbf{W}^* \in \mathbb{R}^{n \times k}$ *be any collection of k vectors. Assume that the loss function of the algorithm is of the form described in Equation (3.3), with cover parameters* $\chi_i$, *and its value is upper bounded,* $\mathbb{I}\left(\mathbf{W}_i; (\mathbf{x}_i, \mathbf{y}_i; \chi_i)\right) \leq A$. *Then the cumulative loss the algorithm suffers is upper bounded by*

$$\sum_{i=1}^{m} \mathbb{I}\left(\mathbf{W}_i; (\mathbf{x}_i, \mathbf{y}_i, \chi_i)\right) \leq 2 \min\{k-1, A\} R^2 \|\mathbf{W}^*\|^2 + 2 \sum_i \mathbb{H}\left(\mathbf{W}^*; (\mathbf{x}_i, \mathbf{y}_i, \chi_i)\right).$$

The theorem compares the total number of mistakes the algorithm makes through a run over a specific-given input sequence with the performance of a *fixed* prediction function parameterized by $\mathbf{W}^*$. Specifically, it bounds the cumulative loss of the algorithm with a constant that only depends on global properties of the sample and covers, and the cumulative *hinge* loss of the fixed prediction-function $\mathbf{W}^*$ over the same sequence. For each example $i$, both the algorithm of Figure 3.7 and the fixed prediction-function are evaluated using the same cover $\chi_i$.

## 3.6 A Generalized Passive–Aggressive Algorithm

We now turn to describe a Passive–Aggressive (PA) algorithm designed for the problem of multicategorization which includes as a special case, the binary PA algorithm

(Crammer *et al.* 2006). We construct the algorithm in a way similar to transforming the Perceptron algorithm into the MCP algorithms.

The general method we use for deriving our online update rule is similar to the PA algorithm (Crammer *et al.* 2006) and thus we refer to it as the MCPA algorithm, for Multicategorization Passive–Aggressive algorithm. We define the new collection of $k$ weight vectors $\mathbf{W}_{i+1}$ as the solution to the following optimization problem

$$\min_{\mathbf{W}} \ \frac{1}{2}\|\mathbf{W} - \mathbf{W}_i\|_2^2 + C\mathbb{H}\left(\mathbf{W}; (\mathbf{x}_i; y_i; \chi_i)\right), \tag{3.8}$$

where $C > 0$ is a tradeoff parameter. As $C \to 0$ the algorithm focuses on finding prediction functions $\mathbf{W}$ close to the current prediction function $\mathbf{W}_i$, even though it might suffer a large hinge loss. On the other hand, when $C \to \infty$ the algorithm focuses on a prediction function which attains small (or even zero) hinge loss, although the prediction function may be extremely different from the current prediction function.

Writing the optimization problem of Equation (3.8) explicitly we obtain

$$\min_{\mathbf{W}} \ \frac{1}{2}\|\mathbf{W} - \mathbf{W}_i\|_2^2 + C \sum_{\chi \in \chi_i} \xi_\chi \tag{3.9}$$

$$\text{subject to: } \forall \chi \in \chi_i, \ \forall (r,s) \in \chi \ : \ (\mathbf{w}_r \cdot \mathbf{x}_i) - (\mathbf{w}_s \cdot \mathbf{x}_i) \geq 1 - \xi_\chi$$

$$\forall \chi \in \chi_i \ : \ \xi_\chi \geq 0,$$

where $\xi_\chi \geq 0$ are slack variables. The pseudo-code describing the online passive–aggressive approach for multicategory problems is given in Figure 3.8. In the following section we re-write the algorithm in terms of the dual variables, which will allow us to extend into nonlinear prediction function via Mercer kernels (Aizerman *et al.* 1964; Mercer 1909).

### 3.6.1  Dual Formulation

We now compute the dual optimization problem for two reasons, both involving the design of the algorithm. First, as in many such algorithms, the dual optimization problem depends only on inner-products and thus a kernel operator can be directly employed. Second, the primal optimization problem has $n \times k$ variables ($\mathbf{w}_1 \ldots \mathbf{w}_k$) which can be quite large if we use many features (large value of $n \gg k$). On the other hand, the dual has at most $k \times k$ variables, which can be significantly small since $k \ll n$.

We use Lagrange theory to obtain the following dual (the details of derivation are deferred to Section 3.9.1):

$$\min_{\{\alpha_{r,s}^i\}} \ \frac{1}{2}\|\mathbf{x}_i\|^2 \sum_{r=1}^{k}\left(\sum_s \alpha_{r,s}^i\right)^2 + \sum_{(r,s) \in y_i} \alpha_{r,s}^i\left((\mathbf{w}_{i,r} \cdot \mathbf{x}_i) - (\mathbf{w}_{i,s} \cdot \mathbf{x}_i) - 1\right)$$

$$\text{subject to: } \begin{cases} \alpha_{r,s}^i \geq 0 & (r,s) \in y_i \\ \alpha_{s,r}^i = -\alpha_{r,s}^i & (r,s) \in y_i \\ \alpha_{s,r}^i = 0 & \text{otherwise} \end{cases}$$

$$\forall \chi \in \chi_i \ : \ \sum_{(r,s) \in \chi} \alpha_{r,s}^i \leq C. \tag{3.10}$$

**Parameter:**

- Kernel function $K(\mathbf{x}_1, \mathbf{x}_2)$
- Clipping: $C$

**Initialize:** Set $\tau_{r,i} = 0$ for $r = 1, \ldots, k, \ i = 1, \ldots, m$.
**Loop:** For $i = 1, 2, \ldots, m$

- Get a new instance: $\mathbf{x}_i \in \mathbb{R}^n$.
- Predict:

$$\hat{\mathbf{y}}_i = \left( \cdots \sum_{j=1}^{i-1} \tau_{j,r} K(\mathbf{x}_i, \mathbf{x}_j) \cdots \right).$$

- Get a new target $\mathbf{y}_i$ and a new cover $\boldsymbol{\chi}_i$ of it.
- Suffer loss: $\mathbb{L}_i = \mathbb{I}(\mathbf{W}_i; (\mathbf{x}_i; \mathbf{y}_i; \boldsymbol{\chi}_i))$.
- Set $\{\alpha_{r,s}^i\}$ to be the solution of the optimization problem of Equation (3.10).
- Set $\tau_{i,r} = \sum_s \alpha_{r,s}^i$ for $r = 1, \ldots, k$.

**Output:** $h(\mathbf{x}) = \left( \cdots \sum_{j=1}^{m} \tau_{j,r} K(\mathbf{x}, \mathbf{x}_j) \cdots \right).$

Figure 3.9 The Passive–Aggressive algorithm for multicategorization problems (dual formulation).

Note how the specific cover loss used in the objective of Equation (3.9) is translated into a set of inequality constraints, $\sum_{(r,s) \in \chi} \alpha_{r,s}^i \leq C$ for all $\chi \in \boldsymbol{\chi}_i$. Each cover element is translated into a single inequality. The pseudo-code describing the dual MCPA appears in Figure 3.9. A performance bound for the PA algorithm (either in its primal form or its dual form), similar to Theorem 3.5.1, can be derived. We refer the reader to Section 3.8 for more information.

The dual version of MCPA (Figure 3.9) algorithm is tightly related to the MCP (Figure 3.7) presented in the previous section. In fact both algorithms perform exactly the same steps, using the same matrix parameters $(\alpha_{r,s}^i)$ and differ only in the way these parameters are updated. MCPs set their value using a closed form set of equations (Equation (3.7)), while the MCPA set its values to be a solution of a quadratic program (QP), either in the primal parameters or their dual.

## 3.7 A Batch Formulation

The passive–aggressive framework is tightly related to SVM. In both cases we use Euclidean norm for regularization and hinge loss to evaluate the performance. One possible view is that the PA algorithms approximate a solution for the QP of SVMs. Below, we present a QP for our problem of multicategorization which can be solved using any QP solver. Given training data $\{(\mathbf{x}_i, \mathbf{y}_i)\}_1^m$ and a choice of cover $\boldsymbol{\chi}_i$ for each training instance, the primal objective is a sum of Euclidean norm of the parameters and the hinge loss over the sample given by

$$\min_{\mathbf{W}} \ \frac{1}{2} \|\mathbf{W}\|_2^2 + C \sum_i \mathbb{H}(\mathbf{W}; (\mathbf{x}_i; \mathbf{y}_i; \boldsymbol{\chi}_i)). \tag{3.11}$$

Writing the optimization problem explicitly we obtain

$$\min_{\mathbf{W},\{\{\xi_{\chi_i}\}_{\chi_i \in \chi_i}\}_i} \frac{1}{2}\|\mathbf{W}\|_2^2 + C \sum_i \sum_{\chi_i \in \chi_i} \xi_{\chi_i} \tag{3.12}$$

$$\text{subject to: } \forall i, \ \forall \chi_i \in \chi_i, \ \forall(r,s) \in \chi_i \ : \quad (\mathbf{w}_r \cdot \mathbf{x}_i) - (\mathbf{w}_s \cdot \mathbf{x}_i) \geq 1 - \xi_{\chi_i}$$

$$\forall i, \ \forall \chi_i \in \chi_i \ : \quad \xi_{\chi_i} \geq 0,$$

where $\xi_{\chi_i} \geq 0$ are slack variables. We note in passing the two well-known versions of multi-class (single-label) SVMs (Crammer and Singer 2001; Guermeur *et al.* 2000; Weston and Watkins 1999) are a special case of our formulation. First, the label relation $\mathbf{y}$ always contains $k-1$ elements of the form $\mathbf{y} = \{(y,r)\}_{r \neq y}$ for a true label $y \in C$. Second, the cover $\chi$ is either composed of singletons (Guermeur *et al.* 2000; Weston and Watkins 1999) or it equals the label relation $\chi = \mathbf{y}$ (Crammer and Singer 2001).

As before, in order to apply Mercer kernels we compute the dual using Lagrange theory (details appear in Section 3.9.2),

$$\max_{\{\alpha_{r,s}^i\}} -\frac{1}{2} \sum_p \sum_{i,j} \sum_{r,s} \alpha_{p,r}^i \alpha_{p,s}^j \left(\mathbf{x}_i \cdot \mathbf{x}_j\right) + \sum_i \sum_{(r,s) \in \mathbf{y}_i} \alpha_{r,s}^i$$

$$\text{subject to: } \forall i \begin{cases} \alpha_{r,s}^i \geq 0 & (r,s) \in \mathbf{y}_i \\ \alpha_{s,r}^i = -\alpha_{r,s}^i & (r,s) \in \mathbf{y}_i \\ \alpha_{s,r}^i = 0 & \text{otherwise} \end{cases}$$

$$\forall i \forall \chi \in \chi_i \ : \quad \sum_{(r,s) \in \chi} \alpha_{r,s}^i \leq C. \tag{3.13}$$

Two comments are in order. First, when using the dual formulation in practice, the number of parameters can be trivially reduced using the constraints on $\alpha_{r,s}^i$ in Equation (3.13), specifically setting $\alpha_{r,s}^i = 0$ if both $(r,s), (s,r) \notin \mathbf{y}_i$ and $\alpha_{r,s}^i = -\alpha_{s,r}^i$ if $(s,r) \notin \mathbf{y}_i$. Second, the dual of the batch formulation (Equation (3.13) can be reduced to the dual of the online formulation (Equation (3.10)) by choosing some $i$, fixing the value of $\alpha_{r,s}^j$ for $j < i$ and setting $\alpha_{r,s}^j = 0$ for $j > i$. Formally, we factor as follows:

$$\mathbf{w}_r = \sum_{j<i} \sum_s \alpha_{r,s}^j \mathbf{x}_j + \sum_s \alpha_{r,s}^i \mathbf{x}_i$$

$$= \sum_{j>i} \sum_s \alpha_{r,s}^j \mathbf{x}_j$$

$$= \mathbf{w}_{i,r} + \sum_s \alpha_{r,s}^i \mathbf{x}_i,$$

where $\mathbf{w}_{i,r} = \sum_{j<i} \sum_s \alpha_{r,s}^j \mathbf{x}_j$ is fixed and depends only on the examples $\mathbf{x}_j$ for $j < i$. In this respect, the online algorithm of Figure 3.9 *approximates* the optimal solution of the batch optimization problem by iterating over the example index $i$, and on each iteration optimizing for $\{\alpha_{r,s}^i\}_{(r,s)}$.

In general one can expect the performance of the batch formulation to be superior to the online algorithm presented in the previous section. This superiority comes with a price of efficiency. The QP of Equation (3.13) has $\sum_i |\mathbf{y}_i| = \mathcal{O}(mk^2)$ variables, all constrained to be non-negative, and $\sum_i |\mathbf{\chi}_i|$ additional inequality constraints. The online formulation (Figure 3.9) is more efficient as it breaks the large problem into a sequence of smaller QPs, each with $|\mathbf{y}_i|$ non-negative variables and $|\mathbf{\chi}_i|$ constraints. Since the cost of solving a QP is super-linear, then solving a sequence of small QPs will be more efficient. A similar tradeoff between performance and efficiency holds between the two online algorithms. The performance of the MCP algorithm (Figure 3.7) is inferior to performance of the MCPA algorithm (Figure 3.9). However, the latter solves a QP on each iteration, while the MCP set its parameters using a closed form rule.

## 3.8 Concluding Remarks

In each of the sections above a choice was made from a few possible alternatives. We now sketch a few of the alternatives. The chapter investigated algorithms for multi-category problems, where some predictions or mistakes are preferable over others. It generalizes binary-prediction problems (see Chapter 1 and Cortes and Vapnik 1995; Cristianini and Shawe-Taylor 2000; Rosenblatt 1958; Schölkopf and Smola 2002) and its standard generalization of multi-class problems, where there are more than two possible labels, but no mistake is preferable over another (Crammer and Singer 2001, 2003). A common approach for such problems is to reduce a single multi-class problem into few binary-classification problems, and then to combine the output of the binary classifiers into a single multi-class prediction. For example, the one-vs-rest reduction builds $k$ binary classifiers, one per class, each of which is trained to predict if the label of a given input is a specific class $r$ or not (e.g. Cortes and Vapnik 1995). Other reductions include the one-vs.-one (Furnkranz 2002) and error correcting output codes (ECOC) (Allwein *et al.* 2000; Crammer and Singer 2002; Dietterich and Bakiri 1991, 1995).

A well investigated approach for complex problems such as automatic speech recognition is to predict the phonemes associated with an entire acoustic model simultaneously, rather than predicting the phoneme of each frame independently. Such approaches often ignore the fact that some mistakes are preferable over others, yet are modeling correlation between temporal-consecutive frames. Two notable examples are Hidden Markov Models (Rabiner and Juang 1993) or Conditional Random Fields (Lafferty *et al.* 2001). See Chapter 5 and Chapter 10 for more details.

Semi-orders and cover loss functions provide a flexible method to state what predictions are preferable to others and how to evaluate various mistakes. On some occasions, the richness of these loss functions is not enough, and other evaluation measures are used (for example the Levenshtein distance). Our algorithms can be extended to incorporate these loss functions at the price of weaker analysis. See Crammer (2004) for more details.

We presented two online algorithms and one batch algorithm, all of which have counterparts designed for binary classification. Specifically, as noted above, MCP generalizes the Perceptron algorithm (Minsky and Papert 1969; Rosenblatt 1958) and MCPA generalizes the PA algorithms (Crammer *et al.* 2006). Specific cases of our general approach include extensions of the Perceptron algorithms for multi-class (Crammer and Singer 2003) and a more detailed treatment can be found elsewhere (Crammer 2004).

One of the first reductions of the Perceptron algorithm to multi-class single-labeled problems is provided in the widely read and cited book of Duda and Hart (1973). The multi-class version in the book is called Kesler's construction. Kesler's construction is attributed to Carl Kesler and was described by Nilsson (1965). Kesler construction was further generalized to category ranking problems by Har-Peled *et al.* (2002).

There are a few extensions of SVM (see Cristianini and Shawe-Taylor 2000; Schölkopf and Smola 2002, and the references therein) for multi-class problems (Crammer and Singer 2001; Guermeur *et al.* 2000; Hsu and Lin 2002; Vapnik 1998; Weston and Watkins 1999). They differ from each other in the way the various labels are combined. Other extensions for multi-class multi-labeled problems also exist (Elisseeff and Weston 2001).

## 3.9 Appendix. Derivations of the Duals of the Passive–Aggressive Algorithm and the Batch Formulation

### 3.9.1 Derivation of the Dual of the Passive–Aggressive Algorithm

The corresponding Lagrangian of the optimization problem in Equation (3.9) is

$$
\mathcal{L}(\mathbf{W}; \alpha) = \frac{1}{2} \sum_{r=1}^{k} \left\| \mathbf{w}_r - \mathbf{w}_{i,r} \right\|^2 + C \sum_{\chi \in \chi_i} \xi_\chi - \sum_{\chi \in \chi_i} \beta_\chi \xi_\chi
$$
$$
+ \sum_{\chi \in \chi_i} \sum_{(r,s) \in \chi} \alpha_{r,s}^i \left( 1 - \xi_\chi - (\mathbf{w}_r \cdot \mathbf{x}_i) + (\mathbf{w}_s \cdot \mathbf{x}_i) \right), \tag{3.14}
$$

where $\alpha_{r,s}^i \geq 0$ (for $(r, s) \in \mathbf{y}$) are Lagrange multipliers and we used the notation that $\mathbf{W}$ is written as a concatenation of the $k$ weight vectors. Furthermore, $\beta_\chi$ are Lagrange multipliers associated with the inequality constraints $\xi_\chi \geq 0$.

To find a saddle point of $\mathcal{L}$ we first differentiate $\mathcal{L}$ with respect to $\xi_\chi$ (for $\chi \in \chi_i$) and set it to zero:

$$
\nabla_{\xi_\chi} \mathcal{L} = C - \beta_\xi - \sum_{(r,s) \in \chi} \alpha_{r,s}^i. \tag{3.15}
$$

Setting the derivative to zero we get

$$
\sum_{(r,s) \in \chi} \alpha_{r,s}^i = C - \beta_\chi \leq C \quad \forall \chi \in \chi_i, \tag{3.16}
$$

where the inequality holds since $\beta_\chi \geq 0$.

We next differentiate $\mathcal{L}$ with respect to $\mathbf{w}_p$ (for $p = 1, \ldots, k$):

$$
\nabla_{\mathbf{w}_p} \mathcal{L} = \mathbf{w}_p - \mathbf{w}_{i,p} - \sum_{s \,:\, (p,s) \in \mathbf{y}_i} \alpha_{p,s}^i \mathbf{x}_i + \sum_{r \,:\, (r,p) \in \mathbf{y}_i} \alpha_{r,p}^i \mathbf{x}_i = 0
$$
$$
\Rightarrow \mathbf{w}_p = \mathbf{w}_{i,p} + \sum_{s \,:\, (p,s) \in \mathbf{y}_i} \alpha_{p,s}^i \mathbf{x}_i - \sum_{r \,:\, (r,p) \in \mathbf{y}_i} \alpha_{r,p}^i \mathbf{x}_i. \tag{3.17}
$$

The Lagrange coefficients $\alpha_{r,s}^i$ are defined only for $(r, s) \in \mathbf{y}_i$. To simplify Equation (3.17) we define a matrix $\alpha$ of coefficients which extends the definition of $\alpha_{r,s}^i$ for all $r, s = 1, \ldots, k$.

The value of the elements indexed by $(r, s) \in \mathbf{y}_i$ is already defined by the optimization problem. We define the matrix to be anti-symmetric, that is $\alpha^i_{s,r} = -\alpha^i_{r,s}$, which automatically sets the values of the diagonal to zero, $\alpha^i_{r,r} = 0$ and the values of the elements $(s, r)$ for $(r, s) \in \mathbf{y}_i$. We set all the other elements to zero. This definition agrees with the definition of the matrix $\alpha$ in Section 3.5.

Under this extension of the Lagrange multipliers Equation (3.17) is rewritten as

$$\mathbf{w}_r = \mathbf{w}_{i,r} + \sum_s \alpha^i_{r,s} \mathbf{x}_i \overset{\text{def}}{=} \tau_r \mathbf{x}_i. \tag{3.18}$$

This presentation of the update step is of the same form used for the MCP given in Section 3.5. Substituting Equation (3.16) and Equation (3.18) back in the Lagrangian of Equation (3.14) we get

$$
\begin{aligned}
\mathcal{L} = {} & \frac{1}{2} \sum_{r=1}^{k} \left\| \sum_s \alpha^i_{r,s} \mathbf{x}_i \right\|^2 + C \sum_{\chi \in \chi_i} \xi_\chi - \sum_{\chi \in \chi_i} \beta_\chi \xi_\chi \\
& + \sum_{\chi \in \chi_i} \sum_{(r,s) \in \chi} \alpha^i_{r,s} \left( 1 - \xi_\chi - \left( \mathbf{w}_{i,r} + \sum_p \alpha^i_{r,p} \mathbf{x}_i \cdot \mathbf{x}_i \right) + \left( \mathbf{w}_{i,s} + \sum_p \alpha^i_{s,p} \mathbf{x}_i \cdot \mathbf{x}_i \right) \right) \\
= {} & \frac{1}{2} \sum_{r=1}^{k} \left\| \sum_s \alpha^i_{r,s} \mathbf{x}_i \right\|^2 + C \sum_{\chi \in \chi_i} \xi_\chi - \sum_{\chi \in \chi_i} \beta_\chi \xi_\chi \\
& + \sum_{(r,s) \in \mathbf{y}_i} \alpha^i_{r,s} \left( 1 - (\mathbf{w}_{i,r} \cdot \mathbf{x}_i) + (\mathbf{w}_{i,s} \cdot \mathbf{x}_i) \right) - \sum_{\chi \in \chi_i} \xi_\chi \sum_{(r,s) \in \chi} \alpha^i_{r,s} \\
& + \sum_{(r,s) \in \mathbf{y}_i} \alpha^i_{r,s} \left( - \sum_p \alpha^i_{r,p} (\mathbf{x}_i \cdot \mathbf{x}_i) + \sum_p \alpha^i_{s,p} (\mathbf{x}_i \cdot \mathbf{x}_i) \right) \\
= {} & \frac{1}{2} \|\mathbf{x}_i\|^2 \sum_{r=1}^{k} \left( \sum_s \alpha^i_{r,s} \right)^2 + \sum_{(r,s) \in \mathbf{y}_i} \alpha^i_{r,s} \left( 1 - (\mathbf{w}_{i,r} \cdot \mathbf{x}_i) + (\mathbf{w}_{i,s} \cdot \mathbf{x}_i) \right) \\
& + \|\mathbf{x}_i\|^2 \sum_{(r,s) \in \mathbf{y}_i} \alpha^i_{r,s} \sum_p (-\alpha^i_{r,p} + \alpha^i_{s,p}) + \sum_{\chi \in \chi_i} \xi_\chi \left( C - \beta_\chi - \sum_{(r,s) \in \chi} \alpha^i_{r,s} \right). \tag{3.19}
\end{aligned}
$$

The last term equals zero because of Equation (3.15) and thus

$$
\begin{aligned}
\mathcal{L} = {} & \frac{1}{2} \|\mathbf{x}_i\|^2 \sum_{r=1}^{k} \left( \sum_s \alpha^i_{r,s} \right)^2 + \sum_{(r,s) \in \mathbf{y}_i} \alpha^i_{r,s} \left( 1 - (\mathbf{w}_{i,r} \cdot \mathbf{x}_i) + (\mathbf{w}_{i,s} \cdot \mathbf{x}_i) \right) \\
& + \|\mathbf{x}_i\|^2 \sum_{(r,s) \in \mathbf{y}_i} \alpha^i_{r,s} \sum_p (-\alpha^i_{r,p} + \alpha^i_{s,p}). \tag{3.20}
\end{aligned}
$$

We now further derive the last term. Using the asymmetry property used to define the matrix $\alpha$ we get, $\alpha^i_{r,s}(-\alpha^i_{r,p} + \alpha^i_{s,p}) = \alpha^i_{s,r}(-\alpha^i_{s,p} + \alpha^i_{r,p})$. Substituting in the last term of

Equation (3.20), together with the fact that $\alpha^i_{r,s} = 0$ for $(r, s)$, $(s, r) \notin \mathbf{y}_i$ we obtain

$$\sum_{(r,s)\in\mathbf{y}_i} \alpha^i_{r,s} \sum_p (-\alpha^i_{r,p} + \alpha^i_{s,p})$$

$$= \frac{1}{2}\left( \sum_{(r,s)\in\mathbf{y}_i} \alpha^i_{r,s} \sum_p (-\alpha^i_{r,p} + \alpha^i_{s,p}) + \sum_{(s,r)\in\mathbf{y}_i} \alpha^i_{r,s} \sum_p (-\alpha^i_{r,p} + \alpha^i_{s,p}) \right)$$

$$= -\frac{1}{2}\sum_{r,s,p} \alpha^i_{r,s}\alpha^i_{r,p} + \frac{1}{2}\sum_{r,s,p} \alpha^i_{r,s}\alpha^i_{s,p}$$

$$= -\frac{1}{2}\sum_{r,s,p} \alpha^i_{r,s}\alpha^i_{r,p} - \frac{1}{2}\sum_{r,s,p} \alpha^i_{s,r}\alpha^i_{s,p}$$

$$= -\sum_{r,s,p} \alpha^i_{r,s}\alpha^i_{r,p}, \tag{3.21}$$

where the last equality is due to change of names of the indices, and the equality preceding it holds by using the asymmetry of the matrix $\alpha$. Substituting Equation (3.21) back in Equation (3.20) we obtain

$$\mathcal{L} = \frac{1}{2}\|\mathbf{x}_i\|^2 \sum_{r=1}^k \left(\sum_s \alpha^i_{r,s}\right)^2 + \sum_{(r,s)\in\mathbf{y}_i} \alpha^i_{r,s}\left(1 - (\mathbf{w}_{i,r}\cdot\mathbf{x}_i) + (\mathbf{w}_{i,s}\cdot\mathbf{x}_i)\right)$$

$$+ \frac{1}{2}\|\mathbf{x}_i\|^2 \sum_{r,s} \alpha^i_{r,s} \sum_p (-\alpha^i_{r,p} + \alpha^i_{s,p})$$

$$= -\frac{1}{2}\|\mathbf{x}_i\|^2 \sum_{r=1}^k \left(\sum_s \alpha^i_{r,s}\right)^2 + \sum_{(r,s)\in\mathbf{y}_i} \alpha^i_{r,s}\left(1 - (\mathbf{w}_{i,r}\cdot\mathbf{x}_i) + (\mathbf{w}_{i,s}\cdot\mathbf{x}_i)\right).$$

To conclude, the dual is given by

$$\min_{\{\alpha^i_{r,s}\}} \frac{1}{2}\|\mathbf{x}_i\|^2 \sum_{r=1}^k \left(\sum_s \alpha^i_{r,s}\right)^2 + \sum_{(r,s)\in\mathbf{y}_i} \alpha^i_{r,s}\left((\mathbf{w}_{i,r}\cdot\mathbf{x}_i) - (\mathbf{w}_{i,s}\cdot\mathbf{x}_i) - 1\right)$$

$$\text{subject to: } \begin{cases} \alpha^i_{r,s} \geq 0 & (r,s)\in\mathbf{y}_i \\ \alpha^i_{s,r} = -\alpha^i_{r,s} & (r,s)\in\mathbf{y}_i \\ \alpha^i_{s,r} = 0 & \text{otherwise} \end{cases}$$

$$\forall \chi \in \chi_i : \sum_{(r,s)\in\chi} \alpha^i_{r,s} \leq C. \tag{3.22}$$

### 3.9.2 Derivation of the Dual of the Batch Formulation

We first write the Lagrangian of Equation (3.12):

$$\mathcal{L}(\mathbf{W}; \alpha) = \frac{1}{2}\sum_{r=1}^k \|\mathbf{w}_r\|^2 + C\sum_i \sum_{\chi_i\in\chi_i} \xi^i_{\chi_i} - \sum_i \sum_{\chi_i\in\chi_i} \beta^i_{\chi_i}\xi^i_{\chi_i} \tag{3.23}$$

$$+ \sum_i \sum_{\chi_i\in\chi_i} \sum_{(r,s)\in\chi_i} \alpha^i_{r,s}(1 - \xi^i_{\chi_i} - (\mathbf{w}_r\cdot\mathbf{x}_i) + (\mathbf{w}_s\cdot\mathbf{x}_i)),$$

To find a saddle point of $\mathcal{L}$ we first differentiate $\mathcal{L}$ with respect to $\xi^i_{\chi_i}$ (for $i = 1, \ldots, m$ and $\chi_i \in \boldsymbol{\chi}_i$) and set it to zero:

$$\nabla_{\xi^i_{\chi_i}} \mathcal{L} = C - \beta^i_{\chi_i} - \sum_{(r,s) \in \chi_i} \alpha^i_{r,s}. \tag{3.24}$$

Setting the derivative to zero we get

$$\sum_{(r,s) \in \chi_i} \alpha^i_{r,s} = C - \beta^i_{\chi_i} \leq C \quad \forall \chi_i \in \boldsymbol{\chi}_i, \tag{3.25}$$

where the inequality holds since $\beta^i_{\chi_i} \geq 0$.

We next differentiate $\mathcal{L}$ with respect to $\mathbf{w}_p$ (for $p = 1, \ldots, k$):

$$\nabla_{\mathbf{w}_p} \mathcal{L} = \mathbf{w}_p - \sum_i \sum_{s \,:\, (p,s) \in \mathbf{y}_i} \alpha^i_{p,s} \mathbf{x}_i + \sum_i \sum_{r \,:\, (r,p) \in \mathbf{y}_i} \alpha^i_{r,p} \mathbf{x}_i = 0$$

$$\Rightarrow \mathbf{w}_p = \sum_i \sum_{s \,:\, (p,s) \in \mathbf{y}_i} \alpha^i_{p,s} \mathbf{x}_i - \sum_i \sum_{r \,:\, (r,p) \in \mathbf{y}_i} \alpha^i_{r,p} \mathbf{x}_i. \tag{3.26}$$

As before, the Lagrange coefficients $\alpha^i_{r,s}$ are defined only for $(r,s) \in \mathbf{y}_i$. To simplify Equation (3.26) we define a set of matrices of coefficients $\alpha^i$, one per example, which extends the definition of $\alpha^i_{r,s}$ for all $r, s = 1, \ldots, k$ similarly to the extension defined in the PA algorithms. Under this extension of the Lagrange multipliers Equation (3.26) is rewritten as

$$\mathbf{w}_r = \sum_i \sum_s \alpha^i_{r,s} \mathbf{x}_i \overset{\text{def}}{=} \sum_i \tau_{r,i} \mathbf{x}_i. \tag{3.27}$$

Substituting Equation (3.27) back in the Lagrangian of Equation (3.23) we get

$$\mathcal{L} = \frac{1}{2} \sum_{p=1}^k \left\| \sum_i \sum_s \alpha^i_{p,s} \mathbf{x}_i \right\|^2 + C \sum_i \sum_{\chi \in \boldsymbol{\chi}_i} \xi_\chi - \sum_i \sum_{\chi_i \in \boldsymbol{\chi}_i} \beta^i_{\chi_i} \xi_\chi$$

$$+ \sum_i \sum_{\chi \in \boldsymbol{\chi}_i} \sum_{(r,s) \in \chi} \alpha^i_{r,s} \left( 1 - \xi_\chi - \sum_j \sum_p \alpha^j_{r,p} (\mathbf{x}_j \cdot \mathbf{x}_i) + \sum_j \sum_p \alpha^j_{s,p} (\mathbf{x}_j \cdot \mathbf{x}_i) \right)$$

$$= \frac{1}{2} \sum_p \sum_{i,j} \sum_{r,s} \alpha^i_{p,r} \alpha^j_{p,s} (\mathbf{x}_i \cdot \mathbf{x}_j) + C \sum_i \sum_{\chi \in \boldsymbol{\chi}_i} \xi_\chi - \sum_i \sum_{\chi_i \in \boldsymbol{\chi}_i} \beta^i_{\chi_i} \xi_\chi$$

$$+ \sum_i \sum_{\chi \in \boldsymbol{\chi}_i} \sum_{(r,s) \in \chi} \alpha^i_{r,s} - \sum_i \sum_{\chi \in \boldsymbol{\chi}_i} \sum_{(r,s) \in \chi} \alpha^i_{r,s} \xi_\chi$$

$$+ \sum_i \sum_{\chi \in \boldsymbol{\chi}_i} \sum_{(r,s) \in \chi} \alpha^i_{r,s} \left( -\sum_j \sum_p \alpha^j_{r,p} (\mathbf{x}_j \cdot \mathbf{x}_i) + \sum_j \sum_p \alpha^j_{s,p} (\mathbf{x}_j \cdot \mathbf{x}_i) \right)$$

$$= \frac{1}{2} \sum_p \sum_{i,j} \sum_{r,s} \alpha^i_{p,r} \alpha^j_{p,s} (\mathbf{x}_i \cdot \mathbf{x}_j) + \sum_i \sum_{\chi \in \boldsymbol{\chi}_i} \sum_{(r,s) \in \chi} \alpha^i_{r,s}$$

$$+ \sum_i \sum_{\chi \in \chi_i} \sum_{(r,s) \in \chi} \alpha^i_{r,s} \left( - \sum_j \sum_p \alpha^j_{r,p} \left( \mathbf{x}_j \cdot \mathbf{x}_i \right) + \sum_j \sum_p \alpha^j_{s,p} \left( \mathbf{x}_j \cdot \mathbf{x}_i \right) \right)$$

$$+ \sum_i \sum_{\chi_i \in \chi_i} \left( C - \beta^i_{\chi_i} - \sum_{(r,s) \in \chi} \alpha^i_{r,s} \right).$$

The last terms equal zero because of Equation (3.25). Rearranging the terms we get

$$\mathcal{L} = \frac{1}{2} \sum_p \sum_{i,j} \sum_{r,s} \alpha^i_{p,r} \alpha^j_{p,s} \left( \mathbf{x}_i \cdot \mathbf{x}_j \right) + \sum_i \sum_{\chi \in \chi_i} \sum_{(r,s) \in \chi} \alpha^i_{r,s}$$

$$+ \sum_{i,j} \left( \mathbf{x}_j \cdot \mathbf{x}_i \right) \sum_{(r,s) \in \mathbf{y}_i} \alpha^i_{r,s} \left( - \sum_p \alpha^j_{r,p} + \sum_p \alpha^j_{s,p} \right)$$

$$= \frac{1}{2} \sum_p \sum_{i,j} \sum_{r,s} \alpha^i_{p,r} \alpha^j_{p,s} \left( \mathbf{x}_i \cdot \mathbf{x}_j \right) + \sum_i \sum_{\chi \in \chi_i} \sum_{(r,s) \in \chi} \alpha^i_{r,s}$$

$$- \sum_{i,j} \left( \mathbf{x}_j \cdot \mathbf{x}_i \right) \sum_{r,s,p} \alpha^i_{r,s} \alpha^j_{r,p},$$

where the last equality is due to Equation (3.21). Rearranging the terms again and adding the constraints of Equation (3.25) we get the dual optimization problem:

$$\max_{\{\alpha^i_{r,s}\}} \ - \frac{1}{2} \sum_p \sum_{i,j} \sum_{r,s} \alpha^i_{p,r} \alpha^j_{p,s} \left( \mathbf{x}_i \cdot \mathbf{x}_j \right) + \sum_i \sum_{(r,s) \in \mathbf{y}_i} \alpha^i_{r,s}$$

$$\text{subject to: } \forall i \begin{cases} \alpha^i_{r,s} \geq 0 & (r,s) \in \mathbf{y}_i \\ \alpha^i_{s,r} = -\alpha^i_{r,s} & (r,s) \in \mathbf{y}_i \\ \alpha^i_{s,r} = 0 & \text{otherwise} \end{cases}$$

$$\forall i \forall \chi \in \chi_i : \ \sum_{(r,s) \in \chi} \alpha^i_{r,s} \leq C. \tag{3.28}$$

# References

Aizerman MA, Braverman EM and Rozonoer LI 1964 Theoretical foundations of the potential function method in pattern recognition learning. *Automation and Remote Control* **25**, 821–837.

Allwein E, Schapire R and Singer Y 2000 Reducing multiclass to binary: A unifying approach for margin classifiers. *Journal of Machine Learning Research* **1**, 113–141.

Cortes C and Vapnik V 1995 Support-vector networks. *Machine Learning* **20**(3), 273–297.

Crammer K 2004 *Online Learning of Complex Categorial Problems* Hebrew University, PhD thesis.

Crammer K and Singer Y 2001 On the algorithmic implementation of multiclass kernel-based vector machines. *Journal of Machine Learning Research* **2**, 265–292.

Crammer K and Singer Y 2002 On the learnability and design of output codes for multiclass problems. *Machine Learning* **47**, 201–233.

Crammer K and Singer Y 2003 Ultraconservative online algorithms for multiclass problems. *Journal of Machine Learning Research* **3**, 951–991.

Crammer K, Dekel O, Keshet J, Shalev-Shwartz S and Singer Y 2006 Online passive–aggressive algorithms. *Journal of Machine Learning Research* **7**, 551–585.

Cristianini N and Shawe-Taylor J 2000 *An Introduction to Support Vector Machines*. Cambridge University Press.

Dekel O, Keshet J and Singer Y 2005 Online algorithm for hierarchical phoneme classification *Workshop on Multimodal Interaction and Related Machine Learning Algorithms (Lecture Notes in Computer Science)*. Springer-Verlag, pp. 146–159.

Dietterich T and Bakiri G 1991 Error-correcting output codes: A general method for improving multiclass inductive learning programs. *Proceedings of the 9th National Conference on Artificial Intelligence*. AAAI Press.

Dietterich T and Bakiri G 1995 Solving multiclass learning problems via error-correcting output codes. *Journal of Artificial Intelligence Research* **2**, 263–286.

Duda RO and Hart PE 1973 *Pattern Classification and Scene Analysis*. John Wiley & Sons.

Elisseeff A and Weston J 2001 A kernel method for multi-labeled classification *Advances in Neural Information Processing Systems 14*.

Furnkranz J 2002 Round robin classification. *Journal of Machine Learning Research* **2**, 721–747.

Guermeur Y, Elisseeff A and Paugam-Moisy H 2000 A new multi-class svm based on a uniform convergence result. *Proceedings of IJCNN-2000*. IEEE Computer Society, Washington, DC.

Har-Peled S, Roth D and Zimak D 2002 Constraint classification for multiclass classification and ranking *Advances in Neural Information Processing Systems 15*.

Hsu CW and Lin CJ 2002 A comparison of methods for multi-class support vector machines. *IEEE Transactions on Neural Networks* **13**, 415–425.

Lafferty J, McCallum A and Pereira F 2001 Conditional random fields: Probabilistic models for segmenting and labeling sequence data. *Proceedings of the 18th International Conference on Machine Learning*. Morgan Kaufmann, San Francisco, CA.

Mercer J 1909 Functions of positive and negative type and their connection with the theory of integral equations. *Philosophical Transactions of the Royal Society of London. Series A: Mathematical and Physical Sciences* **209**, 415–446.

Minsky ML and Papert SA 1969 *Perceptrons*. MIT Press. (Expanded Edition 1990.)

Nilsson NJ 1965 *Learning Machines: Foundations of trainable pattern classification systems*. McGraw-Hill, New York.

Rabiner L and Juang B-H 1993 *Fundamentals of Speech Recognition*. Prentice Hall.

Rosenblatt F 1958 The perceptron: A probabilistic model for information storage and organization in the brain. *Psychological Review* **65**, 386–407. (Reprinted in *Neurocomputing* (MIT Press, 1988).)

Schölkopf B and Smola AJ 2002 *Learning with Kernels: Support Vector Machines, Regularization, Optimization and Beyond*. MIT Press.

Vapnik VN 1998 *Statistical Learning Theory*. John Wiley & Sons.

Weston J and Watkins C 1999 Support vector machines for multi-class pattern recognition. *Proceedings of the 7th European Symposium on Artificial Neural Networks*. Available at: http://www.dice.ucl.ac.be/esann/prcoeedings/papers.php?ann=1999.

# Part II

# Acoustic Modeling

# 4

# A Large Margin Algorithm for Forced Alignment[1]

## Joseph Keshet, Shai Shalev-Shwartz, Yoram Singer and Dan Chazan

We describe and analyze a discriminative algorithm for learning to align a phoneme sequence of a speech utterance with its acoustical signal counterpart by predicting a timing sequence representing the phoneme start times. In contrast to common Hidden Markov Models (HMM)-based approaches, our method employs a discriminative learning procedure in which the learning phase is tightly coupled with the forced alignment task. The alignment function we devise is based on mapping the input acoustic-symbolic representations of the speech utterance along with the target timing sequence into an abstract vector space. We suggest a specific mapping into the abstract vector space which utilizes standard speech features (e.g. spectral distances) as well as confidence outputs of a frame-based phoneme classifier. Generalizing the notion of separation with a margin used in Support Vector Machines (SVMs) for binary classification, we cast the learning task as the problem of finding a vector in an abstract inner-product space. We set the prediction vector to be the solution of a minimization problem with a large set of constraints. Each constraint enforces a gap between the projection of the correct target timing sequence and the projection of an alternative, incorrect, timing sequence onto the vector. Though the number of constraints is very large, we describe a simple iterative algorithm for efficiently learning the vector and analyze the formal properties of the resulting learning algorithm. We report new experimental results comparing the proposed algorithm with previous studies on forced alignment. The new results suggest

---

[1]Portions reprinted, with permission, from Joseph Keshet, Shai Shalev-Shwartz, Yoram Singer and Dan Chazan, Large Margin Algorithm for Speech-to-Phoneme and Music-to-Score Alignment, *IEEE Transactions on Audio, Speech and Language Processing*, **15**(8) 2373–2382, 2007. © 2008 IEEE.

---

*Automatic Speech and Speaker Recognition: Large Margin and Kernel Methods*   Joseph Keshet and Samy Bengio
© 2009 John Wiley & Sons, Ltd

that the discriminative alignment algorithm outperforms the state-of-the-art systems when evaluated on the TIMIT corpus.

## 4.1   Introduction

Forced alignment is the task of proper positioning of a sequence of phonemes in relation to a corresponding continuous speech signal. An accurate and fast forced alignment procedure is a necessary tool for developing speech recognition and text-to-speech systems. Most previous work on forced alignment has focused on a generative model of the speech signal using HMMs. See for example Brugnara *et al.* (1993), Hosom (2002), Toledano *et al.* (2003) and the references therein. In this chapter we present a discriminative supervised algorithm for forced alignment. The work presented in this chapter is based on Keshet *et al.* (2007).

This chapter is organized as follows. In Section 4.2 we formally introduce the forced alignment problem. In our algorithm we use a cost function of predicting incorrect timing sequence. This function is defined in Section 4.3. In Section 4.4 we describe a large margin approach for the alignment problem. Our specific learning algorithm is described and analyzed in Section 4.5. The evaluation of the alignment function and the learning algorithm are both based on an optimization problem for which we give an efficient dynamic programming procedure in Section 4.6. Next, in Section 4.7 we describe the base alignment functions we use. Finally, we present experimental results in which we compare our method with state-of-the-art approaches in Section 4.8.

## 4.2   Problem Setting

In this section we formally describe the forced alignment problem. In the forced alignment problem, we are given a speech utterance along with a phonetic representation of the utterance. Our goal is to generate an alignment between the speech signal and the phonetic representation. We denote the domain of the acoustic feature vectors by $\mathcal{X} \subset \mathbb{R}^d$. The acoustic feature representation of a speech signal is therefore a sequence of vectors $\bar{\mathbf{x}} = (\mathbf{x}_1, \ldots, \mathbf{x}_T)$, where $\mathbf{x}_t \in \mathcal{X}$ for all $1 \leq t \leq T$. A phonetic representation of an utterance is defined as a string of phoneme symbols. Formally, we denote each phoneme by $p \in \mathcal{P}$, where $\mathcal{P}$ is the set of phoneme symbols. Therefore, a phonetic representation of a speech utterance consists of a sequence of phoneme values $\bar{p} = (p_1, \ldots, p_k)$. Note that the number of phonemes clearly varies from one utterance to another and thus $k$ is not fixed. We denote by $\mathcal{P}^\star$ (and similarly $\mathcal{X}^\star$) the set of all finite-length sequences over $\mathcal{P}$. In summary, the input is a pair $(\bar{\mathbf{x}}, \bar{p})$ where $\bar{\mathbf{x}}$ is an acoustic representation of the speech signal and $\bar{p}$ is a phonetic representation of the same signal. An alignment between the acoustic and phonetic representations of a spoken utterance is a timing sequence $\bar{s} = (s_1, \ldots, s_k)$ where $s_i \in \mathbb{N}$ is the start time (measured as frame number) of phoneme $i$ in the acoustic signal. Each phoneme $i$ therefore starts at frame $s_i$ and ends at frame $s_{i+1} - 1$. An example of the notation described above is depicted in Figure 4.1.

Clearly, there are different ways to pronounce the same utterance. Different speakers have different accents and tend to speak at different rates. Our goal is to learn an alignment function that predicts the true start times of the phonemes from the speech signal and the phonetic representation.

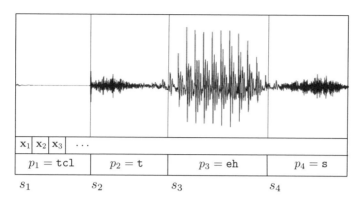

Figure 4.1  A spoken utterance labeled with the sequence of phonemes $/p_1\ p_2\ p_3\ p_4/$ and its corresponding sequence of start-times $(s_1\ s_2\ s_3\ s_4)$.

## 4.3  Cost and Risk

In this section we describe a discriminative supervised learning approach for learning an alignment function $f$ from a training set of examples. Each example in the training set is composed of a speech utterance, $\bar{x}$, a sequence of phonemes, $\bar{p}$, and the true timing sequence, $\bar{s}$, i.e. a sequence of phoneme start times. Our goal is to find an alignment function, $f$, which performs well on the training set as well as on unseen examples. First, we define a quantitative assessment of alignment functions. Let $(\bar{x},\ \bar{p},\ \bar{s})$ be an input example and let $f$ be an alignment function. We denote by $\gamma(\bar{s},\ f(\bar{x},\ \bar{p}))$ the cost of predicting the timing sequence $f(\bar{x},\ \bar{p})$ where the true timing sequence is $\bar{s}$. Formally, $\gamma : \mathbb{N}^* \times \mathbb{N}^* \to \mathbb{R}$ is a function that gets two timing sequences (of the same length) and returns a scalar which is the cost of predicting the second timing sequence where the true timing sequence is the first. We assume that $\gamma(\bar{s},\ \bar{s}') \geq 0$ for any two timing sequences $\bar{s},\ \bar{s}'$ and that $\gamma(\bar{s},\ \bar{s}) = 0$. An example for a cost function is

$$\gamma(\bar{s},\ \bar{s}') = \frac{1}{|\bar{s}|} |\{i : |s_i - s_i'| > \varepsilon\}|. \tag{4.1}$$

In words, the above cost is the average number of times the absolute difference between the predicted timing sequence and the true timing sequence is greater than $\varepsilon$. Recall that our goal is to find an alignment function $f$ that attains small cost on unseen examples. Formally, let $D$ be any (unknown) distribution over the domain of the examples, $\mathcal{X}^* \times \mathcal{P}^* \times \mathbb{N}^*$. The goal of the learning process is to minimize the risk of using the alignment function, defined as the expected cost of $f$ on the examples, where the expectation is taken with respect to the distribution $D$:

$$\text{risk}(f) = \mathbb{E}_{(\bar{x},\bar{p},\bar{s}) \sim D}[\gamma(\bar{s},\ f(\bar{x},\ \bar{p}))].$$

To do so, we assume that the examples of our training set are identically and independently distributed (i.i.d.) according to the distribution $D$. Note that we only observe the training examples but we do not know the distribution $D$. The training set of examples is used as a restricted window throughout which we estimate the quality of alignment functions according to the distribution of unseen examples in the real world, $D$. In the next sections we show how to use the training set in order to find an alignment function, $f$, which achieves a small cost on the training set, and which achieves a small cost on unseen examples with high probability as well.

## 4.4    A Large Margin Approach for Forced Alignment

In this section we describe a large margin approach for learning an alignment function. Recall that a supervised learning algorithm for alignment receives as input a training set $\mathcal{T} = \{(\bar{\mathbf{x}}_1, \bar{p}_1, \bar{s}_1), \ldots, (\bar{\mathbf{x}}_m, \bar{p}_m, \bar{s}_m)\}$ and returns a alignment function $f$. To facilitate an efficient algorithm we confine ourselves to a restricted class of alignment functions. Specifically, we assume the existence of a predefined set of base alignment feature functions, $\{\phi_j\}_{j=1}^n$. Each base alignment feature is a function of the form $\phi_j : \mathcal{X}^* \times \mathcal{P}^* \times \mathbb{N}^* \to \mathbb{R}$. That is, each base alignment feature gets the acoustic representation, $\bar{\mathbf{x}}$, and the sequence of phonemes, $\bar{p}$, together with a candidate timing sequence, $\bar{s}$, and returns a scalar which, intuitively, represents the confidence in the suggested timing sequence $\bar{s}$. We denote by $\boldsymbol{\phi}(\bar{\mathbf{x}}, \bar{p}, \bar{s})$ the vector in $\mathbb{R}^n$ whose $j$th element is $\phi_j(\bar{\mathbf{x}}, \bar{p}, \bar{s})$. The alignment functions we use are of the form

$$f(\bar{\mathbf{x}}, \bar{p}) = \arg\max_{\bar{s}} \mathbf{w} \cdot \boldsymbol{\phi}(\bar{\mathbf{x}}, \bar{p}, \bar{s}), \tag{4.2}$$

where $\mathbf{w} \in \mathbb{R}^n$ is a vector of importance weights that we need to learn. In words, $f$ returns a suggestion for a timing sequence by maximizing a weighted sum of the confidence scores returned by each base alignment function $\phi_j$. Since $f$ is parameterized by $\mathbf{w}$ we use the notation $f_{\mathbf{w}}$ for an alignment function $f$, which is defined as in Equation (4.2). Note that the number of possible timing sequences, $\bar{s}$, is exponentially large. Nevertheless, as we show later, under mild conditions on the form of the base alignment functions, $\{\phi_j\}$, the optimization problem in Equation (4.2) can be efficiently calculated using a dynamic programming procedure.

We now describe a large margin approach for learning the weight vector $\mathbf{w}$, which defines an alignment function as in Equation (4.2), from a training set $\mathcal{T} = \{(\bar{\mathbf{x}}_1, \bar{p}_1, \bar{s}_1), \ldots, (\bar{\mathbf{x}}_m, \bar{p}_m, \bar{s}_m)\}$ of examples. Similar to the SVM algorithm for binary classification, our approach for choosing the weight vector $\mathbf{w}$ is based on the idea of large-margin separation. However, in our case, timing sequences are not merely correct or incorrect. Instead, the cost function $\gamma(\bar{s}, \bar{s}')$ is used for assessing the quality of sequences. Therefore, we do not aim at separating correct timing sequences from incorrect ones but rather try to rank the sequences according to their quality. Theoretically, our approach can be described as a two-step procedure: first, we construct a vector $\boldsymbol{\phi}(\bar{\mathbf{x}}_i, \bar{p}_i, \bar{s}')$ in the vector space $\mathbb{R}^n$ based on each instance $(\bar{\mathbf{x}}_i, \bar{p}_i)$ in the training set $\mathcal{T}$ and each possible timing sequence $\bar{s}'$. Second, we find a vector $\mathbf{w} \in \mathbb{R}^n$, such that the projection of vectors onto $\mathbf{w}$ ranks the vectors constructed in the first step above according to their quality. Ideally, for each instance $(\bar{\mathbf{x}}_i, \bar{p}_i)$ and for each possible suggested timing sequence $\bar{s}'$, we would like the following constraint to hold:

$$\mathbf{w} \cdot \boldsymbol{\phi}(\bar{\mathbf{x}}_i, \bar{p}_i, \bar{s}_i) - \mathbf{w} \cdot \boldsymbol{\phi}(\bar{\mathbf{x}}_i, \bar{p}_i, \bar{s}') \geq \gamma(\bar{s}_i, \bar{s}'). \tag{4.3}$$

That is, $\mathbf{w}$ should rank the correct timing sequence $\bar{s}_i$ above any other possible timing sequence $\bar{s}'$ by at least $\gamma(\bar{s}_i, \bar{s}')$. We refer to the difference $\mathbf{w} \cdot \boldsymbol{\phi}(\bar{\mathbf{x}}_i, \bar{p}_i, \bar{s}_i) - \mathbf{w} \cdot \boldsymbol{\phi}(\bar{\mathbf{x}}_i, \bar{p}_i, \bar{s}')$ as the *margin* of $\mathbf{w}$ with respect to the sequence $\bar{s}'$. Note that if the prediction of $\mathbf{w}$ is incorrect then the margin is negative. The constraints in Equation (4.3) imply that the margin of $\mathbf{w}$ with respect to any possible timing sequence $\bar{s}'$ should be at least the cost of predicting $\bar{s}'$ instead of the true timing sequence $\bar{s}_i$. Naturally, if $\mathbf{w}$ ranks the different possible timing sequences correctly, the margin requirements given in Equation (4.3) can be satisfied by simply multiplying $\mathbf{w}$ by a large scalar. The SVM algorithm solves this problem by minimizing $\frac{1}{2}\|\mathbf{w}\|^2$ subject to the constraints given in Equation (4.3).

In practice, it might be the case that the constraints given in Equation (4.3) cannot be satisfied. To overcome this obstacle, we follow the soft SVM approach and define the following hinge loss function for alignment:

$$\ell(\mathbf{w}; (\bar{\mathbf{x}}_i, \bar{p}_i, \bar{s}_i)) = \max_{\bar{s}'} \left[ \gamma(\bar{s}_i, \bar{s}') - \mathbf{w} \cdot (\boldsymbol{\phi}(\bar{\mathbf{x}}_i, \bar{p}_i, \bar{s}_i) - \boldsymbol{\phi}(\bar{\mathbf{x}}_i, \bar{p}_i, \bar{s}')) \right]_+, \qquad (4.4)$$

where $[a]_+ = \max\{0, a\}$. The hinge loss measures the maximal violation of any of the constraints given in Equation (4.3). The soft SVM approach for forced alignment is to choose the vector $\mathbf{w}^\star$, which minimizes the following optimization problem:

$$\min_{\mathbf{w}, \xi \geq 0} \frac{1}{2}\|\mathbf{w}\|^2 + C \sum_{i=1}^{m} \xi_i \qquad (4.5)$$

$$\text{s.t.} \quad \mathbf{w} \cdot \boldsymbol{\phi}(\bar{\mathbf{x}}_i, \bar{p}_i, \bar{s}_i) - \mathbf{w} \cdot \boldsymbol{\phi}(\bar{\mathbf{x}}_i, \bar{p}_i, \bar{s}') \geq \gamma(\bar{s}_i, \bar{s}') - \xi_i \quad \forall i, \bar{s}',$$

where the parameter $C$ serves as a complexity–accuracy trade-off parameter (see Cristianini and Shawe-Taylor 2000) and each $\xi_i$ is a non-negative slack variable that indicates the loss of the $i$th example.

Solving the quadratic optimization problem given in Equation (4.5) is complicated since the number of constraints is exponentially large. Several authors suggested specific algorithms for manipulating the exponential number of constraints (Taskar *et al.* 2003; Tsochantaridis *et al.* 2004). However, these methods are problematic when the size of the dataset is very large since several passes over the data are required. In the next section, we propose an alternative method, which visits each example only once.

## 4.5 An Iterative Algorithm

In this section we describe an iterative algorithm for learning an alignment function, parameterized by $\mathbf{w}$. Our iterative algorithm first constructs a sequence of weight vectors $\mathbf{w}_1, \ldots, \mathbf{w}_m, \mathbf{w}_{m+1}$. The first weight vector is set to be the zero vector, $\mathbf{w}_1 = \mathbf{0}$. On iteration $i$ of the algorithm, we utilize the $i$th example of the training set along with the previous weight vector $\mathbf{w}_i$, for defining the next weight vector $\mathbf{w}_{i+1}$ as follows. Let $\bar{s}'_i$ be the timing sequence, which corresponds to the highest violated margin constraint of the $i$th example according to $\mathbf{w}_i$, that is:

$$\bar{s}'_i = \arg\max_{\bar{s}} \gamma(\bar{s}, \bar{s}_i) - \mathbf{w}_i \cdot (\boldsymbol{\phi}(\bar{\mathbf{x}}_i, \bar{p}_i, \bar{s}_i) - \boldsymbol{\phi}(\bar{\mathbf{x}}_i, \bar{p}_i, \bar{s})). \qquad (4.6)$$

In Section 4.6 we provide an algorithm that efficiently calculates the above optimization problem using dynamic programming. We set the next weight vector $\mathbf{w}_{i+1}$ to be the minimizer of the following optimization problem:

$$\min_{\mathbf{w} \in \mathbb{R}^n, \xi \geq 0} \frac{1}{2} \|\mathbf{w} - \mathbf{w}_i\|^2 + C\xi \tag{4.7}$$

$$\text{s.t.} \quad \mathbf{w} \cdot \boldsymbol{\phi}(\bar{\mathbf{x}}_i, \bar{p}_i, \bar{s}_i) - \mathbf{w} \cdot \boldsymbol{\phi}(\bar{\mathbf{x}}_i, \bar{p}_i, \bar{s}'_i) \geq \gamma(\bar{s}_i, \bar{s}'_i) - \xi.$$

This optimization problem can be thought of as a relaxed version of the SVM optimization problem with two major differences. First, we replace the exponential number of constraints from Equation (4.5) with a single constraint. This constraint is based on the timing sequence $\bar{s}'_i$ defined in Equation (4.6). Second, we replaced the term $\|\mathbf{w}\|^2$ in the objective function of the SVM with the term $\|\mathbf{w} - \mathbf{w}_i\|^2$. Intuitively, we would like to minimize the loss of $\mathbf{w}$ on the current example, i.e. the slack variable $\xi$, while remaining as close as possible to our previous weight vector $\mathbf{w}_i$.

The solution to the optimization problem in Equation (4.7) has a simple closed form solution:

$$\mathbf{w}_{i+1} = \mathbf{w}_i + \min\left\{ \frac{\ell_i}{\|\Delta\boldsymbol{\phi}_i\|^2}, C \right\} \cdot \Delta\boldsymbol{\phi}_i,$$

where $\Delta\boldsymbol{\phi}_i = \boldsymbol{\phi}(\bar{\mathbf{x}}_i, \bar{p}_i, \bar{s}_i) - \boldsymbol{\phi}(\bar{\mathbf{x}}_i, \bar{p}_i, \bar{s}'_i)$ and $\ell_i = \ell(\mathbf{w}_i; (\mathbf{x}_i, \bar{p}_i, \bar{s}_i))$. We now show how this update is derived using standard tools from convex analysis (see for instance Boyd and Vandenberghe (2004)). If $\ell_i = 0$ then $\mathbf{w}_i$ itself satisfies the constraint in Equation (4.7) and is clearly the optimal solution. If $\ell_i > 0$ we derive these updates by defining the Lagrangian of the respective optimization problem and satisfying the Karush–Khun–Tucker (KKT) conditions (Boyd and Vandenberghe 2004). The Lagrangian of the optimization problem is

$$\mathcal{L}(\mathbf{w}, \xi, \tau, \lambda) = \frac{1}{2}\|\mathbf{w} - \mathbf{w}_i\|^2 + C\xi + \tau(\gamma(\bar{s}_i, \bar{s}'_i) - \xi - \mathbf{w} \cdot \Delta\boldsymbol{\phi}_i) - \lambda\xi, \tag{4.8}$$

where $\tau \geq 0$ and $\lambda \geq 0$ are Lagrange multipliers. Setting the partial derivatives of $\mathcal{L}$ with respect to the elements of $\mathbf{w}$ to zero gives

$$0 = \nabla_{\mathbf{w}}\mathcal{L}(\mathbf{w}, \xi, \tau, \lambda) = \mathbf{w} - \mathbf{w}_i - \tau\Delta\boldsymbol{\phi}_i \quad \Longrightarrow \quad \mathbf{w} = \mathbf{w}_i + \tau\Delta\boldsymbol{\phi}_i. \tag{4.9}$$

Differentiating the Lagrangian with respect to $\xi$ and setting that partial derivative to zero gives

$$0 = \frac{\partial\mathcal{L}(\mathbf{w}, \xi, \tau, \lambda)}{\partial\xi} = C - \tau - \lambda \quad \Longrightarrow \quad C = \tau + \lambda. \tag{4.10}$$

The KKT conditions require $\lambda$ to be non-negative so we conclude that $\tau \leq C$. We now discuss two possible cases: if $\ell_i / \|\Delta\boldsymbol{\phi}_i\|^2 \leq C$ then we can plug Equation (4.10) back into Equation (4.8) and get $\tau = \ell_i / \|\Delta\boldsymbol{\phi}_i\|^2$. The other case is when $\ell_i / \|\Delta\boldsymbol{\phi}_i\|^2 > C$. This condition can be rewritten as

$$C\|\Delta\boldsymbol{\phi}_i\|^2 < \gamma(\bar{s}_i, \bar{s}'_i) - \xi - \mathbf{w} \cdot \Delta\boldsymbol{\phi}_i. \tag{4.11}$$

We also know that the constraint in Equation (4.7) must hold at the optimum, so $\mathbf{w} \cdot \Delta\boldsymbol{\phi}_i \geq \gamma(\bar{s}_i, \bar{s}'_i) - \xi$. Using the explicit form of $\mathbf{w}$ given in Equation (4.9), we can rewrite this constraint as $\mathbf{w}_i \cdot \Delta\boldsymbol{\phi}_i + \tau\|\Delta\boldsymbol{\phi}_i\|^2 \geq \gamma(\bar{s}_i, \bar{s}'_i) - \xi$. Combining this inequality with the inequality in Equation (4.11) gives

$$C\|\Delta\boldsymbol{\phi}_i\|^2 - \tau\|\Delta\boldsymbol{\phi}_i\| < \xi.$$

We now use our earlier conclusion that $\tau \leq C$ to obtain $0 < \xi$. Turning to the KKT complementarity condition, we know that $\xi\lambda = 0$ at the optimum. Having concluded that $\xi$ is strictly positive, we get that $\lambda$ must equal zero. Plugging $\lambda = 0$ into Equation (4.10) gives $\tau = C$. Summing up, we used the KKT conditions to show that in the case where $\ell_i/\|\Delta\phi_i\|^2 > C$, it is optimal to select $\tau = C$. Folding all of the possible cases into a single equation, we get

$$\tau = \min\{C, \ell_i/\|\Delta\phi_i\|^2\}. \tag{4.12}$$

The above iterative procedure gives us a sequence of $m + 1$ weight vectors, $\mathbf{w}_1, \ldots, \mathbf{w}_{m+1}$. In the sequel we prove that the average performance of this sequence of vectors is comparable to the performance of the SVM solution. Formally, let $\mathbf{w}^\star$ be the optimum of the SVM problem given in Equation (4.5). Then, we show in the sequel that setting $C = 1/\sqrt{m}$ gives

$$\frac{1}{m}\sum_{i=1}^{m} \ell(\mathbf{w}_i; (\bar{\mathbf{x}}_i, \bar{p}_i, \bar{s}_i)) \leq \frac{1}{m}\sum_{i=1}^{m} \ell(\mathbf{w}^\star; (\bar{\mathbf{x}}_i, \bar{p}_i, \bar{s}_i)) + \frac{1}{\sqrt{m}}\left(\|\mathbf{w}^\star\|^2 + \frac{1}{2}\right). \tag{4.13}$$

That is, the average loss of our iterative procedure is upper bounded by the average loss of the SVM solution plus a factor that decays to zero. However, while each prediction of our iterative procedure is calculated using a different weight vector, our learning algorithm is required to output a *single* weight vector, which defines the output alignment function. To overcome this obstacle, we calculate the average cost of each of the weight vectors $\mathbf{w}_1, \ldots, \mathbf{w}_{m+1}$ on a validation set, denoted $\mathcal{T}_{\text{valid}}$, and choose the one achieving the lowest average cost. We show in the sequel that with high probability, the weight vector which achieves the lowest cost on the validation set also generalizes well. A pseudo-code of our algorithm is given in Figure 4.2.

---

**Input:** training set $\mathcal{T} = \{(\bar{\mathbf{x}}_i, \bar{p}_i, \bar{s}_i)\}_{i=1}^{m}$; validation set $\mathcal{T}_{\text{valid}}$; parameter $C$

**Initialize:** $\mathbf{w}_1 = \mathbf{0}$

**For** $i = 1, \ldots, m$

**Predict:** $\bar{s}_i' = \arg\max_{\bar{s}} \gamma(\bar{s}_i, \bar{s}) - \mathbf{w}_i \cdot \boldsymbol{\phi}(\bar{\mathbf{x}}_i, \bar{p}_i, \bar{s}_i) + \mathbf{w}_i \cdot \boldsymbol{\phi}(\bar{\mathbf{x}}_i, \bar{p}_i, \bar{s})$

**Set:** $\Delta\boldsymbol{\phi}_i = \boldsymbol{\phi}(\bar{\mathbf{x}}_i, \bar{p}_i, \bar{s}_i) - \boldsymbol{\phi}(\bar{\mathbf{x}}_i, \bar{p}_i, \bar{s}_i')$

**Set:** $\ell_i = \max\{\gamma(\bar{s}_i, \bar{s}_i') - \mathbf{w} \cdot \Delta\boldsymbol{\phi}_i, 0\}$

**Update:** $\mathbf{w}_{i+1} = \mathbf{w}_i + \min\{\ell_i/\|\Delta\boldsymbol{\phi}_i\|^2, C\}\,\Delta\boldsymbol{\phi}_i$

**Output:** The weight vector which achieves the lowest average cost on the validation set $\mathcal{T}_{\text{valid}}$.

---

Figure 4.2 The forced alignment algorithm.

We now analyze our alignment algorithm from Figure 4.2. Our first theorem shows that the average loss of our alignment algorithm is comparable to the average loss of the SVM solution for the alignment problem defined in Equation (4.5).

**Theorem 4.5.1** *Let* $T = \{(\bar{\mathbf{x}}_1, \bar{p}_1, \bar{s}_1), \ldots, (\bar{\mathbf{x}}_m, \bar{p}_m, \bar{s}_m)\}$ *be a set of training examples and assume that for all $i$ and $\bar{s}'$ we have that $\|\boldsymbol{\phi}(\bar{\mathbf{x}}_i, \bar{p}_i, \bar{s}')\| \leq 1/2$. Let $\mathbf{w}^\star$ be the optimum of the SVM problem given in Equation (4.5). Let $\mathbf{w}_1, \ldots, \mathbf{w}_m$ be the sequence of weight vectors obtained by the algorithm in Figure 4.2 given the training set $T$. Then,*

$$\frac{1}{m}\sum_{i=1}^{m}\ell(\mathbf{w}_i; (\bar{\mathbf{x}}_i, \bar{p}_i, \bar{s}_i)) \leq \frac{1}{m}\sum_{i=1}^{m}\ell(\mathbf{w}^\star; (\bar{\mathbf{x}}_i, \bar{p}_i, \bar{s}_i)) + \frac{1}{Cm}\|\mathbf{w}^\star\|^2 + \frac{1}{2}C. \qquad (4.14)$$

*In particular, if $C = 1/\sqrt{m}$ then*

$$\frac{1}{m}\sum_{i=1}^{m}\ell(\mathbf{w}_i; (\bar{\mathbf{x}}_i, \bar{p}_i, \bar{s}_i)) \leq \frac{1}{m}\sum_{i=1}^{m}\ell(\mathbf{w}^\star; (\bar{\mathbf{x}}_i, \bar{p}_i, \bar{s}_i)) + \frac{1}{\sqrt{m}}\left(\|\mathbf{w}^\star\|^2 + \frac{1}{2}\right). \qquad (4.15)$$

*Proof.* Our proof relies on Theorem 2 in Shalev-Shwartz and Singer (2006). We first construct a sequence of binary classification examples, $(\Delta\boldsymbol{\phi}_1, +1), \ldots, (\Delta\boldsymbol{\phi}_m, +1)$. For all $i$ and for all $\mathbf{w} \in \mathbb{R}^n$, define the following classification hinge loss:

$$\ell_i^c(\mathbf{w}) = \max\{\gamma(\bar{s}_i, \bar{s}_i') - \mathbf{w}\cdot\Delta\boldsymbol{\phi}_i, 0\}.$$

Theorem 2 in Shalev-Shwartz and Singer (2006) implies that the following bound holds for all $\mathbf{w} \in \mathbb{R}^n$:

$$\sum_{i=1}^{m}\mu(\ell_i^c(\mathbf{w}_i)) \leq \frac{1}{C}\|\mathbf{w}\|^2 + \sum_{i=1}^{m}\ell_i^c(\mathbf{w}), \qquad (4.16)$$

where

$$\mu(a) = \frac{1}{C}\left(\min\{a, C\}\left(a - \frac{1}{2}\min\{a, C\}\right)\right).$$

Let $\mathbf{w}^\star$ denote the optimum of the alignment problem given by Equation (4.5). The bound of Equation (4.16) holds for any $\mathbf{w}$ and in particular for the optimal solution $\mathbf{w}^\star$. Furthermore, the definition of $\ell_i^c$ implies that $\ell_i^c(\mathbf{w}^\star) \leq \ell(\mathbf{w}^\star; (\bar{\mathbf{x}}_i, \bar{p}_i, \bar{s}_i))$ and $\ell_i^c(\mathbf{w}_i) = \ell(\mathbf{w}_i; (\bar{\mathbf{x}}_i, \bar{p}_i, \bar{s}_i))$ for all $i$. Using the latter two facts in Equation (4.16) gives that

$$\sum_{i=1}^{m}\mu(\ell(\mathbf{w}_i; (\bar{\mathbf{x}}_i, \bar{p}_i, \bar{s}_i))) \leq \frac{1}{C}\|\mathbf{w}^\star\|^2 + \sum_{i=1}^{m}\ell(\mathbf{w}^\star; (\bar{\mathbf{x}}_i, \bar{p}_i, \bar{s}_i)). \qquad (4.17)$$

By definition, the function $\mu$ is bounded below by a linear function, that is, for any $a > 0$,

$$\mu(a) \geq a - \tfrac{1}{2}C.$$

Using the lower bound with the argument $\ell(\mathbf{w}_i; (\bar{\mathbf{x}}_i, \bar{p}_i, \bar{s}_i))$ and summing over $i$ we obtain

$$\sum_{i=1}^{m}\ell(\mathbf{w}_i; (\bar{\mathbf{x}}_i, \bar{p}_i, \bar{s}_i)) - \frac{1}{2}Cm \leq \sum_{i=1}^{m}\mu(\ell(\mathbf{w}_i; (\bar{\mathbf{x}}_i, \bar{p}_i, \bar{s}_i))).$$

Combining the above inequality with Equation (4.17) and rearranging terms gives the bound stated in the theorem and concludes our proof. $\qquad \square$

The next theorem tells us that the output alignment function of our algorithm is likely to have good generalization properties.

**Theorem 4.5.2** *Under the same conditions of Theorem 4.5.1. assume that the training set $T$ and the validation set $T_{\text{valid}}$ are both sampled i.i.d. from a distribution D. Denote by $m_{\text{valid}}$ the size of the validation set. Assume in addition that $\gamma(\bar{s}, \bar{s}') \leq 1$ for all $\bar{s}$ and $\bar{s}'$. Let $\mathbf{w}$ be the output weight vector of the algorithm in Figure 4.2 and let $f_{\mathbf{w}}$ be the corresponding alignment function. Then, with probability of at least $1 - \delta$ we have that*

$$\text{risk}(f_{\mathbf{w}}) \leq \frac{1}{m}\sum_{i=1}^{m} \ell(\mathbf{w}^{\star}; (\bar{\mathbf{x}}_i, \bar{p}_i, \bar{s}_i)) + \frac{\|\mathbf{w}^{\star}\|^2 + \frac{1}{2} + \sqrt{2\ln(2/\delta)}}{\sqrt{m}} + \frac{\sqrt{2\ln(2m/\delta)}}{\sqrt{m_{\text{valid}}}}. \quad (4.18)$$

*Proof.* Denote by $f_1, \ldots, f_m$ the alignment prediction functions corresponding to the weight vectors $\mathbf{w}_1, \ldots, \mathbf{w}_m$ that are found by the alignment algorithm. Proposition 1 in Cesa-Bianchi *et al.* (2004) implies that with probability of at least $1 - \delta_1$ the following bound holds:

$$\frac{1}{m}\sum_{i=1}^{m}\text{risk}(f_i) \leq \frac{1}{m}\sum_{i=1}^{m}\gamma(\bar{s}_i, f_i(\bar{\mathbf{x}}_i, \bar{p}_i)) + \frac{\sqrt{2\ln(1/\delta_1)}}{\sqrt{m}}.$$

By definition, the hinge loss $\ell(\mathbf{w}_i; (\bar{\mathbf{x}}_i, \bar{p}_i, \bar{s}_i))$ bounds from above the loss $\gamma(\bar{s}_i, f_i(\bar{\mathbf{x}}_i, \bar{p}_i))$. Combining this fact with Theorem 4.5.1 we obtain that

$$\frac{1}{m}\sum_{i=1}^{m}\text{risk}(f_i) \leq \frac{1}{m}\sum_{i=1}^{m}\ell(\mathbf{w}^{\star}; (\bar{\mathbf{x}}_i, \bar{p}_i, \bar{s}_i)) + \frac{\|\mathbf{w}^{\star}\|^2 + \frac{1}{2} + \sqrt{2\ln(1/\delta_1)}}{\sqrt{m}}. \quad (4.19)$$

The left-hand side of the above inequality upper bounds $\text{risk}(f_b)$, where $b = \arg\min_i \text{risk}(f_i)$. Therefore, among the finite set of alignment functions, $F = \{f_1, \ldots, f_m\}$, there exists at least one alignment function (for instance the function $f_b$) whose true risk is bounded above by the right-hand side of Equation (4.19). Recall that the output of our algorithm is the alignment function $f_{\mathbf{w}} \in F$, which minimizes the average cost over the validation set $T_{\text{valid}}$. Applying Hoeffding inequality together with the union bound over $F$ we conclude that with probability of at least $1 - \delta_2$,

$$\text{risk}(f_{\mathbf{w}}) \leq \text{risk}(f_b) + \sqrt{\frac{2\ln(m/\delta_2)}{m_{\text{valid}}}},$$

where to remind the reader $m_{\text{valid}} = |T_{\text{valid}}|$. We have therefore shown that with probability of at least $1 - \delta_1 - \delta_2$ the following inequality holds:

$$\text{risk}(f_{\mathbf{w}}) \leq \frac{1}{m}\sum_{i=1}^{m}\ell(\mathbf{w}^{\star}; (\bar{\mathbf{x}}_i, \bar{p}_i, \bar{s}_i)) + \frac{\|\mathbf{w}^{\star}\|^2 + \frac{1}{2} + \sqrt{2\ln(1/\delta_1)}}{\sqrt{m}} + \frac{\sqrt{2\ln(m/\delta_2)}}{\sqrt{m_{\text{valid}}}}.$$

Setting $\delta_1 = \delta_2 = \delta/2$ concludes our proof. $\square$

As mentioned before, the learning algorithm we present in this chapter share similarities with the SVM method for structured output prediction (Taskar *et al.* 2003; Tsochantaridis *et al.* 2004). Yet, the weight vector resulted by our method is not identical to the one obtained by directly solving the SVM optimization problem. We would like to note in passing that our generalization bound from Theorem 4.5.2 is comparable to generalization bounds derived for the SVM method (see for example Taskar *et al.* (2003)). The major advantage of our method over directly solving the SVM problem is its simplicity and efficiency.

## 4.6    Efficient Evaluation of the Alignment Function

So far we have put aside the problem of evaluation time of the function $f$ given in
Equation (4.2). Recall that calculating $f$ requires solving the following optimization problem:

$$f(\bar{\mathbf{x}}, \bar{p}) = \arg \max_{\bar{s}} \mathbf{w} \cdot \boldsymbol{\phi}(\bar{\mathbf{x}}, \bar{p}, \bar{s}).$$

Similarly, we need to find an efficient way for solving the maximization problem given in
Equation (4.6). A direct search for the maximizer is not feasible since the number of possible
timing sequences, $\bar{s}$, is exponential in the number of events. Fortunately, as we show below,
by imposing a few mild conditions on the structure of the alignment feature functions and on
the cost function, $\gamma$, both problems can be solved in polynomial time.

   We start with the problem of calculating the prediction given in Equation (4.2). For
simplicity, we assume that each base feature function, $\phi_j$, can be decomposed as follows.
Let $\hat{\phi}_j$ be any function from $\mathcal{X}^* \times \mathcal{P}^* \times \mathbb{N}^3$ into the reals, which can be computed in a
constant time. That is, $\hat{\phi}_j$ receives as input the signal, $\bar{\mathbf{x}}$, the sequence of events, $\bar{p}$, and three
time points. Additionally, we use the convention $s_0 = 0$ and $s_{|\bar{p}|+1} = T + 1$. Using the above
notation, we assume that each $\phi_j$ can be decomposed to be

$$\phi_j(\bar{\mathbf{x}}, \bar{p}, \bar{s}) = \sum_{i=2}^{|\bar{s}|-1} \hat{\phi}_j(\bar{\mathbf{x}}, \bar{p}, s_{i-1}, s_i, s_{i+1}). \tag{4.20}$$

The base alignment functions we derive in the next section can be decomposed as in
Equation (4.20).

   We now describe an efficient algorithm for calculating the best timing sequence assuming
that $\phi_j$ can be decomposed as in Equation (4.20). Similar algorithms can be constructed
for any base feature functions that can be described as a dynamic Bayesian network (Dean
and Kanazawa 1989; Taskar *et al.* 2003). Given $i \in \{1, \ldots, |\bar{p}|\}$ and two time indices $t, t' \in
\{1, \ldots, T\}$, denote by $\rho(i, t, t')$ the score for the prefix of the events sequence $1, \ldots, i$,
assuming that their actual start times are $s_1, \ldots, s_i$, where $s_i = t'$ and assuming that $s_{i+1} = t$.
This variable can be computed efficiently in a similar fashion to the forward variables
calculated by the Viterbi procedure in HMMs (see for instance Rabiner and Juang (1993)).
The pseudo-code for computing $\rho(i, t, t')$ recursively is shown in Figure 4.3. The best
sequence of actual start times, $\bar{s}'$, is obtained from the algorithm by saving the intermediate
values that maximize each expression in the recursion step. The complexity of the algorithm
is $O(|\bar{p}| |\bar{\mathbf{x}}|^3)$. However, in practice, we can use the assumption that the maximal length of
an event is bounded, $t - t' \leq L$. This assumption reduces the complexity of the algorithm to
be $O(|\bar{p}| |\bar{\mathbf{x}}| L^2)$.

   Solving the maximization problem given in Equation (4.6) can be performed in a similar
manner as we now briefly describe. Assume that $\gamma(\bar{s}, \bar{s}')$ can be decomposed as follows:

$$\gamma(\bar{s}, \bar{s}') = \sum_{i=1}^{|\bar{s}|} \hat{\gamma}(s_i, s_i'),$$

where $\hat{\gamma}$ is any computable function. For example, for the definition of $\gamma$ given in
Equation (4.1) we can set $\hat{\gamma}(s_i, s_i')$ to be zero if $|s_i - s_i'| \leq \epsilon$ and otherwise $\hat{\gamma}(s_i, s_i') = 1/|\bar{s}|$.

**Input:** audio signal $\bar{\mathbf{x}}$; sequence of phonemes $\bar{p}$ ; weight vector $\mathbf{w}$ ; maximum phoneme duration $L$

**Initialize:**

$\rho(0, t, 0) = 0, \quad 1 \leq t \leq L$

$\varsigma(0, t, 0) = 0, \quad 1 \leq t \leq L$

**Recursion:**

For $i = 1, \ldots, |\bar{p}|$

  For $t = 1, \ldots, |\bar{\mathbf{x}}|$

    For $t' = t - L, \ldots, t - 1$

      If $t - L \geq 0$

$$\rho(i, t, t') = \max_{t''} \ \rho(i-1, t', t'') + \mathbf{w} \cdot \hat{\boldsymbol{\phi}}(\bar{\mathbf{x}}, \bar{p}, t'', t', t)$$

$$\varsigma(i, t, t') = \arg\max_{t''} \ \rho(i-1, t', t'') + \mathbf{w} \cdot \hat{\boldsymbol{\phi}}(\bar{\mathbf{x}}, \bar{p}, t'', t', t)$$

**Termination:**

$\rho^\star = \max_{t'} \rho(|\bar{p}|, T, t')$

$s^\star_{|\bar{p}|+1} = T$

$s^\star_{|\bar{p}|} = \arg\max_{t'} \rho(|\bar{p}|, T, t')$

**Path backtracking:**

$s^\star_i = \varsigma(i + 1, s^\star_{i+2}, s^\star_{i+1}), \quad i = |\bar{p}| - 2, |\bar{p}| - 3, \ldots, 1$

Figure 4.3 An efficient procedure for evaluating the alignment function.

A dynamic programming procedure for calculating Equation (4.6) can be obtained from Figure 4.3 by replacing the recursion definition of $\rho(i, t, t')$ to

$$\rho(i, t, t') = \max_{t''} \rho(i - 1, t', t'') + \hat{\gamma}(s_{i+1}, t) + \mathbf{w} \cdot \hat{\boldsymbol{\phi}}(\bar{\mathbf{x}}, \bar{p}, t'', t', t). \qquad (4.21)$$

To conclude this section we discuss the global complexity of our proposed method. In the training phase, our algorithm performs $m$ iterations, one iteration per each training example. At each iteration the algorithm evaluates the alignment function once, updates the alignment function, if needed, and evaluates the new alignment function on a validation set of size $m_{\text{valid}}$. Each evaluation of the alignment function takes an order of $O(|\bar{p}| \, |\bar{\mathbf{x}}| \, L^2)$ operations. Therefore the total complexity of our method becomes $O(m \, m_{\text{valid}} \, |\bar{p}| \, |\bar{\mathbf{x}}| \, L^2)$. In practice, however, we can evaluate the updated alignment function only for the last 50 iterations or so, which reduces the global complexity of the algorithm to $O(m \, |\bar{p}| \, |\bar{\mathbf{x}}| \, L^2)$. In all of our experiments, evaluating the alignment function only for the last 50 iterations was found empirically to give sufficient results. Finally, we compare the complexity of our method to the complexity of other algorithms which directly solve the SVM optimization problem given in Equation (4.5). The algorithm given in Taskar *et al.* (2003) is based on the SMO algorithm for solving SVM problems. While there is no direct complexity analysis for this algorithm, in practice it usually required at least $m^2$ iterations, which results in a total complexity of the order $O(m^2 \, |\bar{p}| \, |\bar{\mathbf{x}}| \, L^2)$. The complexity of the algorithm presented in Tsochantaridis *et al.* (2004) depends on the choice of several parameters. For reasonable choice of these parameters the total complexity is also of the order $O(m^2 \, |\bar{p}| \, |\bar{\mathbf{x}}| \, L^2)$.

## 4.7   Base Alignment Functions

Recall that our construction is based on a set of base alignment functions, $\{\phi_j\}_{j=1}^n$, which maps an acoustic–phonetic representation of a speech utterance as well as a suggested phoneme start time sequence into an abstract vector-space. All of our base alignment functions are decomposable as in Equation (4.20) and therefore it suffices to describe the functions $\{\hat{\phi}_j\}$. We start the section by introducing a specific set of base functions, which is highly adequate for the forced alignment problem. Next, we report experimental results comparing our algorithm with alternative state-of-the-art approaches.

We utilize seven different base alignment functions ($n = 7$). These base functions are used for defining our alignment function $f(\bar{\mathbf{x}}, \bar{p})$ as in Equation (4.2).

Our first four base functions aim at capturing transitions between phonemes. These base functions are based on the distance between frames of the acoustical signal at two sides of phoneme boundaries as suggested by a phoneme start time sequence $\bar{s}$. The distance measure we employ, denoted by $d$, is the Euclidean distance between feature vectors. Our underlying assumption is that if two frames, $\mathbf{x}_t$ and $\mathbf{x}_{t'}$, are derived from the same phoneme then the distance $d(\mathbf{x}_t, \mathbf{x}_{t'})$ should be smaller than if the two frames are derived from different phonemes. Formally, our first four base functions are defined as

$$\hat{\phi}_j(\bar{\mathbf{x}}, \bar{p}, s_{i-1}, s_i, s_{i+1}) = d(\mathbf{x}_{s_i-j}, \mathbf{x}_{s_i+j}), \quad j \in \{1, 2, 3, 4\}. \tag{4.22}$$

If $\bar{s}$ is the correct timing sequence then distances between frames across the phoneme change points are likely to be large. In contrast, an incorrect phoneme start time sequence is likely to compare frames from the same phoneme, often resulting in small distances. Note that the first four base functions described above only use the start time of the $i$th phoneme and do not use the values of $s_{i-1}$ and $s_{i+1}$. A schematic illustration of this base function form is depicted in Figure 4.4.

The fifth base function we use is based on the framewise phoneme classifier described in Dekel *et al.* (2004). Formally, for each phoneme $p \in \mathcal{P}$ and frame $\mathbf{x} \in \mathcal{X}$, there is a confidence, denoted $g_p(\mathbf{x})$, that the phoneme $p$ is pronounced in the frame $\mathbf{x}$. The resulting

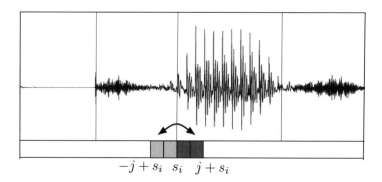

Figure 4.4   A schematic description of one of the first four base functions. The depicted base function is the sum of the Euclidean distances between the sum of two frames before and after any presumed boundary $s_i$.

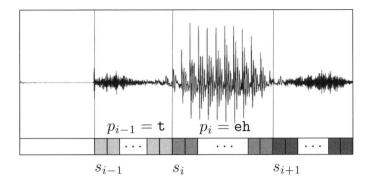

Figure 4.5 A schematic description of the fifth base function. This function is the sum of all the scores obtained from a large margin classifier, given a sequence of phonemes and a presumed sequence of start times.

base function measures the cumulative confidence of the complete speech signal given the phoneme sequence and their start-times:

$$\hat{\phi}_5(\bar{\mathbf{x}}, \bar{p}, s_{i-1}, s_i, s_{i+1}) = \sum_{t=s_i}^{s_{i+1}-1} g_{p_i}(\mathbf{x}_t). \tag{4.23}$$

The fifth base function use both the start time of the $i$th phoneme and the $(i + 1)$th phoneme but ignores $s_{i-1}$. A schematic illustration of the fifth base function is depicted in Figure 4.5.

Our next base function scores timing sequences based on phoneme durations. Unlike the previous base functions, the sixth base function is oblivious to the speech signal itself. It merely examines the length of each phoneme, as suggested by $\bar{s}$, compared with the typical length required to pronounce this phoneme. Formally:

$$\hat{\phi}_6(\bar{\mathbf{x}}, \bar{p}, s_{i-1}, s_i, s_{i+1}) = \log \mathcal{N}(s_{i+1} - s_i; \hat{\mu}_{p_i}, \hat{\sigma}_{p_i}), \tag{4.24}$$

when $\mathcal{N}$ is a Normal probability density function with mean $\hat{\mu}_p$ and standard deviation $\hat{\sigma}_p$, representing the average and standard deviation of the duration of phoneme $p$, respectively. In our experiments, we estimated $\hat{\mu}_p$ and $\hat{\sigma}_p$ from the entire TIMIT training set, excluding SA1 and SA2 utterances. A schematic illustration of the sixth base function is depicted in Figure 4.6.

Our last base function exploits assumptions on the speaking rate of a speaker. Intuitively, people usually speak at an almost steady rate and therefore a timing sequence in which speech rate is changed abruptly is probably incorrect. Formally, let $\hat{\mu}_p$ be the average duration required to pronounce the $p$th phoneme. We denote by $r_i$ the relative speech rate, $r_i = (s_{i+1} - s_i)/\hat{\mu}_p$. That is, $r_i$ is the ratio between the actual length of phoneme $p_i$ as suggested by $\bar{s}$ and its average length. The relative speech rate presumably changes slowly over time. In practice the speaking rate ratios often differ from speaker to speaker and within a given utterance. We measure the local change in the speaking rate as $(r_i - r_{i-1})^2$ and we define the base function $\hat{\phi}_7$ as the local change in the speaking rate:

$$\hat{\phi}_7(\bar{\mathbf{x}}, \bar{p}, s_{i-1}, s_i, s_{i+1}) = (r_i - r_{i-1})^2. \tag{4.25}$$

Note that $\hat{\phi}_7$ relies on all three start times it receives as an input, $s_{i-1}, s_i, s_{i+1}$.

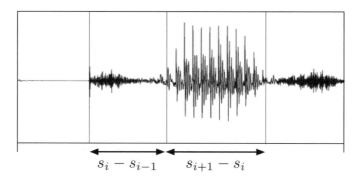

$$s_i - s_{i-1} \qquad s_{i+1} - s_i$$

**Figure 4.6** A schematic description of the sixth base function. This function is the sum of the confidences in the duration of each phoneme given its presumed start time.


## 4.8 Experimental Results

In this section we present experimental results that demonstrate the accuracy and robustness of our algorithm. We performed experiments on read speech using the TIMIT corpus (originally sampled at 16 kHz), TIMIT corpus sampled at 8 kHz, and on HTIMIT. In all the experiments, the baseline discriminative system and HMM system were trained on the clean read-speech TIMIT corpus with a set of 48 phonemes defined by Lee and Hon (1989). We first divided the training portion of TIMIT (excluding the SA1 and SA2 utterances) into three disjoint parts containing 500, 100 and 3093 utterances, respectively. The first part of the training set was used for learning the functions $g_p$ (Equation (4.23)), which defines the base function $\psi_5$. The second and the third sets of 100 and 3096 utterances formed the validation set and training set for the forced alignment algorithm, respectively.

First, the functions $g_p$ were trained using the algorithm described in Dekel *et al.* (2004) with MFCC $+ \Delta + \Delta\Delta$ acoustic features and a Gaussian kernel (the parameters $\sigma = 4.3$ and $C = 1.0$ were chosen on the validation set). Note that the functions $g_p$ are far from being a fine phoneme classifier, since using the functions $g_p$ as a frame-based phoneme classifier resulted in classification accuracy of 55% per frame on the TIMIT core test set, comparing with almost 70% using Multilayer Perception (MLP) (Pinto and Hermansky 2008).

Second, we ran our iterative alignment algorithm as described in Section 4.5 on the training and validation sets with value of $\varepsilon$ in the definition of $\gamma$ equals to 1, i.e. 10 ms.

We evaluated the trained forced aligner on the core test set of TIMIT. We compare our results with the results reported by Brugnara *et al.* (1993) and the results reported by Hosom (2002), which are considered to be the state-of-the-art forced aligners. The results are summarized in Table 4.1. For each tolerance value $\tau \in \{10 \text{ ms}, 20 \text{ ms}, 30 \text{ ms}, 40 \text{ ms}\}$, we counted the number of predictions whose distance to the true boundary, $t = |y_i - y_i'|$, is less than $\tau$. As can be seen in the table our discriminative large margin algorithm outperforms the best results reported on TIMIT so far. Furthermore, we found out in our experiments that the algorithm converges after the first 50 utterances.

The TIMIT is originally sampled at 16 kHz, but we also trained and checked the discriminative model on TIMIT sub-sampled at 8 kHz. This was done to evaluate the

Table 4.1 Percentage of correctly positioned phoneme boundaries, given a predefined tolerance on the TIMIT corpus

|  | $t \leq 10$ ms | $t \leq 20$ ms | $t \leq 30$ ms | $t \leq 40$ ms |
|---|---|---|---|---|
| **TIMIT core test-set (16 kHz)** | | | | |
| Discriminative alignment | **84.2** | **93.3** | **96.7** | **98.4** |
| Brugnara *et al.* (1993) | 75.3 | 88.9 | 94.4 | 97.1 |
| Hosom (2002) | | 92.6 | | |
| **TIMIT core test-set (8 kHz)** | | | | |
| Discriminative alignment | 84.0 | 93.1 | 96.7 | 98.3 |
| **HTIMIT CB1 834 utterances** | | | | |
| Discriminative alignment | 71.2 | 89.5 | 94.8 | 96.9 |

performance of the algorithm on a more realistic sampling rate and to use the model trained on the sub-sampled TIMIT on other corpora sampled at 8 kHz. The results are presented in Table 4.1. We can see that the results for TIMIT sub-sampled at 8 kHz still outperforms the results reported by Brugnara *et al.* (1993) and by Hosom (2002).

In the next experiments we examined the robustness of the proposed algorithm to different environments. We used the model trained on TIMIT but we tested it on different corpora without any further training or adaptation. First we checked the discriminative model on the HTIMIT corpus (Reynolds 1997). The HTIMIT corpus was generated by playing the TIMIT speech through a loudspeaker into a different set of phone handsets. We used the model trained on TIMIT sampled at 8 kHz and tested it on the CB1 portion of the HTIMIT corpus. We see that there is a degradation in the results compared with the clean speech as expected, especially for $\tau \leq 10$ ms. The overall results, however, are reasonable given that we did not re-train or adapt the model to the new environment.

## 4.9 Discussion

We described and experimented with a discriminative method for phoneme forced alignment. The proposed approach is based on recent advances in large margin classifiers. Our training algorithm is simple to implement and entertains convergence guarantees. In contrast to HMM training procedures, which are prone to local maxima variabilities, our proposed algorithm is guaranteed to converge to a solution which has good generalization properties under mild conditions. Indeed, the experiments reported above suggest that the discriminative training requires fewer training examples than an HMM-based alignment procedure while achieving the best reported results for this task.

## References

Boyd S and Vandenberghe L 2004 *Convex Optimization*. Cambridge University Press.

Brugnara F, Falavigna D and Omologo M 1993 Automatic segmentation and labeling of speech based on hidden Markov models. *Speech Communication* **12**, 357–370.

Cesa-Bianchi N, Conconi A and Gentile C 2004 On the generalization ability of on-line learning algorithms. *IEEE Transactions on Information Theory* **50**(9), 2050–2057.

Cristianini N and Shawe-Taylor J 2000 *An Introduction to Support Vector Machines*. Cambridge University Press.

Dean T and Kanazawa K 1989 A model for reasoning about persistence and causation. *Computational Intelligence* **5**(3), 142–150.

Dekel O, Keshet J and Singer Y 2004 Online algorithm for hierarchical phoneme classification *Workshop on Multimodal Interaction and Related Machine Learning Algorithms (Lecture Notes in Computer Science)*. Springer-Verlag, pp. 146–159.

Hosom JP 2002 Automatic phoneme alignment based on acoustic-phonetic modeling. *Proceedings of the 7th International Conference on Spoken Language Processing (ICSLP2002–INTERSPEECH2002)*, Denver, CO, September 16–20, 2002 (eds. J H L Hansen and B Pellom), pp. 357–360.

Keshet J, Shalev-Shwartz S, Singer Y and Chazan D 2007 A large margin algorithm for speech and audio segmentation. *IEEE Transactions on Audio, Speech and Language Processing* **15**(8), 2373–2382.

Lee KF and Hon HW 1989 Speaker independent phone recognition using hidden Markov models. *IEEE Transactions on Acoustic, Speech and Signal Processing* **37**(2), 1641–1648.

Pinto J and Hermansky H 2008 Combining evidence from a generative and a discriminative model in phoneme recognition. *Proceedings of the 10th International Conference on Spoken Language Processing (ICSLP2008–INTERSPEECH2008)*, Brisbane, Australia, September 22–26, 2008.

Rabiner L and Juang B-H 1993 *Fundamentals of Speech Recognition*. Prentice Hall.

Reynolds D 1997 HTIMIT and LLHDB: speech corpora for the study of handset transducer effects. *Proceedings of the International Conference on Audio, Speech and Signal Processing*, Munich, Germany, April 21–24, 1997, pp. 1535–1538.

Shalev-Shwartz S and Singer Y 2006 Online learning meets optimization in the dual. *Proceedings of the 19th Annual Conference on Computational Learning Theory*.

Taskar B, Guestrin C and Koller D 2003 Max-margin Markov networks *Advances in Neural Information Processing Systems 17*.

Toledano D, Gomez L and Grande L 2003 Automatic phoneme segmentation. *IEEE Transactions on Speech and Audio Processing* **11**(6), 617–625.

Tsochantaridis I, Hofmann T, Joachims T and Altun Y 2004 Support vector machine learning for interdependent and structured output spaces. *Proceedings of the 21st International Conference on Machine Learning*, Banff, Alberta, Canada, July 4–8, 2004.

# 5

# A Kernel Wrapper for Phoneme Sequence Recognition

**Joseph Keshet and Dan Chazan**

We describe a kernel wrapper, a Mercer kernel for the task of phoneme sequence recognition which is based on operations with the Gaussian kernel, and suitable for any sequence kernel classifier. We start by presenting a kernel based algorithm for phoneme sequence recognition, which aims at minimizing the Levenshtein distance (edit distance) between the predicted phoneme sequence and the true phoneme sequence. Motivated by the good results of frame based phoneme classification using Support Vector Machines (SVMs) with Gaussian kernel, we devised a kernel for speech utterances and phoneme sequences which generalizes the kernel function for phoneme frame based classification and adds timing constraints in the form of transition and duration constraints. The kernel function has three parts corresponding to phoneme acoustic model, phoneme duration model and phoneme transition model. We present initial encouraging experimental results with the TIMIT corpus.

## 5.1 Introduction

Kernel based algorithms enable to exploit a set of nonlinear feature functions which can replace the traditional probability function estimations, such as those used in Hidden Markov Models (HMMs). Moreover, they enable the use of phoneme specific feature functions to define the special properties of each phoneme. For example, a special time-frequency resolution can be used for the acoustic feature of the stop consonants, and a generic time-frequency resolution can be used for the rest of the phonemes. The feature functions do not have to be probabilities and can be nonlinear by using Mercer kernels (Vapnik 1998).

In this chapter we devise a new set of Mercer kernels for phoneme sequence recognition. The kernel functions are specifically designed for prediction of phoneme sequences given

*Automatic Speech and Speaker Recognition: Large Margin and Kernel Methods*   Joseph Keshet and Samy Bengio
© 2009 John Wiley & Sons, Ltd

acoustic feature vectors, and can be used with any sequence prediction kernel based classifier, such as the structured SVM algorithm (Taskar *et al.* 2003; Tsochantaridis *et al.* 2005). In parallel to the HMM phoneme recognizers, the kernel function has three parts corresponding to phoneme acoustic model, phoneme duration model and phoneme transition model.

Although our kernel functions can be used with any sequence prediction kernel based algorithm, we used them here with a kernel based algorithm which aims at minimizing the Levenshtein distance between the model based predicted phoneme sequence and the correct one (Keshet *et al.* 2006). We show that this algorithm has good performance on the test data both theoretically and empirically. Most previous work on phoneme sequence recognition has focused on HMMs. See for example Chapter 7 and the many references therein. These models are all based on probability estimations and maximization of the sequence likelihood and they are not trained to minimize the Levenshtein distance, the cost usually used to assess the quality of phoneme sequence prediction.

The chapter is organized as follows. In Section 5.2 we formally introduce the phoneme sequence recognition problem. Our kernel wrapper is designed to extend the frame based phoneme classifier which is presented in Section 5.3. Next, a kernel based iterative algorithm which minimizes the Levenshtein distance is given in Section 5.4. In Section 5.5 we present the kernel functions used for phoneme recognition. Preliminary experimental results are given in Section 5.6. We conclude the chapter with a discussion in Section 5.7.

## 5.2   Problem Setting

In the problem of phoneme sequence recognition, we are given a speech utterance and our goal is to predict the phoneme sequence corresponding to it. We represent a speech signal as a sequence of acoustic feature-vectors $\bar{\mathbf{x}} = (\mathbf{x}_1, \ldots, \mathbf{x}_T)$, where $\mathbf{x}_t \in \mathcal{X}$ for all $1 \leq t \leq T$. Each utterance corresponds to a sequence of phoneme symbols. Formally, we denote each phoneme symbol by $p \in \mathcal{P}$, where $\mathcal{P}$ is a set of phoneme symbols, and we denote the sequence of phoneme symbols by $\bar{p} = (p_1, \ldots, p_L)$. Furthermore, we denote by $s_l \in \mathbb{N}$ the start time of phoneme $p_l$ (in frame units) and we denote by $\bar{s} = (s_1, \ldots, s_L)$ the sequence of phoneme start-times, which corresponds to the phoneme sequence $\bar{p}$. Naturally, the length of the speech signal and hence the number of phonemes varies from one utterance to another and thus $T$ and $L$ are not fixed. We denote by $\mathcal{P}^*$ (and similarly $\mathcal{X}^*$ and $\mathbb{N}^*$) the set of all finite-length sequences over $\mathcal{P}$. Our goal is to learn a function $f$ that predicts the correct phoneme sequence given an acoustic sequence. That is, $f$ is a function from $\mathcal{X}^*$ to the set of finite-length sequences over the domain of phoneme symbols, $\mathcal{P}^*$. We also refer to $f$ as a phoneme sequence recognizer or predictor.

The ultimate goal of the phoneme sequence prediction is usually to minimize the Levenshtein distance between the predicted sequence and the correct one. The Levenshtein distance (also called edit distance) is a string metric. The Levenshtein distance between two strings is given by the minimum number of operations needed to transform one string into the other, where the allowed operations are insertion, deletion, or substitution of a single character. For example, the Levenshtein distance between *kitten* and *sitting* is three, since these three edits change one into the other, and there is no way to do it with fewer than three edits (two substitutions and one insertion). The best known use of the Levenshtein distance is in the application of spellcheck. However, it is also used to assess the quality of predicting a phoneme sequence (Lee and Hon 1989) or to compute the word error rate in speech recognition. Throughout this chapter we denote by $\gamma(\bar{p}, \bar{p}')$ the Levenshtein distance between the predicted phoneme sequence $\bar{p}'$ and the true phoneme sequence $\bar{p}$.

## 5.3 Frame-based Phoneme Classifier

Before describing the kernel based phoneme *recognizer* let us take a brief detour and describe a kernel based phoneme *classifier*. A phoneme classifier is a function $g : \mathcal{X} \times \mathcal{P} \to \mathbb{R}$, which, given a vector of acoustic features $\mathbf{x}_t \in \mathcal{X}$ extracted from the $t$th frame and a phoneme symbol $p_t$, returns a score indicating the confidence that the phoneme $p_t$ is uttered at the $t$th frame. Note that throughout this chapter we denote by $p_l$ the $l$th phoneme is the sequence $\bar{p}$, and by $p_t$ the phoneme uttered at the $t$th frame.

In hybrid Artificial Neural Network (ANN) and HMM systems (Bourlard and Morgan 1994) the front-end estimates the *posterior* phoneme conditional probability density functions over the phoneme set on a frame-by-frame basis using Multilayer Perceptron (MLP), that is, $g$ is defined to return a score in the simplex $[0, 1]$ rather than in $\mathbb{R}$. Those posteriors are then decoded into a word sequence using the HMM Viterbi decoding. Analogously, Salomon *et al.* (2002) used an SVM to classify speech frames with the *one-versus-one* scheme, where one classifier is trained for every possible pair of phonemes, resulting in $|\mathcal{P}|(|\mathcal{P}| - 1)/2$ classifiers. The reported results for framewise accuracy were around 71%.

A more modern approach for *multiclass* SVM classification (Crammer and Singer 2001) uses the following classification scheme instead of one-versus-one (or one-versus-rest):

$$\hat{p}_t = \arg \max_p \; g(\mathbf{x}_t, p) = \arg \max_p \; \mathbf{w}^p \cdot \mathbf{x}_t, \qquad (5.1)$$

where $\mathbf{w}^p$ will be referred to as the prototype of phoneme $p$. This type of classifier finds the projection of the speech frame instance $\mathbf{x}_t$ over the prototypes of all phonemes $p \in \mathcal{P}$, and predicts the one with the highest score $\mathbf{w}^p \cdot \mathbf{x}_t$. The difference between this method and other methods which reduce the multiclass problem into a set of binary problems is that in this method all the prototypes $\mathbf{w}^p$ are trained jointly to minimize the multiclass error. Motivated by the good accuracy of these frame based methods, we describe a kernel wrapper for a whole phoneme sequence prediction. Before describing the kernel wrapper we describe briefly an efficient kernel based algorithm for phoneme recognition.

## 5.4 Kernel-based Iterative Algorithm for Phoneme Recognition

In this section we describe a discriminative supervised learning algorithm for learning a phoneme sequence recognizer $f$ from a training set of examples. This algorithm was published elsewhere (Keshet *et al.* 2006), and it is given here only for completeness. Each example in the training set is composed of an acoustic signal, $\bar{\mathbf{x}}$, a sequence of phonemes, $\bar{p}$, and a sequence of phoneme start-times, $\bar{s}$.

Our construction is based on a predefined vector feature function $\boldsymbol{\phi} : \mathcal{X}^\star \times (\mathcal{P} \times \mathbb{N})^\star \to \mathcal{H}$, where $\mathcal{H}$ is a reproducing kernel Hilbert space (RKHS). Thus, the input of this function is an acoustic representation, $\bar{\mathbf{x}}$, together with a candidate phoneme symbol sequence $\bar{p}$ and a candidate phoneme start time sequence $\bar{s}$. The feature function returns a vector in $\mathcal{H}$, where, intuitively, each element of the vector represents the confidence in the suggested phoneme sequence. For example, one element of the feature function can sum the number of times phoneme $p$ comes after phoneme $p'$, while other elements of the feature function may extract properties of each acoustic feature vector $\mathbf{x}_t$ provided that phoneme $p$ was pronounced at time $t$. The description of the concrete form of the feature function is deferred to Section 5.5.

Our goal is to learn a phoneme sequence recognizer $f$, which takes as input a sequence of acoustic features $\bar{\mathbf{x}}$ and returns a sequence of phoneme symbols $\bar{p}$. The form of the function $f$ we use is

$$f(\bar{\mathbf{x}}) = \arg\max_{\bar{p}} \left( \max_{\bar{s}} \ \mathbf{w} \cdot \boldsymbol{\phi}(\bar{\mathbf{x}}, \bar{p}, \bar{s}) \right), \tag{5.2}$$

where $\mathbf{w} \in \mathcal{H}$ is a vector of importance weights that should be learned. In words, $f$ returns a suggestion for a phoneme sequence by maximizing a weighted sum of the scores returned by the feature function elements. Learning the weight vector $\mathbf{w}$ is analogous to the estimation of the parameters of the local probability functions in HMMs. Our approach, however, does not require $\mathbf{w}$ to take a probabilistic form. The maximization defined by Equation (5.2) is over an exponentially large number of all possible phoneme sequences. Nevertheless, as in HMMs, if the feature function, $\boldsymbol{\phi}$, is decomposable, the optimization in Equation (5.2) can be efficiently calculated using a dynamic programming procedure.

We now describe a simple iterative algorithm for learning the weight vector $\mathbf{w}$. The algorithm receives as input a training set $\mathcal{T} = \{(\bar{\mathbf{x}}_1, \bar{p}_1, \bar{s}_1), \dots, (\bar{\mathbf{x}}_m, \bar{p}_m, \bar{s}_m)\}$ of examples. Initially we set $\mathbf{w} = \mathbf{0}$. At each iteration the algorithm updates $\mathbf{w}$ according to the $i$th example in $\mathcal{T}$ as we now describe. Denote by $\mathbf{w}_{i-1}$ the value of the weight vector before the $i$th iteration. Let $(\bar{p}'_i, \bar{s}'_i)$ be the predicted phoneme sequence for the $i$th example according to $\mathbf{w}_{i-1}$:

$$(\bar{p}'_i, \bar{s}'_i) = \arg\max_{(\bar{p}, \bar{s})} \ \mathbf{w}_{i-1} \cdot \boldsymbol{\phi}(\bar{\mathbf{x}}_i, \bar{p}, \bar{s}). \tag{5.3}$$

We set the next weight vector $\mathbf{w}_i$ to be the minimizer of the following optimization problem:

$$\mathbf{w}_i = \arg\min_{\mathbf{w} \in \mathcal{H}, \xi \geq 0} \frac{1}{2}\|\mathbf{w} - \mathbf{w}_{i-1}\|^2 + C\xi$$

$$\text{s.t. } \mathbf{w} \cdot \boldsymbol{\phi}(\bar{\mathbf{x}}_i, \bar{p}_i, \bar{s}_i) - \mathbf{w} \cdot \boldsymbol{\phi}(\bar{\mathbf{x}}_i, \bar{p}'_i, \bar{s}'_i) \geq \gamma(\bar{p}_i, \bar{p}'_i) - \xi, \tag{5.4}$$

where $C$ serves as a complexity–accuracy trade-off parameter as in the SVM algorithm (see Cristianini and Shawe-Taylor (2000)) and $\xi$ is a non-negative slack variable, which indicates the loss of the $i$th example. Intuitively, we would like to minimize the loss of the current example, i.e. the slack variable $\xi$, while keeping the weight vector $\mathbf{w}$ as close as possible to our previous weight vector $\mathbf{w}_{i-1}$. The constraint makes the projection of the correct phoneme sequence $(\bar{p}_i, \bar{s}_i)$ onto $\mathbf{w}$ higher than the projection of the predicted phoneme sequence $(\bar{p}'_i, \bar{s}'_i)$ onto $\mathbf{w}$ by at least the Levenshtein distance between them. Using similar derivation as in Keshet *et al.* (2007), it can be shown that the solution to the above optimization problem is

$$\mathbf{w}_i = \mathbf{w}_{i-1} + \alpha_i \Delta \boldsymbol{\phi}_i, \tag{5.5}$$

where $\Delta \boldsymbol{\phi}_i = \boldsymbol{\phi}(\bar{\mathbf{x}}_i, \bar{p}_i, \bar{s}_i) - \boldsymbol{\phi}(\bar{\mathbf{x}}_i, \bar{p}'_i, \bar{s}'_i)$. The value of the scalar $\alpha_i$ is based on the Levenshtein distance $\gamma(\bar{p}_i, \bar{p}'_i)$, the different scores that $\bar{p}_i$ and $\bar{p}'_i$ received according to

---

**Input:** training set $\mathcal{T} = \{(\bar{\mathbf{x}}_i, \bar{p}_i, \bar{s}_i)\}_{i=1}^m$; validation set $\mathcal{T}_{\text{valid}} = \{(\bar{\mathbf{x}}_i, \bar{p}_i, \bar{s}_i)\}_{i=1}^{m_{\text{valid}}}$; parameter $C$

**Initialize:** $\mathbf{w}_0 = \mathbf{0}$

**For** $i = 1, \ldots, m$

  **Predict:** $(\bar{p}_i', \bar{s}_i') = \arg \max_{\bar{p}, \bar{s}} \mathbf{w}_{i-1} \cdot \boldsymbol{\phi}(\bar{\mathbf{x}}_i, \bar{p}, \bar{s})$

  **Set:** $\Delta\boldsymbol{\phi}_i = \boldsymbol{\phi}(\bar{\mathbf{x}}_i, \bar{p}_i, \bar{s}_i) - \boldsymbol{\phi}(\bar{\mathbf{x}}_i, \bar{p}_i', \bar{s}_i')$

  **Set:** $\ell(\mathbf{w}_{i-1}; \bar{\mathbf{x}}_i, \bar{p}_i, \bar{s}_i) = \max\{\gamma(\bar{p}_i, \bar{p}_i') - \mathbf{w}_{i-1} \cdot \Delta\boldsymbol{\phi}_i, 0\}$

  **If** $\ell(\mathbf{w}_{i-1}; \bar{\mathbf{x}}_i, \bar{p}_i, \bar{s}_i) > 0$

    **Set:** $\alpha_i = \min\left\{C, \dfrac{\ell(\mathbf{w}_{i-1}; \bar{\mathbf{x}}_i, \bar{p}_i, \bar{s}_i)}{\|\Delta\boldsymbol{\phi}_i\|^2}\right\}$

  **Update:** $\mathbf{w}_i = \mathbf{w}_{i-1} + \alpha_i \cdot \Delta\boldsymbol{\phi}_i$

**Output:** The weight vector $\mathbf{w}^*$ which achieves best performance on a validation set $\mathcal{T}_{\text{valid}}$:

$$\mathbf{w}^* = \arg \min_{\mathbf{w} \in \{\mathbf{w}_1, \ldots, \mathbf{w}_m\}} \sum_{j=1}^{m_{\text{valid}}} \gamma(\bar{p}_j, f(\bar{\mathbf{x}}_j))$$

---

Figure 5.1 An iterative algorithm for phoneme recognition.

$\mathbf{w}_{i-1}$ and a parameter $C$. Formally,

$$\alpha_i = \min\left\{C, \frac{\max\{\gamma(\bar{p}_i, \bar{p}_i') - \mathbf{w}_{i-1} \cdot \Delta\boldsymbol{\phi}_i, 0\}}{\|\Delta\boldsymbol{\phi}_i\|^2}\right\}. \tag{5.6}$$

A pseudo-code of our algorithm is given in Figure 5.1.

Before analyzing the algorithm we define the following hinge loss function for phoneme sequence recognition:

$$\ell(\mathbf{w}; (\bar{\mathbf{x}}_i, \bar{p}_i, \bar{s}_i)) = \max_{\bar{p}', \bar{s}'}[\gamma(\bar{p}_i, \bar{p}') - \mathbf{w} \cdot (\boldsymbol{\phi}(\bar{\mathbf{x}}_i, \bar{p}_i, \bar{s}_i) - \boldsymbol{\phi}(\bar{\mathbf{x}}_i, \bar{p}', \bar{s}'))]_+, \tag{5.7}$$

where $[a]_+ = \max\{0, a\}$. The phoneme sequence loss function differs from the alignment loss function defined in Chapter 4 only by the cost function $\gamma$. Here, the cost is the Levenshtein distance, while in the alignment case the cost is defined in Equation (4.1).

We now analyze our algorithm from Figure 5.1. Our first theorem shows that under some mild technical conditions, the cumulative Levenshtein distance of the iterative procedure, $\sum_{i=1}^m \gamma(\bar{p}_i, \bar{p}_i')$, is likely to be small.

**Theorem 5.4.1** *Let* $\mathcal{T} = \{(\bar{\mathbf{x}}_1, \bar{p}_1, \bar{s}_1), \ldots, (\bar{\mathbf{x}}_m, \bar{p}_m, \bar{s}_m)\}$ *be a set of training examples and assume that for all* $i$, $\bar{p}'$ *and* $\bar{s}'$ *we have that* $\|\boldsymbol{\phi}(\bar{\mathbf{x}}_i, \bar{p}', \bar{s}')\| \leq 1/2$. *Assume there exists a weight vector* $\mathbf{w}^*$ *that satisfies*

$$\mathbf{w} \cdot \boldsymbol{\phi}(\bar{\mathbf{x}}_i, \bar{p}_i, \bar{s}_i) - \mathbf{w} \cdot \boldsymbol{\phi}(\bar{\mathbf{x}}_i, \bar{p}', \bar{s}') \geq \gamma(\bar{p}_i, \bar{p}')$$

*for all $1 \leq i \leq m$ and $\bar{p}'$. Let $\mathbf{w}_1, \ldots, \mathbf{w}_m$ be the sequence of weight vectors obtained by the algorithm in Figure 5.1 given the training set $\mathcal{T}$. Then,*

$$\frac{1}{m} \sum_{i=1}^{m} \ell(\mathbf{w}_i; (\bar{\mathbf{x}}_i, \bar{p}_i, \bar{s}_i)) \leq \frac{1}{m} \sum_{i=1}^{m} \ell(\mathbf{w}^\star; (\bar{\mathbf{x}}_i, \bar{p}_i, \bar{s}_i)) + \frac{1}{Cm} \|\mathbf{w}^\star\|^2 + \frac{1}{2} C. \tag{5.8}$$

*In particular, if $C = 1/\sqrt{m}$ then,*

$$\frac{1}{m} \sum_{i=1}^{m} \ell(\mathbf{w}_i; (\bar{\mathbf{x}}_i, \bar{p}_i, \bar{s}_i)) \leq \frac{1}{m} \sum_{i=1}^{m} \ell(\mathbf{w}^\star; (\bar{\mathbf{x}}_i, \bar{p}_i, \bar{s}_i)) + \frac{1}{\sqrt{m}} \left( \|\mathbf{w}^\star\|^2 + \frac{1}{2} \right). \tag{5.9}$$

The proof of the theorem follows the line of reasoning of Theorem 4.5.1.

The loss bound of Theorem 5.4.1 can be translated into a bound on the Levenshtein distance error as follows. Note that the hinge loss defined by Equation (5.7) is always greater than $\gamma(\bar{p}_i, \bar{p}'_i)$:

$$\gamma(\bar{p}_i, \bar{p}') \leq \ell(\mathbf{w}_i; (\bar{\mathbf{x}}_i, \bar{p}_i, \bar{s}_i)).$$

Therefore, we get the following corollary.

**Corollary 5.4.2** *Under the conditions of Theorem 5.4.1 the following bound on the cumulative Levenshtein distance holds:*

$$\sum_{i=1}^{m} \gamma(\bar{p}_i, \bar{p}'_i) \leq \frac{1}{m} \sum_{i=1}^{m} \ell(\mathbf{w}^\star; (\bar{\mathbf{x}}_i, \bar{p}_i, \bar{s}_i)) + \frac{1}{\sqrt{m}} \left( \|\mathbf{w}^\star\|^2 + \frac{1}{2} \right). \tag{5.10}$$

The next theorem tells us that if the cumulative Levenshtein distance of the iterative procedure is small, there exists at least one weight vector among the vectors $\{\mathbf{w}_1, \ldots, \mathbf{w}_m\}$ which attains small averaged Levenshtein distance on unseen examples as well. To find this weight vector we simply calculate the averaged Levenshtein distance attained by each of the weight vectors on a validation set. The average Levenshtein distance in defined as

$$\mathbb{E}[\gamma(\bar{p}, \bar{p}')] = \mathbb{E}_{(\bar{\mathbf{x}}, \bar{p}, \bar{s}) \sim \mathcal{D}}[\gamma(\bar{p}, f(\bar{\mathbf{x}}))],$$

where $\bar{p}'$ is the phoneme sequence predicted by the function $f$ defined in Equation (5.2), and the expectation is taken over a fixed and unknown distribution $\mathcal{D}$ over the domain of the examples, $\mathcal{X}^* \times \mathcal{P}^* \times \mathbb{N}^*$.

**Theorem 5.4.3** *Under the same conditions of Theorem 5.4.1 assume that the training set $\mathcal{T}$ and the validation set $\mathcal{T}_{\text{valid}}$ are both sampled i.i.d. from a distribution $\mathcal{D}$. Denote by $m_{\text{valid}}$ the size of the validation set. Assume in addition that $\gamma(\bar{p}, \bar{p}') \leq 1$ for all $\bar{p}$ and $\bar{p}'$. Then, with probability of at least $1 - \delta$ we have that*

$$\mathbb{E}[\gamma(\bar{p}, \bar{p}')] \leq \frac{1}{m} \sum_{i=1}^{m} \ell(\mathbf{w}^\star; (\bar{\mathbf{x}}_i, \bar{p}_i, \bar{s}_i)) + \frac{\|\mathbf{w}^\star\|^2 + \frac{1}{2} + \sqrt{2 \ln(2/\delta)}}{\sqrt{m}} + \frac{\sqrt{2 \ln(2m/\delta)}}{\sqrt{m_{\text{valid}}}}. \tag{5.11}$$

The proof of the theorem goes along the lines of the proof of Theorem 4.5.2. Note that the Levenshtein distance is generally not upper bounded by 1. The theorem presented here is only to show a qualitative result, and can be changed to the case where $\gamma(\bar{p}, \bar{p}') \leq \gamma_0$, for $\gamma_0 \in \mathbb{R}$ a constant.

## 5.5 Nonlinear Feature Functions

Recall that the kernel based phoneme recognizer is not formulated with probabilities, but rather as a dot product of the vector feature function $\phi$ and a weight vector $\mathbf{w} \in \mathcal{H}$:

$$\bar{p}' = \arg\max_{\bar{p}} \left( \max_{\bar{s}} \ \mathbf{w} \cdot \phi(\bar{\mathbf{x}}, \bar{p}, \bar{s}) \right).$$

In the spirit of a generative HMM-based speech recognizer, this function can be written as a sum of the phoneme acoustic model, phoneme duration model and phoneme transition model:

$$\bar{p}' = \arg\max_{\bar{p}} \Big( \max_{\bar{s}} \ \mathbf{w}^{\text{acoustic}} \cdot \phi^{\text{acoustic}}(\bar{\mathbf{x}}, \bar{p}, \bar{s})$$

$$+ \mathbf{w}^{\text{duration}} \cdot \phi^{\text{duration}}(\bar{p}, \bar{s}) + \mathbf{w}^{\text{transition}} \cdot \phi^{\text{transition}}(\bar{p}) \Big). \tag{5.12}$$

where

$$\mathbf{w} = (\mathbf{w}^{\text{acoustic}}, \mathbf{w}^{\text{duration}}, \mathbf{w}^{\text{transition}})$$

and

$$\phi(\bar{\mathbf{x}}, \bar{p}, \bar{s}) = \left( \phi^{\text{acoustic}}(\bar{\mathbf{x}}, \bar{p}, \bar{s}), \phi^{\text{duration}}(\bar{p}, \bar{s}), \phi^{\text{transition}}(\bar{p}) \right).$$

Let us now present each of the models in detail.

### 5.5.1 Acoustic Modeling

Assume that we would like to use the multiclass phoneme classifier described in Section 5.3 to classify a whole speech utterance. Let us also assume that we already have the set of prototypes $\{\mathbf{w}^1, \ldots, \mathbf{w}^{|\mathcal{P}|}\}$, which defines the multiclass phoneme classifier as in Equation (5.1). The predicted phoneme sequence given the input speech utterance $\bar{\mathbf{x}} = (\mathbf{x}_1, \ldots, \mathbf{x}_T)$ using the phoneme classifier is

$$(p_1, \ldots, p_T) = \arg\max_{(p_1, \ldots, p_T)} \sum_{t=1}^{T} \mathbf{w}^{p_t} \cdot \mathbf{x}_t. \tag{5.13}$$

The advantage of a phoneme recognizer over a phoneme classifier is that it can incorporate temporal relations into the recognition process. Such a temporal relation can be stated in terms of minimum and maximum phoneme duration or in terms of the allowable transition between phonemes. We base our phoneme recognition kernel on the sum of the scores:

$$f(\bar{\mathbf{x}}, \bar{p}, \bar{s}) = \sum_{t=1}^{T} \mathbf{w}^{p_t} \cdot \mathbf{x}_t \tag{5.14}$$

$$= \sum_{l=1}^{|\bar{p}|} \sum_{t=s_l}^{s_{l+1}-1} \mathbf{w}^{p_l} \cdot \mathbf{x}_t \tag{5.15}$$

$$= \sum_{p \in \mathcal{P}} \mathbf{w}^p \cdot \sum_{l=1}^{|\bar{p}|} \sum_{t=s_l}^{s_{l+1}-1} \mathbb{1}_{\{p=p_l\}} \mathbf{x}_t. \tag{5.16}$$

In the transition from Equation (5.14) to Equation (5.15) we change the summation over frames to summation over phonemes of a phoneme sequence. Note that the score $f(\bar{\mathbf{x}}, \bar{p}, \bar{s})$ can be thought of as the non-probabilistic equivalent of the likelihood of the acoustic sequence $\bar{\mathbf{x}}$ given the phoneme sequence $\bar{p}$ and the timing sequence $\bar{s}$.

Denote by $\mathbf{w}^{\text{acoustic}}$ the weight vector composed of a concatenation of the prototypes $\mathbf{w}^p$, $1 \le p \le |\mathcal{P}|$, namely, $\mathbf{w}^{\text{acoustic}} = (\mathbf{w}^1, \ldots, \mathbf{w}^{|\mathcal{P}|})$. We can now define the acoustic feature function of phoneme $p$ as

$$\boldsymbol{\phi}^p(\bar{\mathbf{x}}, \bar{p}, \bar{s}) = \sum_{l=1}^{|\bar{p}|} \sum_{t=s_l}^{s_{l+1}-1} \mathbb{1}_{\{p=p_l\}} \mathbf{x}_t. \tag{5.17}$$

Hence we have

$$f(\bar{\mathbf{x}}, \bar{p}, \bar{s}) = \sum_{p \in \mathcal{P}} \mathbf{w}^p \boldsymbol{\phi}^p(\bar{\mathbf{x}}, \bar{p}, \bar{s}) = \mathbf{w}^{\text{acoustic}} \cdot \boldsymbol{\phi}^{\text{acoustic}}(\bar{\mathbf{x}}, \bar{p}, \bar{s}), \tag{5.18}$$

where $\boldsymbol{\phi}^{\text{acoustic}}(\bar{\mathbf{x}}, \bar{p}, \bar{s}) = (\boldsymbol{\phi}^1(\bar{\mathbf{x}}, \bar{p}, \bar{s}), \ldots, \boldsymbol{\phi}^{|\mathcal{P}|}(\bar{\mathbf{x}}, \bar{p}, \bar{s}))$.

So far we have described the acoustic feature function $\boldsymbol{\phi}^{\text{acoustic}}(\bar{\mathbf{x}}, \bar{p}, \bar{s})$ as a linear function of the acoustic feature vectors $\{\mathbf{x}_t\}$. We now turn to describe the nonlinear version of the same feature function, based on Mercer kernels. First, we show that the weight vector $\mathbf{w}$ can always be written as a linear function of $\boldsymbol{\phi}(\bar{\mathbf{x}}, \bar{p}, \bar{s})$. Then we replace the linear dependency of $\boldsymbol{\phi}^{\text{acoustic}}$ in the acoustic vectors with a nonlinear dependency, and get a nonlinear phoneme recognizer.

Note that in the kernel based algorithm presented in Section 5.4, the weight vector $\mathbf{w}_i$ can be written as (see Equation (5.5))

$$\mathbf{w}_i = \mathbf{w}_{i-1} + \alpha_i \Delta \boldsymbol{\phi}_i = \mathbf{w}_{i-2} + \alpha_{i-1} \Delta \boldsymbol{\phi}_{i-1} + \alpha_i \Delta \boldsymbol{\phi}_i = \cdots = \sum_{j=1}^{i} \alpha_j \Delta \boldsymbol{\phi}_j. \tag{5.19}$$

This way of presenting $\mathbf{w}$ is common to all kernel based algorithms and it is supported by the *representer theorem* (Kimeldorf and Wahba 1971). Focusing on the acoustic part of $\mathbf{w}$ and expanding $\Delta \boldsymbol{\phi}_j$ we have

$$\mathbf{w}^{\text{acoustic}} = \sum_{j=1}^{i} \alpha_j \left( \boldsymbol{\phi}^{\text{acoustic}}(\bar{\mathbf{x}}_j, \bar{p}_j, \bar{s}_j) - \boldsymbol{\phi}^{\text{acoustic}}(\bar{\mathbf{x}}_j, \bar{p}'_j, \bar{s}'_j) \right). \tag{5.20}$$

Plugging Equation (5.17) into the last equation we have

$$\mathbf{w}^{\text{acoustic}} = \sum_{p \in \mathcal{P}} \sum_{j=1}^{i} \alpha_j^p (\boldsymbol{\phi}^p(\bar{\mathbf{x}}_j, \bar{p}_j, \bar{s}_j) - \boldsymbol{\phi}^p(\bar{\mathbf{x}}_j, \bar{p}'_j, \bar{s}'_j))$$

$$= \sum_{p \in \mathcal{P}} \sum_{j=1}^{i} \alpha_j^p \left( \sum_{l=1}^{|\bar{p}_j|} \sum_{t=s_{j_l}}^{s_{j_{l+1}}-1} \mathbb{1}_{\{p=p_{j_l}\}} \mathbf{x}_{j_t} - \sum_{l=1}^{|\bar{p}'_j|} \sum_{t=s'_{j_l}}^{s'_{j_{l+1}}-1} \mathbb{1}_{\{p=p'_{j_l}\}} \mathbf{x}_{j_t} \right)$$

$$= \sum_{p \in \mathcal{P}} \sum_{j=1}^{i} \alpha_j^p \sum_{t=1}^{|\bar{\mathbf{x}}_j|} (\mathbb{1}_{\{p=p_{j_t} \vee p \neq p'_{j_t}\}} \mathbf{x}_{j_t} - \mathbb{1}_{\{p \neq p_{j_t} \vee p = p'_{j_t}\}} \mathbf{x}_{j_t}), \tag{5.21}$$

Table 5.1 Calculation of phoneme sequence support patterns: only the bold lettered phonemes are added to the support set

| Frame $t$ | 1 | 2 | 3 | 4 | 5 | 6 | 7 | 8 | 9 | 10 |
|---|---|---|---|---|---|---|---|---|---|---|
| True phoneme seq. $\bar{p}$ | f | f | f | **aa** | aa | aa | aa | r | r | r |
| Predicted phoneme seq. $\bar{p}'$ | f | f | f | **f** | ih | ih | ih | r | r | r |

where $p_{j_t}$ is the phoneme uttered in the $t$th frame of the $j$th phoneme sequence. Plugging the last equation and Equation (5.17) into Equation (5.18) we get

$$
\mathbf{w}^{\text{acoustic}} \cdot \boldsymbol{\phi}^{\text{acoustic}} = \sum_{p \in \mathcal{P}} \sum_{j=1}^{i} \alpha_j^p \sum_{t=1}^{|\bar{\mathbf{x}}_j|} \left( \mathbb{1}_{\{p = p_{j_t} \vee\, p \neq p'_{j_t}\}} \sum_{l=1}^{|\bar{p}|} \sum_{t=s_l}^{s_{l+1}-1} \mathbb{1}_{\{p = p_l\}} \mathbf{x}_{j_t} \cdot \mathbf{x}_t \right.
$$

$$
\left. - \mathbb{1}_{\{p \neq p_{j_t} \vee\, p = p'_{j_t}\}} \sum_{l=1}^{|\bar{p}|} \sum_{t=s_l}^{s_{l+1}-1} \mathbb{1}_{\{p = p_l\}} \mathbf{x}_{j_t} \cdot \mathbf{x}_t \right). \tag{5.22}
$$

We can now replace the acoustic features $\{\mathbf{x}_t\}$ with a mapped version of it, $\psi(\mathbf{x}_t)$, where $\psi : \mathcal{X} \to \mathcal{H}$, $\mathcal{X}$ is the acoustic feature vector space and $\mathcal{H}$ is a Hilbert space. The mapping $\psi$ can be expressed implicitly using Mercer kernel which defines the inner product as $k(\mathbf{x}_{t_1}, \mathbf{x}_{t_2}) = \psi(\mathbf{x}_{t_1}) \cdot \psi(\mathbf{x}_{t_2})$.

Let us take a deeper look at Equation (5.22). The first sum is over all phonemes in the phoneme set, and it can be thought of as the sum over feature functions for each phoneme. The second sum is over the support set. The support vectors defining this set can be better understood by looking at Equation (5.21). Usually the support set is composed of those vectors which the iterative classifier predicted incorrectly during training, or which define the boundaries of the decision hyperplane. In a sequence-based classification, as we have here, the support is not composed of the whole sequence which the recognizer does not predict correctly but only the *misclassified frames* in these sequences. Suppose the frame $t$ was not predicted correctly, then the corresponding acoustic feature $\mathbf{x}_t$ becomes the $j$th support, with positive constant $\alpha_j^p$ for the correct label $p_{j_t}$ and negative constant $-\alpha_j^p$ for the mis-predicted label $p'_{j_t}$. This is depicted in Table 5.1.

### 5.5.2 Duration Modeling

The duration modeling is oblivious of the speech signal itself and merely examines the duration of each phoneme. Let $D = (d_1, \ldots, d_N)$ denote a set of $N$ predefined possible phoneme duration values. For each phoneme in our phoneme set $\rho \in \mathcal{P}$ and $d \in D$ let $\phi_{\rho,d}^{\text{duration}}(\bar{p}, \bar{s})$ denote the number of times the phoneme $\rho$ appeared in $\bar{p}$ while its duration was at least $d$; that is,

$$
\phi_{\rho,d}^{\text{duration}}(\bar{p}, \bar{s}) = |\{k : p_l = \rho \wedge (s_{l+1} - s_l) \geq d\}|. \tag{5.23}
$$

The overall duration modeling is a vector which builds as a concatenation of the terms in Equation (5.23), namely

$$
\boldsymbol{\phi}^{\text{duration}}(\bar{p}, \bar{s}) = \left( \phi_{\rho_1,d_1}^{\text{duration}}(\bar{p}, \bar{s}), \phi_{\rho_1,d_2}^{\text{duration}}(\bar{p}, \bar{s}), \ldots \right). \tag{5.24}
$$

### 5.5.3  Transition Modeling

Finally, we present the phoneme transition modeling. Let $A(p', p)$ be an estimated transition probability matrix from phoneme $p'$ to phoneme $p$. That is,

$$A(p', p) = \frac{\text{expected number of transitions from } p' \text{ to } p}{\text{expected number of transitions from } p'}.$$

Additionally, let $\Theta = (\theta_1, \ldots, \theta_M)$ be a set of threshold values. For each $\theta_i \in \Theta$ let $\phi^{\text{transition}}_{\theta, p_{l-1}, p_l}$ be the number of times a switch from phoneme $p_{l-1}$ to phoneme $p_l$ occurs in $\bar{p}$ such that $A(p_{l-1}, p_l)$ is at least $\theta$. That is,

$$\phi^{\text{transition}}_{\theta, p_{l-1}, p_l} (\bar{p}) = |\{k : A(p_{l-1}, p_l) \geq \theta\}|.$$

The overall transition modeling is a vector which builds as a concatenation of the terms.

## 5.6  Preliminary Experimental Results

To validate the effectiveness of the proposed approach we performed experiments with the TIMIT corpus. All the experiments described here have followed the same methodology. We divided the training portion of TIMIT (excluding the SA1 and SA2 utterances) into two disjoint parts containing 3600 and 96 utterances. The first part is used as a training set and the second part is used as a validation set. Mel-frequency cepstrum coefficients (MFCC) along with their first and second derivatives were extracted from the speech waveform in a standard way along with cepstral mean subtraction (CMS). The TIMIT original phoneme set of 61 phonemes was mapped to a 39 phoneme set as proposed by Lee and Hon (1989). Performance was evaluated over the TIMIT core test set by calculating the Levenshtein distance between the predicted phoneme sequence and the correct one. This test set was proposed in the TIMIT documentation and includes representative 192 utterances from all TIMIT test utterances.

We applied our method as discussed in Section 5.4 and Section 5.5 where the hyper-parameters were chosen on the validation set ($\sigma^2 = 14$, $C = 1$). We compared the results of our method to the *monophone* HMM approach, where each phoneme was represented by a simple left-to-right HMM of five emitting states with 40 diagonal Gaussians. These models were enrolled as follows: first the HMMs were initialized using $K$-means, and then enrolled independently using EM. The second step, often called embedded training, re-enrolls all the models by relaxing the segmentation constraints using a forced alignment. Minimum values of the variances for each Gaussian were set to 20% of the global variance of the data. All HMM experiments were done using the *Torch* package (Collobert *et al.* 2002). All hyper-parameters including number of states, number of Gaussians per state, variance flooring factor were tuned using the validation set. The overall results are given in Table 5.2. We report the number of insertions, deletions and substitutions, as calculated by the Levenshtein distance. The Levenshtein distance is defined as the sum of insertions, deletions, and substitutions. 'Accuracy' stands for 100% minus the Levenshtein distance and 'correct' stands for accuracy plus insertions. As can be seen, the HMM method outperforms our method in terms of accuracy, mainly due to the high level of insertions of our method, suggesting that a better duration model should be explored. Nevertheless, we believe that the potential of our method is larger than the results reported.

Table 5.2 Phoneme recognition results on TIMIT core test set. The label 'Kernel based' refers to the discriminative algorithm using the kernel wrapper with Gaussian kernel, while the label 'HMM' refers to a monophone (no context) Hidden Marko Model

|  | Correct | Accuracy | Insertions | Deletions | Substitutions |
|---|---|---|---|---|---|
| Kernel based | 64.6 | 54.6 | 9.9 | 9.2 | 26.3 |
| HMM | 62.7 | 59.1 | 3.6 | 10.5 | 26.8 |

## 5.7 Discussion: Can we Hope for Better Results?

In this chapter we showed a kernel wrapper for phoneme recognition. The kernel wrapper was built so as to generalize the SVM frame based phoneme classifier and to improve its performance. However, the performance of the kernel wrapper with the discriminative iterative algorithm was found to be inferior to the performance of a monophone HMM. We assume there are several reasons for that.

**The Algorithm.** While the kernel wrapper presented here is suitable for any structured SVM optimizer for sequences, we experimented it with the iterative algorithm. The iterative algorithm is shown theoretically to be competitive with the optimal classifier of the form of Equation (5.2), which attains small Levenshtein distance error. Theorem 5.4.3 promises that when the number of training utterances, $m$, goes to infinity, the iterative algorithm will converge. It might be the case that the convergence rate of the algorithm is not fast enough and the training set is too small. This assumption is indeed verified while looking at the learning curve of the algorithm depicted in Figure 5.2. In the figure we see the cumulative iterative accuracy (also called online error) as a function of the iteration number, that is, the accuracy in predicting the next training utterance before receiving its phonetic labeling. This graph suggests that there is room for improvements of accuracy if there were more training examples drawn from the same distribution, since the graph does not reach pick performance asymptote. A potential remedy to this problem might be to use a different algorithm, such as a batch structured SVM algorithm with Hildreth optimization (Hildreth 1957). Such an algorithm visits each utterance many times, and converges to the SVM solution. Unfortunately, this is also a problem. The number of support patterns in the iterative algorithm is around half a million, which is more than half the size of the training set. This is more or less the expected number of support patterns in the batch algorithm, which means a very heavy computing load.

**Incorrect Assumptions.** In Section 5.3 we assumed that the joint training of the multiclass problem is better than the one-versus-one training scheme. It may not be the case for the task of phoneme recognition, although it was shown in Crammer and Singer (2001) that joint training outperforms other multiclass methods for SVM. Another hidden assumption is that the maximization over the sequence is better than the maximization over the sum of each frame as given in Equation (5.14). Comparing with other domains, like vision and text, we believe that these assumptions are reasonable and not the major source of the accuracy problem.

**Efficiency.** One of the most acute problems of the algorithm is the training and test computational load in terms of both time and space. The support set includes more than

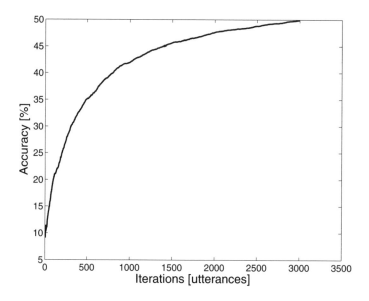

Figure 5.2 Cumulative accuracy during the iterative training of phoneme recognition.

half of the training set. The prediction itself depends linearly on the size of the support, and inference of each frame involved in applying the kernel operation for every single support pattern.

Future work will be focused on more accurate algorithms which converge faster to the SVM solution and on algorithms with reduced support set such as the *Forgetron* (Dekel *et al.* 2008) and the *Projectron* (Orabona *et al.* 2008).

# References

Bourlard H and Morgan N 1994 *Connectionist Speech Recognition: A Hybrid Approach*. Kluwer Academic Publishers.

Collobert R, Bengio S and Mariéthoz J 2002 Torch: a modular machine learning software library. IDIAP-RR 46, IDIAP.

Crammer K and Singer Y 2001 On the algorithmic implementation of multiclass kernel-based vector machines. *Journal of Machine Learning Research* **2**, 265–292.

Cristianini N and Shawe-Taylor J 2000 *An Introduction to Support Vector Machines*. Cambridge University Press.

Dekel O, Shalev-Shwartz S and Singer Y 2008 The Forgetron: A kernel-based perceptron on a budget. *SIAM Journal of Computing* **34**, 1342–1372.

Hildreth C 1957 A quadratic programming procedure. *Naval Research Logistics Quarterly* **4**, 79–85. Erratum, ibidem, p. 361.

Keshet J, Shalev-Shwartz S, Bengio S, Singer Y and Chazan D 2006 Discriminative kernel-based phoneme sequence recognition. *Proceedings of the 9th International Conference on Spoken Language Processing*, Pittsburgh, PA, September 16–21, 2006, paper 1284-Mon3BuP.2.

Keshet J, Shalev-Shwartz S, Singer Y and Chazan D 2007 A large margin algorithm for speech and audio segmentation. *IEEE Transactions on Audio, Speech and Language Processing* **15**(8), 2373–2382.

Kimeldorf G and Wahba G 1971 Some results on tchebycheffian spline functions. *Journal of Mathematical Analysis and Applications* **33**, 82–95.

Lee KF and Hon HW 1989 Speaker independent phone recognition using hidden Markov models. *IEEE Transactions on Acoustic, Speech and Signal Processing* **37**(2), 1641–1648.

Orabona F, Keshet J and Caputo B 2008 The Projectron: a bounded kernel-based Perceptron. *Proceedings of the 25th International Conference on Machine Learning*, Helsinki, Finland, July 5–9, 2008, pp. 720–727.

Salomon J, King S and Osborne M 2002 Framewise phone classification using support vector machines. *Proceedings of the 7th International Conference on Spoken Language Processing (ICSLP2002–INTERSPEECH2002)*, Denver, CO, September 16–20, 2002 (eds. J H L Hansen and B Pellom), pp. 2645–2648.

Taskar B, Guestrin C and Koller D 2003 Max-margin Markov networks. *Advances in Neural Information Processing Systems 17*.

Tsochantaridis I, Joachims T, Hofmann T and Altun Y 2005 Large margin methods for structured and interdependent output variables. *Journal of Machine Learning Research* **6**, 1453–1484.

Vapnik VN 1998 *Statistical Learning Theory*. John Wiley & Sons.

# 6

# Augmented Statistical Models: Using Dynamic Kernels for Acoustic Models

Mark J. F. Gales

There has been significant interest in developing alternatives to Hidden Markov Models (HMMs) as the underlying acoustic model for speech recognition. In particular, statistical models which improve the weak temporal dependency representations in HMMs have been popular. A number of schemes have been proposed based on additional observation, or latent variable, dependencies. These approaches will be briefly reviewed in this chapter followed by a detailed look at an alternative approach related to dynamic, or sequence, kernels. The feature-spaces associated with dynamic kernels are one scheme for extracting information from data sequences such as speech. These features, and the resulting kernels, may be directly used in a classifier such as a Support Vector Machine (SVM). However, they may also be viewed as features for training statistical models. One appropriate form of dynamic kernel for this problem is the generative kernel, a generalization of the Fisher kernel. This kernel uses a 'base' generative model, here an HMM, to extract features from the data. These features do not have the same conditional independence assumptions as the original base generative model used to derive them. Thus, they yield a systematic approach to adding temporal dependencies into statistical models. These models are referred to as Augmented Statistical Models. Two forms of Augmented Statistical Model are discussed in this chapter. The first is a generative form, for which a large margin approach to training the parameters is described. The second form of model is a discriminative one. Here the model parameters are trained using conditional maximum likelihood training. The performance of these two models is evaluated on a TIMIT phone classification task.

*Automatic Speech and Speaker Recognition: Large Margin and Kernel Methods*    Joseph Keshet and Samy Bengio
© 2009 John Wiley & Sons, Ltd

# 6.1   Introduction

There has been a wide variety of acoustic models applied to the speech recognition task. These range from the standard HMM, to segmental models (Ostendorf *et al.* 1996), switching linear dynamical systems (LDSs) (Rosti and Gales 2003), Buried Markov Models (BMMs) (Bilmes 2003a) and mixed memory models (Nock 2001). Many of these models can be viewed as graphical models (Rosti and Gales 2003). The underlying aspect of all these models is how to appropriately model the dependencies (and complexities) of the speech signal. For example, observations may be assumed to be conditionally independent of all other observations given the state that generated them, as in an HMM (Rabiner 1989). Though HMMs have one of the simplest forms of temporal correlation modeling, they are currently the dominant form of acoustic model for speech recognition. When considering alternatives, the fundamental questions that must be answered are: what latent variables should be used? what dependencies between observations should be modelled? and how are the distributions of the observations altered by these dependencies?

In this chapter a structured approach to obtaining the statistics to model the dependencies between observations is described. Dynamic kernels, sometimes called sequence kernels, are used to specify the statistics. These kernels have been applied in a range of speech processing tasks: speaker verification (Campbell *et al.* 2005; Wan and Renals 2004), speech recognition (Smith and Gales 2001) and spoken language understanding (Cortes *et al.* 2003). The interesting attribute of many of these dynamic kernels is that the feature-spaces associated with them contain temporal information about the data sequences. One standard scheme is to use these kernels directly in a classifier, such as a SVM (Vapnik 1998). An interesting alternative is described in this chapter. Here the feature-spaces from these kernels are used as features in a statistical model. This type of model will be referred to as an *Augmented Statistical Model* (Layton 2006; Smith 2003), since the approach increases the modeling ability of a standard initial statistical model. A particular form of dynamic kernel is examined here for use in Augmented Statistical Models, the *generative kernel* (Layton and Gales 2004), which is closely related to the Fisher kernel (Jaakkola and Haussler 1999). This class of kernel has some nice attributes when used in this form. It may be viewed as a Taylor series expansion of standard generative models (Smith 2003; Smith and Gales 2001), as well as being related to *constrained exponential models* (Amari and Nagaoka 2000).

The next section briefly describes some of the standard approaches used in speech recognition for temporal correlation modeling. This is followed by a discussion on kernels and specifically dynamic kernels. Two forms of Augmented Statistical Model are then described, one generative, the other discriminative. A description of some initial experiments using both of these forms of augmented model is then given.

# 6.2   Temporal Correlation Modeling

One of the challenges, and interests, in speech processing applications is that the training and test data comprises sequences of observations. For example in speech recognition it is common to use a fixed frame-rate of about 10 ms to extract feature vectors from the digitized waveform. With a fixed frame-rate the number of feature vectors for a series of utterances of even the same word, for example the digit *one*, will vary. Any form of classifier for

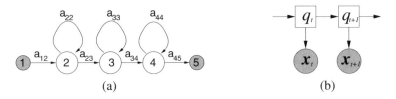

(a)                                    (b)

Figure 6.1  Two representations of Hidden Markov Models. (a) Typical phone topology. (b) Dynamic Bayesian Network.

speech recognition must be able to handle this variability in the length of the data sequences. Furthermore, the feature-vectors will be heavily correlated in time due to the nature of speech production. It is also necessary to model this temporal correlation in some fashion. Attempting to explicitly model whole sequences has proved to be impractical to date. Thus current acoustic models make use of conditional independence assumptions to handle both these problems.

The most common form of acoustic model for speech recognition, and also used in a range of other sequence classification and labeling problems, is the HMM (Rabiner 1989). An HMM consists of a set of discrete states, with probabilities of transitions between the states. Associated with each state is an output distribution which may be discrete or continuous. The observations generated by each state are seen (these are the extracted feature vectors), whereas the state sequence that generates a particular observation sequence is hidden. A typical topology used in speech recognition is shown in Figure 6.1(a). There are three emitting states and two non-emitting states, the latter shown shaded. A strict left-to-right topology, as shown, is often enforced on the state transitions.

The HMM can also be described in terms of conditional independence assumptions. This form may be viewed as being complementary to the topological constraints described above. A standard way to represent these assumptions is to use a *Dynamic Bayesian Network* (DBN) (Bilmes 2003b). The DBN describes the conditional dependencies of the model. The DBN for an HMM is also shown in Figure 6.1(b). In the DBN notation used here, squares denote discrete variables; circles continuous variables; shading indicates an observed variable and no shading an unobserved variable. The lack of an arc between variables shows conditional independence. Thus Figure 6.1 indicates that the observations generated by an HMM are conditionally independent given the unobserved, hidden, state that generated it. To reduce the impact of this conditional independence assumption first and second derivatives (delta and delta–delta) parameters (Furui 1986) are often used in speech recognition systems. The likelihood of an observation sequence, $\bar{\mathbf{x}}_{1:T} = \mathbf{x}_1, \ldots, \mathbf{x}_T$, being generated by an HMM with parameters $\lambda$ can be expressed as

$$p(\bar{\mathbf{x}}_{1:T}; \lambda) = \sum_{\bar{q} \in Q_T} \left( \prod_{t=1}^{T} P(q_t|q_{t-1}) p(\mathbf{x}_t|q_t; \lambda) \right), \qquad (6.1)$$

where $Q_T$ is the set of all valid state sequences of length $T$ and $q_t$ is the state at time $t$ in sequence $\bar{q}$. Though the form of temporal modeling in HMMs is quite weak, it has proved to be remarkably successful for speech recognition.

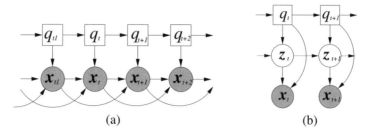

(a)                                                    (b)

Figure 6.2  Dynamic Bayesian Networks for Hidden Markov Model extensions. (a) Buried Markov models. (b) Switching linear dynamical system.

One of the desirable attributes of the DBN representation is that it is simple to show extensions to HMMs in terms of how they modify the conditional independence assumptions of the model. There are two approaches, which may be combined, to extend the HMM structure in Figure 6.1 to improve temporal modeling: adding unobserved, latent, variables; and adding dependency arcs between variables. An example of adding arcs, in this case between observations, is shown in Figure 6.2(a). Here the observation distribution is dependent on the previous two observations in addition to the state that generated it. This form describes various extensions to HMMs: explicit temporal correlation modeling (Wellekens 1987); vector predictors (Woodland 1992); and BMM (Bilmes 2003a). Although an interesting direction for refining an HMM, which yields large increases in log-likelihood values, this approach has not shown significant performance gains. One of the fundamental issues with this approach is to determine which dependencies should be added. Though discriminative approaches can be used for this selection (Bilmes 2003a), the choice is made from only a restricted subset of all possible dependencies, normally based on a few frames either side of the current frame. Augmented statistical models offer a more systematic way of adding longer-term dependencies to an existing statistical model.

Instead of adding arcs, more unobserved, latent or hidden, variables may be added. To compute probabilities, these hidden variables must then be marginalized out in the same fashion as the unobserved state sequence for HMMs. For continuous variables this requires integration over all values of the hidden variable and for discrete variables a summation over all values. In Figure 6.2(b), the DBN for a switching LDS (Digalakis 1992) is shown. Here another continuous latent variable is added that is conditionally independent of all other variables given the discrete state and previous continuous state value. The observations are dependent on this continuous state variable in addition to the standard HMM dependencies. Unfortunately to date performance with this form of model has been poor (Rosti and Gales 2003). An issue with adding latent variables is that it is not possible to precisely determine what variables should be added, and how they should influence the state output distribution. Augmented statistical models do not alter the form of latent variables in the acoustic models, relying instead on standard forms of latent variable models, such as HMMs and Gaussian Mixture Models (GMMs).

## 6.3  Dynamic Kernels

This section briefly summarizes some of the standard dynamic kernels that have been used in the speech processing and related fields. The general area of kernels is briefly discussed, with an emphasis on forms of dynamic kernels. Then generative kernels are described in detail including the nature of the feature-space. A simple example illustrating the advantages of using these generative kernels, and associated feature-space, for sequence classification is detailed.

## 6.3.1 Static and Dynamic Kernels

Kernel approaches are becoming increasingly popular in a wide range of machine learning applications. The basic concept is that the input space, the observation, is mapped to some higher dimensional feature-space via a nonlinear transformation $\boldsymbol{\phi}()$. The kernel is then the inner product, with an appropriate metric, of two observations in this high-dimensional space. When a fixed dimensional observation, $\mathbf{x}$, is used this will be referred to as a static kernel. The static feature-space mapping, $\boldsymbol{\phi}^{\mathrm{s}}()$, and associated kernel may be expressed as

$$k(\mathbf{x}_i, \mathbf{x}_j) = \langle \boldsymbol{\phi}^{\mathrm{s}}(\mathbf{x}_i), \boldsymbol{\phi}^{\mathrm{s}}(\mathbf{x}_j) \rangle, \tag{6.2}$$

where $\langle \cdot, \cdot \rangle$ indicates the inner-product. Standard static kernels are polynomial kernels and Gaussian kernels (Cristianini and Shawe-Taylor 2000). For these standard, static, kernels it is not necessary to explicitly map the input-space to the feature-space as the inner product can be expressed in terms of the original input-space. For example an in-homogenous polynomial kernel of order-2 can be expressed as

$$k(\mathbf{x}_i, \mathbf{x}_j) = (1 + \langle \mathbf{x}_i, \mathbf{x}_j \rangle)^2. \tag{6.3}$$

The choice of feature-space and resulting kernel becomes more interesting when sequence data is being used. Here the dynamic kernel operation is of the form

$$k(\bar{\mathbf{x}}^{(i)}, \bar{\mathbf{x}}^{(j)}) = \langle \boldsymbol{\phi}(\bar{\mathbf{x}}^{(i)}), \boldsymbol{\phi}(\bar{\mathbf{x}}^{(j)}) \rangle, \tag{6.4}$$

where $\bar{\mathbf{x}}^{(i)}$ is the $i$th sequence. The mapping from the input-space to the feature-space, $\boldsymbol{\phi}()$, is additionally required to map from the variable length sequences to a fixed dimension in which the inner product is calculated. A number of kernels have been proposed for handling sequence data (see Chapter 5), including marginalized count kernels (Tsuda *et al.* 2002), Fisher kernels (Jaakkola and Haussler 1999), string kernels (Saunders *et al.* 2003) and rational kernels (Cortes *et al.* 2003). Originally these kernels were examined for discrete sequences. For speech data they need to be able to handle sequences of continuous observations. One such kernel is the generative kernel (Layton and Gales 2004), which is related to Fisher kernels and is discussed in detail in the next section.

It is possible to combine static and dynamic kernels in two distinct fashions. First the input-space defined by the dynamic kernel can be kernelized (Layton 2006). This has the general form

$$k(\bar{\mathbf{x}}^{(i)}, \bar{\mathbf{x}}^{(j)}) = \langle \boldsymbol{\phi}^{\mathrm{s}}(\boldsymbol{\phi}(\bar{\mathbf{x}}^{(i)})), \boldsymbol{\phi}^{\mathrm{s}}(\boldsymbol{\phi}(\bar{\mathbf{x}}^{(j)})) \rangle. \tag{6.5}$$

An alternative is to kernelize the observation prior to the application of the dynamic kernel. Here

$$\boldsymbol{\phi}^{\mathrm{s}}(\bar{\mathbf{x}}_{1:T}) = \boldsymbol{\phi}^{\mathrm{s}}(\mathbf{x}_1), \ldots, \boldsymbol{\phi}^{\mathrm{s}}(\mathbf{x}_T). \tag{6.6}$$

One of the problems with using this form of static and dynamic kernel combination is that possible forms of dynamic kernel are limited unless the static input space is explicitly produced. For this reason the sequence kernel (Campbell *et al.* 2005) and related kernels (Mariéthoz and Bengio 2007) simply use an average, effectively a single component generative model. Some approximations to allow this form of kernel with more complex dynamic kernels are possible (Longworth and Gales 2008).

An interesting contrast to dynamic kernels for acoustic modeling, where additional information about the temporal correlations may be useful, are the kernels used for text-independent speaker verification (see Chapter 12). Here the emphasis is on mapping the variable length sequences to a fixed dimensionality rather than temporal correlation modeling, as in text independent verification there is little consistent temporal information between frames. A range of dynamic kernels and feature-spaces have been examined, which can be roughly partitioned into parametric and derivative kernels (Longworth and Gales 2007). Examples of parametric kernels are mean super-vector kernel (Campbell *et al.* 2006; Stolcke *et al.* 2005) and derivative kernels (Longworth and Gales 2007; Wan and Renals 2004) based on GMMs. One form of derivative kernel is the generative kernel described in the next section. In contrast to the GMMs used in speaker verification they will be based on HMMs, thus allowing additional temporal information to be incorporated.

## 6.3.2  Generative Kernels

An interesting class of dynamic kernels are based on feature-spaces obtained from derivatives with respect to the model parameters of the log-likelihood from a generative model. Both Fisher kernels (Jaakkola and Haussler 1999) and generative kernels (Layton and Gales 2004) fall in this category. Since the number of model parameters is fixed, the first, or higher, order derivatives of the log-likelihood with respect to the model parameters is a fixed size. Thus the feature-space associated with these kernels will be a fixed dimension. In generative kernels the feature space used has the form

$$
\phi(\bar{\mathbf{x}}_{1:T}; \lambda) = \begin{bmatrix} \log(p(\bar{\mathbf{x}}_{1:T}; \lambda)) \\ \dfrac{\partial}{\partial \lambda} \log(p(\bar{\mathbf{x}}_{1:T}; \lambda)) \\ \vdots \\ \dfrac{\partial^\rho}{\partial \lambda^\rho} \log(p(\bar{\mathbf{x}}_{1:T}; \lambda)) \end{bmatrix}, \tag{6.7}
$$

where $\rho$ is the order of the kernel and $\lambda$ is the parameters of the generative model. When specifying a generative kernel the generative model and its parameters must be given along with the order of the derivatives. The form in Equation (6.7) becomes the feature-space for a Fisher kernel when $\rho = 1$ and the log-likelihood term is removed.

It is interesting to look at the general attributes of this form of feature-space. It will be assumed that the observations are conditional independent given the latent variable sequence $\bar{q}$ and that the output distribution given the latent variable is a member of the exponential family. For an HMM, $\bar{q}$ specifies the state-sequence. An individual state (or state-component pair) will be specified by, for example, $s_j$. The log-likelihood may be expressed as

$$\log(p(\bar{\mathbf{x}}_{1:T}; \lambda)) = \log\left(\sum_{\bar{q} \in Q_T} P(\bar{q})\left(\prod_{t=1}^{T} p(\mathbf{x}_t|q_t; \lambda)\right)\right), \tag{6.8}$$

where $p(\mathbf{x}_t|s_j; \lambda)$ is a member of the exponential family associated with latent variable $s_j$ with natural parameters $\lambda_j$ and $Q_T$ is the set of latent variable sequences of length $T$. The member of the exponential family will be written as[1]

$$p(\mathbf{x}_t|s_j; \lambda_j) = \frac{1}{Z(\lambda_j)} \exp(\lambda_j^{\mathsf{T}} \mathbf{T}(\mathbf{x}_t)), \tag{6.9}$$

where $Z(\lambda_j)$ is the normalization term required to ensure the distribution is a valid PDF and $\mathbf{T}(\mathbf{x}_t)$ are the sufficient statistics. The derivative with respect to the parameters of the latent variable $s_j$ can be expressed as Layton (2006)

$$\frac{\partial}{\partial \lambda_j} \log(p(\bar{\mathbf{x}}_{1:T}; \lambda)) = \sum_{t=1}^{T} \gamma_j(t)\left(\mathbf{T}(\mathbf{x}_t) - \frac{\partial}{\partial \lambda_j} \log(Z(\lambda_j))\right), \tag{6.10}$$

where $\gamma_j(t)$ is the posterior probability of the latent variable state, $s_j$, generating the observation at time $t$ given the complete observation sequence

$$\gamma_j(t) = P(q_t = s_j|\bar{\mathbf{x}}_{1:T}; \lambda). \tag{6.11}$$

The second term in Equation (6.10) is not a function of the observations, so depending on the metric it can often be ignored. Thus the feature-space is a weighted version of the exponential model sufficient statistics, where the weighting is determined by the posterior probability. For the example of an HMM with Gaussian state output distributions, the derivative with respect to the mean of state $s_j$, $\boldsymbol{\mu}_j$, is

$$\frac{\partial}{\partial \boldsymbol{\mu}_j} \log(p(\bar{\mathbf{x}}_{1:T}; \lambda)) = \sum_{t=1}^{T} \gamma_j(t)\boldsymbol{\Sigma}^{-1}(\mathbf{x}_t - \boldsymbol{\mu}_j). \tag{6.12}$$

Depending on the metric used the offset from the mean and transformation of the inverse covariance matrix may be ignored as this may be subsumed with the metric.

It is informative to contrast the nature of the temporal dependencies in the generative kernel feature-space to those in the generative model used to derive this feature-space. Independence assumptions in the initial, base, statistical model are maintained in the augmented model. Thus for the case of using a GMM to model the sequence (the latent variables now represent the component generating the observation) there is no additional temporal information. However, this is not the case for systems with conditional independence, such as HMMs. The posterior, $\gamma_j(t)$, in Equation (6.11) will be a function of all the observations, $\bar{\mathbf{x}}_{1:T}$. Thus the statistics associated with the derivative feature-space will contain temporal correlation information from many observations. For more information about higher-order derivatives and relationships to other sequence kernels see Layton (2006).

---

[1]The reference distribution has been bundled into the score-space in the generative model. This will be the case for all the versions considered here.

One of the problems with using generative kernels, compared with using static kernels such as a polynomial kernel, is that the kernel requires the feature-space to be explicitly calculated. However, if a static kernel is to be applied afterwards, it may be efficiently implemented in the standard fashion as a kernel on the dynamic feature-space. In practice this is not a major issue as for most of the situations of interest the number of model parameters is only up to a few thousand.

An important aspect of the kernel is the metric to be used with the inner-product. The kernel between two sequences, $\bar{\mathbf{x}}^{(i)}$ and $\bar{\mathbf{x}}^{(j)}$, given the feature-space $\boldsymbol{\phi}()$ is often expressed as

$$k(\bar{\mathbf{x}}^{(i)}, \bar{\mathbf{x}}^{(j)}; \boldsymbol{\lambda}) = \boldsymbol{\phi}(\bar{\mathbf{x}}^{(i)}; \boldsymbol{\lambda})^{\mathsf{T}} \mathbf{G}^{-1} \boldsymbol{\phi}(\bar{\mathbf{x}}^{(j)}; \boldsymbol{\lambda}), \tag{6.13}$$

where $\mathbf{G}$ defines the metric and should be positive semi-definite. If only the first order derivatives are used (as in the Fisher kernel), then an appropriate metric for calculating distances (as used for example in an SVM) is the Fisher Information matrix defined as (Jaakkola and Haussler 1999)

$$\mathbf{G} = \mathbb{E}\left\{ \left( \frac{\partial}{\partial \boldsymbol{\lambda}} \log(p(\bar{\mathbf{x}}; \boldsymbol{\lambda})) \right) \left( \frac{\partial}{\partial \boldsymbol{\lambda}} \log(p(\bar{\mathbf{x}}; \boldsymbol{\lambda})) \right)^{\mathsf{T}} \right\}, \tag{6.14}$$

where the expectation is over all sequences $\bar{\mathbf{x}}$. For most latent variable models there is no closed-form solution, so the matrix may be approximated by the identity matrix (Jaakkola and Haussler 1999), or by using an estimate based on the training sequences. In the experiments described later the latter approach was used and also applied to all elements of the feature-space.

## 6.3.3 Simple Example

The generative kernel differs from the Fisher kernel in a number of ways. One of the more interesting differences is the possible use of higher-order derivatives in the feature-space. To illustrate the advantage of using these higher order derivatives, as well as the usefulness of derivative-based feature-spaces in general, consider the simple example of a two class problem with two discrete output symbols {A, B}. The training examples (equally distributed) are: Class $\omega_1$: AAAA, BBBB; Class $\omega_2$: AABB, BBAA. If a large amount of training data is available, the discrete two-emitting state HMM shown in Figure 6.3 would be trained using ML for both class $\omega_1$ and $\omega_2$. These HMMs are clearly unable to distinguish between the sequences from class $\omega_1$ and $\omega_2$.

The generative score-space for this discrete case is similar to the component priors in a continuous HMM system. Initially consider the differential with respect to bin $m$ of state $s_j$ (equivalent in the continuous case to the prior of component $m$ in state $s_j$). The first order-derivative is given by

$$\frac{\partial}{\partial c_{jm}} \log(p(\bar{\mathbf{x}}_{1:T}; \boldsymbol{\lambda})) = \sum_{t=1}^{T} \left( \gamma_{jm}(t)/c_{jm} - \sum_{m=1}^{M} \gamma_{jm}(t) \right). \tag{6.15}$$

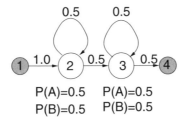

Figure 6.3 Example discrete Hidden Markov Model topology and state probabilities and associated scores.

For the second-order derivative, considering two different bins $m$ and $n$ from states $s_j$ and $s_k$ respectively:

$$\frac{\partial^2}{\partial c_{kn}\partial c_{jm}}\log(p(\bar{\mathbf{x}}_{1:T};\lambda)) = \frac{1}{c_{jm}c_{kn}}\sum_{t=1}^{T}\sum_{\tau=1}^{T}(D(q_t^{(jm)}, q_\tau^{(kn)}) - c_{jm}D(q_t^{(j)}, q_t^{(jm)})$$

$$- c_{kn}D(q_t^{(jm)}, q_\tau^{(k)}) + c_{jm}c_{kn}D(q_t^{(j)}, q_\tau^{(k)})), \qquad (6.16)$$

where

$$D(q_t^{(jm)}, q_\tau^{(kn)}) = P(q_t = s_{jm}, q_\tau = s_{kn}|\bar{\mathbf{x}}_{1:T};\lambda) - \gamma_{jm}(t)\gamma_{kn}(\tau).$$

As with the continuous HMM case, these derivative features are functions of the complete observation sequence as they depend on the state/component posterior $\gamma_{jm}(t)$, which is a function of $\bar{\mathbf{x}}_{1:T}$. Thus long-term dependencies may be represented by these forms of features.

Table 6.1 Feature vector values for a generative kernel

| | Class $\omega_1$ | | Class $\omega_2$ | |
|---|---|---|---|---|
| Feature | AAAA | BBBB | AABB | BBAA |
| $\log(p(\bar{\mathbf{x}}_{1:T};\lambda))$ | −1.11 | −1.11 | −1.11 | −1.11 |
| $\frac{\partial}{\partial c_{2A}}\log(p(\bar{\mathbf{x}}_{1:T};\lambda))$ | 0.50 | −0.50 | 0.33 | −0.33 |
| $\frac{\partial^2}{\partial c_{2A}^2}\log(p(\bar{\mathbf{x}}_{1:T};\lambda))$ | −3.83 | 0.17 | −3.28 | −0.61 |
| $\frac{\partial^2}{\partial c_{2A}\partial c_{3A}}\log(p(\bar{\mathbf{x}}_{1:T};\lambda))$ | −0.17 | −0.17 | −0.06 | −0.06 |

Table 6.1 shows the values of some elements of the feature vector associated with a generative kernel for each of the two classes. It is clear that using the first and higher order derivatives of the log-likelihood allow the two classes to be separated, in some cases using a simple linear classifier. From the last row, $(\partial^2/(\partial c_{2A}\partial c_{3A}))\log(p(\bar{\mathbf{x}}_{1:T};\lambda))$, of the table the feature captures the obvious difference between the two classes that the label changes part way through.

## 6.4 Augmented Statistical Models

One of the standard ways to use dynamic kernels, or their associated feature-spaces, is directly as features in a classifier such as an SVM (Smith and Gales 2001). However, an interesting alternative is to use the feature-space as statistics in, for example, a member of the exponential family. As feature-spaces of these kernels incorporate additional temporal information, with higher-order derivatives including more information, this yields a systematic approach to adding temporal dependencies. These new models will be referred to as Augmented Statistical Models and may either be generative in nature, or discriminative. Both these forms are discussed in this section.

### 6.4.1 Generative Augmented Models

The form of exponential model given in Equation (6.9) considers only features from an individual observation. A more general case is to consider features from the complete sequence $\bar{\mathbf{x}}_{1:T}$. Here

$$p(\bar{\mathbf{x}}_{1:T}; \boldsymbol{\alpha}) = \frac{1}{Z(\boldsymbol{\alpha})} \exp(\boldsymbol{\alpha}^T \mathbf{T}(\bar{\mathbf{x}}_{1:T})), \tag{6.17}$$

where $\boldsymbol{\alpha}$ is now the natural parameters of the exponential model. As previously discussed, the options for sufficient statistics to extract from the sequence $\bar{\mathbf{x}}_{1:T}$ are very broad and will have an important influence on the nature of the acoustic model.

An interesting subset of the set of all possible members of the exponential family is the *constrained exponential family* (Amari and Nagaoka 2000). Here, a local exponential approximation to the reference distribution is used as the statistical model, where the local approximation replicates some of the properties of the reference distribution. A slightly more general form of statistical model than the constrained exponential family is possible. In addition to the values of the local exponential model, the distribution parameters of the model to derive the local exponential distribution may also be trained from the data. This is the generative form of the Augmented Statistical Model (Smith 2003).

Generative Augmented Statistical Models (Aug) are an attractive approach to building class-conditional probability density functions, since they yield a mathematically consistent formulation to add dependencies into the model. First a base statistical model, $p(\bar{\mathbf{x}}; \boldsymbol{\lambda})$, is defined. A member of the exponential family that locally approximates this base distribution for a particular set of parameters $\boldsymbol{\lambda}$ is then used as the final statistical model. The general form of Augmented Statistical Model for a base statistical model can be expressed as[2]

$$p(\bar{\mathbf{x}}_{1:T}; \boldsymbol{\lambda}, \boldsymbol{\alpha}) = \frac{1}{Z(\boldsymbol{\lambda}, \boldsymbol{\alpha})} \exp(\boldsymbol{\alpha}^T \boldsymbol{\phi}(\bar{\mathbf{x}}_{1:T}; \boldsymbol{\lambda})), \tag{6.18}$$

where $\boldsymbol{\phi}(\bar{\mathbf{x}}_{1:T}; \boldsymbol{\lambda})$ is the score-space shown in Equation (6.7) and $Z(\boldsymbol{\lambda}, \boldsymbol{\alpha})$ is the appropriate normalization term given by

$$Z(\boldsymbol{\lambda}, \boldsymbol{\alpha}) = \int \exp(\boldsymbol{\alpha}^T \boldsymbol{\phi}(\bar{\mathbf{x}}; \boldsymbol{\lambda})) d\bar{\mathbf{x}}. \tag{6.19}$$

---

[2]For simplicity, in this work the *natural basis* and higher order derivatives are assumed to yield a set of orthogonal bases. Given this assumption it is not necessary to distinguish between covariant and contravariant bases and components (Amari and Nagaoka 2000).

The integral is over all observation sequences. Unfortunately this normalization term can be solved for only a small number of possible models. The base models of interest, such as HMMs, cannot be simply solved.

The normalization issue can be dealt with for binary problems by directly considering Bayes' decision rule for generative models (Gales and Layton 2006). For the two classes $\omega_1$ and $\omega_2$, with parameters $\alpha^{(1)}$, $\lambda^{(1)}$ and $\alpha^{(2)}$, $\lambda^{(2)}$ respectively, the decision rule is

$$\log\left( \frac{P(\omega_1)Z(\lambda^{(2)}, \alpha^{(2)}) \exp(\alpha^{(1)\mathsf{T}}\phi(\bar{\mathbf{x}}_{1:T}; \lambda^{(1)}))}{P(\omega_2)Z(\lambda^{(1)}, \alpha^{(1)}) \exp(\alpha^{(2)\mathsf{T}}\phi(\bar{\mathbf{x}}_{1:T}; \lambda^{(2)}))} \right) \mathop{\gtrless}_{\omega_2}^{\omega_1} 0 \qquad (6.20)$$

where $P(\omega_1)$ and $P(\omega_2)$ are priors for the two classes. Note this form of discriminative training of generative models is related to discriminative training of HMMs; see for example Gales (2007). The decision rule can be re-expressed as

$$\begin{bmatrix} \alpha^{(1)} \\ \alpha^{(2)} \end{bmatrix}^{\mathsf{T}} \begin{bmatrix} \phi(\bar{\mathbf{x}}_{1:T}; \lambda^{(1)}) \\ -\phi(\bar{\mathbf{x}}_{1:T}; \lambda^{(2)}) \end{bmatrix} + \log\left( \frac{P(\omega_1)Z(\lambda^{(2)}, \alpha^{(2)})}{P(\omega_2)Z(\lambda^{(1)}, \alpha^{(1)})} \right) \mathop{\gtrless}_{\omega_2}^{\omega_1} 0. \qquad (6.21)$$

This is now in the form of a linear decision boundary, where the parameters of the decision boundary, $\alpha^{(1)}$ and $\alpha^{(2)}$, can be estimated using any of the standard approaches. The intractable normalization term is part of the bias term and does not need to be computed explicitly. If for example an SVM is used, which is the scheme used in the results presented later, this is the equivalent of using the following dynamic kernel:

$$k(\bar{\mathbf{x}}^{(i)}, \bar{\mathbf{x}}^{(j)}; \lambda) = \begin{bmatrix} \phi(\bar{\mathbf{x}}^{(i)}; \lambda^{(1)}) \\ -\phi(\bar{\mathbf{x}}^{(i)}; \lambda^{(2)}) \end{bmatrix}^{\mathsf{T}} \mathbf{G}^{-1} \begin{bmatrix} \phi(\bar{\mathbf{x}}^{(j)}; \lambda^{(1)}) \\ -\phi(\bar{\mathbf{x}}^{(j)}; \lambda^{(2)}) \end{bmatrix}. \qquad (6.22)$$

The natural parameters $\alpha^{(1)}$ and $\alpha^{(2)}$ found using this kernel and an SVM will be estimated in a Maximum Margin (MM) fashion.

There are a number of constraints that must be satisfied for this scheme to yield a valid generative model; in particular that $Z(\alpha, \lambda)$ is bounded. This restricts the possible values of $\alpha$ when using a score-space that includes, for example, second-order statistics. For a general discussion of restrictions see Gales and Layton (2006) and Layton (2006). An interesting side-effect of this approach is that the decision boundary obtained from an SVM has a direct probabilistic interpretation, purely as a result of the nature of the feature-space.

Though this form of training allows the natural parameters of a generative augmented model to be trained it is only appropriate for binary classification tasks. The majority of speech processing applications (with the exception of speaker verification) are not binary. To address this problem the *acoustic code-breaking* framework (Venkataramani *et al.* 2003) may be used (Gales and Layton 2006). In this work code-breaking based on confusion networks (Mangu *et al.* 1999) is used. This process is illustrated in Figure 6.4. First word lattices are generated and then converted into confusion networks. These are then pruned so that there is a maximum of two arcs between any nodes. Since classification is based on these arcs, the possibly large vocabulary speech recognition task has been converted into a series of binary classification tasks.

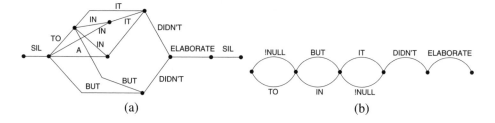

(a)                                                    (b)

Figure 6.4   Multi-class to binary classification for large vocabulary continuous speech recognition. (a) Word lattice. (b) Pruned confusion network.


## 6.4.2   Conditional Augmented Models

Instead of using generative models, discriminative models have become increasingly popular for classifying speech data (Gales 2007), for example using Conditional Random Fields (CRFs) (Lafferty *et al.* 2001) and Hidden CRFs (HCRFs) (Gunawardana *et al.* 2005). Conditional Augmented (C-Aug) (Layton and Gales 2006) Models are a discriminative form of augmented model. In a similar fashion to a number of other models they use a maximum entropy style model. However, rather than restricting the statistics to be the same as those of a standard HMM, which is the starting point for the HCRF implementation, and thus yielding systems very similar to discriminatively trained HMMs (Heigold *et al.* 2007), the feature-space used for the generative kernels is used. This gives an elegant way of combining generative and discriminative models.[3]

The basic form for the C-Aug model is

$$P(\mathbf{w}|\bar{\mathbf{x}}_{1:T}; \boldsymbol{\alpha}, \boldsymbol{\lambda}) = \frac{1}{Z(\boldsymbol{\lambda}, \boldsymbol{\alpha}, \bar{\mathbf{x}}_{1:T})} \exp(\boldsymbol{\alpha}^{(\mathbf{w})\mathsf{T}}\boldsymbol{\phi}(\bar{\mathbf{x}}_{1:T}; \boldsymbol{\lambda}^{(\mathbf{w})})), \tag{6.23}$$

where the observation sequence dependent normalization term is given by

$$Z(\boldsymbol{\lambda}, \boldsymbol{\alpha}, \bar{\mathbf{x}}_{1:T}) = \sum_{\tilde{\mathbf{w}}} \exp(\boldsymbol{\alpha}^{(\tilde{\mathbf{w}})\mathsf{T}}\boldsymbol{\phi}(\bar{\mathbf{x}}_{1:T}; \boldsymbol{\lambda}^{(\tilde{\mathbf{w}})})). \tag{6.24}$$

It is no longer necessary to compute the normalization over all possible sequences; the normalization term is now based on the classes and the specific data sequence of interest. This is tractable so allows standard optimization schemes to be used to train the model. The criterion used in this work is the conditional maximum likelihood. For further details of how these models may be trained see Layton (2006) and Layton and Gales (2006).

In theory these models can be directly used to model the complete word sequences, or utterances, **w**. However, for some tasks, such as large vocabulary speech recognition, this is impractical as the number of possible classes (and model parameters) is vast. The standard approach to addressing this is to combine sub-utterance, either word or phone, models together. It is possible to use sub-utterance C-Aug models (Gales 2007; Layton 2006). For example if the temporal features are assumed to be conditionally independent given, for

---

[3]The dependence of the dynamic kernel on the word sequence, indicated using $\boldsymbol{\lambda}^{(\mathbf{w})}$, is now explicitly marked.

example, the word-level segmentation, then

$$P(\mathbf{w}|\bar{\mathbf{x}}_{1:T}; \boldsymbol{\alpha}, \boldsymbol{\lambda}, \boldsymbol{\theta}) = \prod_{i=1}^{L} \frac{1}{Z(\boldsymbol{\lambda}, \boldsymbol{\alpha}, \bar{\mathbf{x}}_{\theta}^{(i)})} \exp(\boldsymbol{\alpha}^{(w_i)\mathsf{T}} \boldsymbol{\phi}(\bar{\mathbf{x}}_{\theta}^{(i)}; \boldsymbol{\lambda}^{(w_i)})), \tag{6.25}$$

where $\bar{\mathbf{x}}_{\theta}^{(i)}$ is the observations associated with the $i$th word $w_i$ using the segmentation $\boldsymbol{\theta}$ and the word sequence $\mathbf{w} = \{w_1, \ldots, w_L\}$. Preliminary results and discussion for this form of model are given in Layton (2006).

# 6.5 Experimental Results

This section describes initial experiments with Augmented Statistical Models. The first experiments examine the performance of both generative and discriminative augmented models on a TIMIT phone classification task. The second experiment examines the issue with applying the Generative Augmented Statistical Model to a large vocabulary speech recognition task. For further details of the experimental configuration see Layton (2006) and Layton and Gales (2006).

TIMIT is a standard phone classification, or recognition, task. Here TIMIT was used in a phone-classification mode. For these experiments all the training data was used to train the HMM system for initial recognition. All the HMMs had the same topology three-states, strict left-to-right topology, with ten Gaussian components per state. Monophone models were built to allow simple phone classification to be performed. The first set of experiments examined the performance of acoustic code-breaking with augmented models. As this is a classification task, the selection of the two most confusable labels for each phone was simple, based on the initial HMM. Training data for the most confusable pairs were then extracted and used to train augmented models.

Table 6.2 shows a comparison of Maximum Likelihood (ML) and Maximum Mutual Information (MMI) (Gopalakrishnan *et al.* 1991) trained HMMs with an Augmented Statistical Model for three phone pairs. The number of test phones in each of the classes was balanced to remove any biases. Two clear trends can be observed. First, that MMI training, a discriminative training criterion often used for generative models, outperforms ML trained HMMs for all the phone pairings. The use of an Aug model outperformed both forms of standard HMM. The score-space used was a first-order derivative with respect to all the model parameters.

Table 6.2  Classification error using a 10-component Hidden Markov Model on the confusable pairs from the TIMIT core test set

| Classifier | Training | | Phone pair | | |
|---|---|---|---|---|---|
| | $\lambda$ | $\alpha$ | s/z | er/r | m/n |
| HMM | ML | — | 21.9 | 25.2 | 26.7 |
| HMM | MMI | — | 20.7 | 18.6 | 21.0 |
| Aug | ML | MM | 16.4 | 14.8 | 20.0 |

Table 6.3    Classification error using a Maximum Likelihood and Maximum Mutual Information trained Hidden Markov Models and Conditional Augmented models on the TIMIT core test set and the number of components per state (# Comp.)

| Classifier | Training | | # Comp. | |
| --- | --- | --- | --- | --- |
| | $\lambda$ | $\alpha$ | 10 | 20 |
| HMM | ML | — | 29.4 | 27.3 |
| C-Aug | ML | CML | 24.0 | — |
| HMM | MMI | — | 25.3 | 24.8 |
| C-Aug | MMI | CML | 23.4 | — |

An interesting question is the number of binary classifier pairs that are required to achieve significant reductions in the error rate of the complete core test set. Using ten phone-pair classifiers, based on the most confusable phone pairs, about 21% of the TIMIT core test data was rescored, with an absolute reduction of 5.2% in the error rate on those phones rescored. This corresponds to an absolute reduction of 1.1% over the complete core test set.

One of the issues with using the generative form of the augmented model is that a large number of classifiers are required to rescore all the data. This is not an issue with the C-Aug models as it naturally handles multi-class classification. Table 6.3 shows the contrast of ML and MMI trained HMMs along with C-Aug versions of the models trained using Conditional Maximum Likelihood (CML) training. Ten Gaussian components per state were used for the baseline HMM, as well as the augmented model. In addition, 20 Gaussian components per state HMM were also built. The C-Aug model consistently outperforms the standard HMM with both ML and MMI training. As the number of parameters in the C-Aug is approximately double that of the standard model for the same number of components, first-order derivatives of all the model parameters are being used, it is interesting that the 10-component C-Aug model is better than the 20-component HMM. Though significant gains in performance were obtained with the C-Aug model, they are not as good as the results for the HCRF in Gunawardana *et al.* (2005). This is felt to be because the score-space is defined using the standard HMM, the parameters of which are currently not refined (which they are in the HCRF). Furthermore for these experiments no language model scores were used.

It is possible to apply the acoustic code-breaking framework to a large vocabulary continuous speech recognition (LVCSR). A conversational telephone speech task, SwitchBoard, was used. These LVCSR experiments are based on an ML-trained HMM triphone system built on 400 hours of acoustic training data (fsh2004sub). Six thousand distinct decision-tree state-clustered states were used. A trigram language model was used for initial decoding. For further details of the training and language models see Evermann *et al.* (2005). The lattices generated in the initial decoding were converted into confusion networks. The most confusable pairs were then found and data to train the SVM obtained by building confusion networks on all the training.

Compared with the TIMIT phone classification task where there are relatively few phone pairs, the number of possible word pairs for a 58 000 word decoding vocabulary, as used in these experiments, is vast. Which introduces an additional problem. In a similar fashion to the TIMIT task, only the ten most confusable pairs were used. This resulted in approximately

Table 6.4 Recognition results using Maximum Likelihood-trained Hidden Markov Models with confusion network and code-breaking using ten Support Vector Machines on the `eval03` Conversation Telephone Speech Task

| Decoding | WER (%) |
|---|---|
| Viterbi | 30.8 |
| Confusion network | 30.1 |
| + code-breaking | 30.0 |

1.6% of the words being rescored, compared with 21% for the TIMIT classification task. This limits the possible gains on this task. Another issue is that for LVCSR systems language models are essential to achieve good performance. Language models can be incorporated into the acoustic code-breaking framework by adding, for example, the word-posteriors into the score-space (Gales and Layton 2006) or by interpolating word-posteriors with the SVM score (Venkataramani *et al.* 2003). The latter approach is adopted here. Of the words that were rescored the absolute reduction in error rate was 4.0%, comparable to the gains observed on the TIMIT classification task. However, as only 1.6% of the words were rescored the absolute reduction in error rate was less than 0.1% – not statistically significant.

Table 6.4 summarizes these performance numbers. Confusion network decoding outperforms standard Viterbi decoding as the decoding process is tuned to yielding the minimum expected word error rate. The use of code-breaking only marginally improves the performance of the confusion network decoding system.

## 6.6 Conclusions

One of the fundamental issues with using HMMs for speech recognition is that the temporal correlation modeling is poor. Standard approaches to handling this problem are to introduce explicit dependencies between observations, or additional latent variables. However, these approaches have not yielded significant performance gains to date. This chapter has described an alternative approach that uses dynamic kernels and the feature-spaces associated with these kernels to construct acoustic models that have longer term dependencies than standard HMMs. The feature-spaces from generative kernels, which are based on derivatives of the log-likelihood of a generative model with respect to the model parameters, were discussed in detail. Generative kernels are an interesting class to consider as they yield a feature-space with, for example, different temporal dependencies than those of the original generative model used to derive them. These features can then be used in a generative model based on the exponential family and a maximum entropy style discriminative model. Evaluation of these two forms of model on a TIMIT phone-classification task show that these Augmented Statistical Models are useful for speech data classification.

## Acknowledgements

The author would like to thank Nathan Smith, Martin Layton and Chris Longworth, who have all contributed to the theory behind Augmented Statistical Models. In addition the results presented in this chapter are based in the work of Martin Layton described in detail in his thesis (Layton 2006).

# References

Amari S and Nagaoka H 2000 *Methods of Information Geometry*. Oxford University Press. (Translations of *Mathematical Monographs*, vol. 191. American Mathematical Society.)

Bilmes J 2003a Buried Markov models: A graphical-modelling approach to automatic speech recognition. *Computer, Speech and Language* **17**(2–3).

Bilmes J 2003b Graphical models and automatic speech recognition *Mathematical Foundations of Speech and Language Processing (Institute of Mathematical Analysis Volumes in Mathematics Series)*. Springer.

Campbell W, Campbell J, Reynolds D, Singer E and Torres-Carraquillo P 2005 Support vector machines for speaker and language recognition. *Computer Speech and Language* **20**, 210–229.

Campbell W, Sturim D, Reynolds D and Solomonoff A 2006 SVM based speaker verification using a GMM supervector kernel and NAP variability compensation. *Proceedings of the IEEE Conference on Acoustics, Speech and Signal Processing*, vol. 1, pp. 97–100.

Cortes C, Haffner P and Mohri M 2003 Weighted automata kernels – general framework and algorithms. *Proceedings of EuroSpeech*, pp. 989–992.

Cristianini N and Shawe-Taylor J 2000 *Support Vector Machines and other Kernel-based Learning Methods*. Cambridge University Press.

Digalakis V 1992 *Segment-Based Stochastic Models of Spectral Dynamics for Continuous Speech Recognition*. Boston University, PhD thesis.

Evermann G, Chan H, Gales M, Jia B, Mrva D, Woodland P and Yu K 2005 Training LVCSR systems on thousands of hours of data. *Proceedings of the IEEE Conference on Acoustics, Speech and Signal Processing*.

Furui S 1986 Speaker independent isolated word recognition using dynamic features of speech spectrum. *IEEE Transactions Acoustics, Speech and Signal Processing* **34**, 52–59.

Gales M 2007 Discriminative models for speech recognition. *Information Theory and Applications Workshop*, UCSD, CA.

Gales M and Layton M 2006 Training augmented models using SVMs. *IEICE Special Issue on Statistical Modelling for Speech Recognition*.

Gopalakrishnan P, Kanevsky D, Nádas A and Nahamoo D 1991 An inequality for rational functions with applications to some statistical estimation problems. *IEEE Transactions on Information Theory* **37**, 107–113.

Gunawardana A, Mahajan M, Acero A and Platt J 2005 Hidden conditional random fields for phone classification. *Proceedings of InterSpeech*, pp. 1117–1120.

Heigold G, Schlüter R and Ney H 2007 On the equivalence of Gaussian HMM and Gaussian HMM-like hidden conditional random fields. *Proceedings of InterSpeech*, pp. 1721–1724.

Jaakkola T and Haussler D 1999 Exploiting generative models in discriminative classifiers. *Advances in Neural Information Processing Systems 11* (eds. S Solla and D Cohn). MIT Press, pp. 487–493.

Lafferty J, McCallum A and Pereira F 2001 Condition random fields: Probabilistic models for segmenting and labeling sequence data. *Proceedings of the International Conference on Machine Learning*, pp. 591–598.

Layton M 2006 *Augmented Statistical Models for Classifying Sequence Data*. University of Cambridge, PhD thesis.

Layton M and Gales M 2004 Maximum margin training of generative kernels. Department of Engineering, University of Cambridge. *Technical Report CUED/F-INFENG/TR.484*.

Layton M and Gales M 2006 Augmented statistical models for speech recognition. *Proceedings of the IEEE Conference on Acoustics, Speech and Signal Processing*, vol. 1, pp. 129–132.

Longworth C and Gales M 2007 Derivative and parametric kernels for speaker verification. *Proceedings of InterSpeech*, pp. 310–313.

Longworth C and Gales M 2008 A generalised derivative kernel for speaker verification. *Proceedings of InterSpeech*, pp. 1381–1384.

Mangu L, Brill E and Stolcke A 1999 Finding consensus among words: Lattice-based word error minimization. *Proceedings of EuroSpeech*, vol. 1, pp. 495–498.

Mariéthoz J and Bengio S 2007 A kernel trick for sequences applied to text-independent speaker verification systems. *Pattern Recognition* **40**, 2315–2324.

Nock H 2001 *Techniques for modelling Phonological Processes in Automatic Speech Recognition*. University of Cambridge, PhD thesis.

Ostendorf M, Digalakis VV and Kimball OA 1996 From HMMs to segment models: A unified view of stochastic modelling for speech recognition. *IEEE Transactions on Speech and Audio Processing* **4**, 360–378.

Rabiner L 1989 A tutorial on hidden Markov models and selective applications in speech recognition. *Proceedings of the IEEE*, vol. 77, pp. 257–286.

Rosti AVI and Gales MJF 2003 Switching linear dynamical systems for speech recognition. *Technical Report CUED/F-INFENG/TR461*, University of Cambridge. Available from: svr-www.eng.cam.ac.uk/˜mjfg.

Saunders C, Shawe-Taylor J and Vinokourov A 2003 String kernels, fisher kernels and finite state automata. *Advances in Neural Information Processing Systems 15* (eds. S Becker, S Thrun and K Obermayer). MIT Press, pp. 633–640.

Smith N 2003 *Using Augmented Statistical Models and Score Spaces for Classification*. University of Cambridge, PhD thesis.

Smith N and Gales M 2001 Speech recognition using SVMs. *Advances in Neural Information Processing Systems*, vol. 13. MIT Press.

Stolcke A, Ferrer L, Kajarekar S, Shriberg E and Venkataramam A 2005 MLLR transforms as features in speaker recognition. *Proceedings of InterSpeech*, pp. 2425–2428.

Tsuda K, Kin T and Asai K 2002 Marginalized kernels for biological sequences. *Bioinformatics* **18**, S268–S275.

Vapnik V 1998 *Statistical Learning Theory*. John Wiley & Sons.

Venkataramani V, Chakrabartty S and Byrne W 2003 Support vector machines for segmental minimum Bayes risk decoding of continuous speech. *Proceedings of the IEEE Automatic Speech Recognition and Understanding Workshop*, pp. 13–16.

Wan V and Renals S 2004 Speaker verification using sequence discriminant support vector machines. *IEEE Transactions on Speech and Audio Processing* **13**(2), 203–210.

Wellekens C 1987 Explicit time correlation in hidden Markov models for speech recognition. *Proceedings of the IEEE Conference on Acoustics, Speech and Signal Processing*, vol. 1, pp. 384–386.

Woodland P 1992 Hidden Markov models using vector linear predictors and discriminative output distributions. *Proceedings of the IEEE Conference on Acoustics, Speech and Signal Processing*, pp. 509–512.

# 7

# Large Margin Training of Continuous Density Hidden Markov Models

## Fei Sha and Lawrence K. Saul

Continuous Density Hidden Markov Models (CD-HMMs) are an essential component of modern systems for automatic speech recognition (ASR). These models assign probabilities to the sequences of acoustic feature vectors extracted by signal processing of speech waveforms. In this chapter, we investigate a new framework for parameter estimation in CD-HMMs. Our framework is inspired by recent parallel trends in the fields of ASR and machine learning. In ASR, significant improvements in performance have been obtained by discriminative training of acoustic models. In machine learning, significant improvements in performance have been obtained by discriminative training of large margin classifiers. Building on both of these lines of work, we show how to train CD-HMMs by maximizing an appropriately defined margin between correct and incorrect decodings of speech waveforms. We start by defining an objective function over a transformed parameter space for CD-HMMs, then describe how it can be optimized efficiently by simple gradient-based methods. Within this framework, we obtain highly competitive results for phonetic recognition on the TIMIT speech corpus. We also compare our framework for large margin training with other popular frameworks for discriminative training of CD-HMMs.

## 7.1 Introduction

Most modern speech recognizers are built from CD-HMMs. The hidden states in these CD-HMMs are used to model different phonemes or sub-phonetic elements, while the observed outputs are used to model acoustic feature vectors. The accuracy of the speech recognition

*Automatic Speech and Speaker Recognition: Large Margin and Kernel Methods*    Joseph Keshet and Samy Bengio
© 2009 John Wiley & Sons, Ltd

in CD-HMMs depends critically on the careful parameter estimation of their transition and emission probabilities.

The simplest method for parameter estimation in CD-HMMs is the expectation-maximization (EM) algorithm. The EM algorithm is based on maximizing the *joint likelihood* of observed feature vectors and label sequences. It is widely used due to its simplicity and scalability to large data sets, which are common in ASR. However, this approach has the weakness that the model parameters of CD-HMMs are not optimized for sequential classification: in general, maximizing the joint likelihood does not minimize the phoneme or word error rates, which are more relevant metrics for ASR.

Noting this weakness, researchers in ASR have studied alternative frameworks for parameter estimation based on conditional maximum likelihood (Nádas 1983), minimum classification error (Juang and Katagiri 1992) and maximum mutual information (Bahl *et al.* 1986). The learning algorithms in these frameworks optimize *discriminative* criteria that more closely track actual error rates. These algorithms do not enjoy the simple update rules and relatively fast convergence of the EM algorithm, which maximizes the joint likelihood of hidden state sequences and acoustic feature vectors. However, carefully and skillfully implemented, they lead to lower error rates (McDermott *et al.* 2006; Roux and McDermott 2005; Woodland and Povey 2002).

Recently, in a new approach to discriminative acoustic modeling, we proposed a framework for large margin training of CD-HMMs. Our framework explicitly penalizes incorrect decodings by an amount proportional to the number of mislabeled hidden states. It also gives rise to a convex optimization over the parameter space of CD-HMMs, thus avoiding the problem of spurious local minima. Our framework builds on ideas from many previous studies in machine learning and ASR. It has similar motivation as recent frameworks for sequential classification in the machine learning community (Altun *et al.* 2003; Lafferty *et al.* 2001; Taskar *et al.* 2004), but differs in its focus on the real-valued acoustic feature representations used in ASR. It has similar motivation as other discriminative paradigms in ASR (Bahl *et al.* 1986; Gopalakrishnan *et al.* 1991; Juang and Katagiri 1992; Kapadia *et al.* 1993; Nádas 1983; Roux and McDermott 2005; Woodland and Povey 2002), but differs in its goal of margin maximization and its formulation of the learning problem as a convex optimization. The recent margin-based approaches of Jiang and Li (2007), Jiang *et al.* (2006), Li *et al.* (2007, 2005) and Yu *et al.* (2006) are closest in terms of their goals, but entirely different in their mechanics.

In this chapter, we describe our framework for large margin training and present a systematic comparison with other leading frameworks for parameter estimation in CD-HMMs. We compare large margin training not only with maximum likelihood (ML) estimation, but also with popular discriminative methods based on conditional maximum likelihood (CML) (Bahl *et al.* 1986; Nádas 1983) and minimum classification error (MCE) (Juang and Katagiri 1992). We investigate salient differences between CML, MCE, and large margin training through carefully designed experiments on the TIMIT speech corpus (Lamel *et al.* 1986). In particular, we compare the results from multiple phonetic recognizers trained with different parameterizations, initial conditions and learning algorithms. In all other aspects, though, these systems were held fixed: they employed exactly the same acoustic front end and model architectures (e.g. monophone CD-HMMs with full Gaussian covariance matrices). Our experimental results illuminate the significant factors that differentiate competing methods for discriminative training of CD-HMMs.

The chapter is organized as follows. In Section 7.2, we review CD-HMMs and current popular methods for parameter estimation. In Section 7.3, we describe our framework for large margin training of CD-HMMs and give a brief survey of closely related work. In Section 7.4, we compare the performance of phonetic recognizers trained in various different ways. Finally, in Section 7.5, we review our main findings and conclude with a brief discussion of future directions for research.

## 7.2   Background

CD-HMMs are used to specify a joint probability distribution over hidden state sequences $\bar{q} = \{q_1, q_2, \ldots, q_T\}$ and observed output sequences $\bar{\mathbf{x}} = \{\mathbf{x}_1, \mathbf{x}_2, \ldots, \mathbf{x}_T\}$. The logarithm of this joint distribution is given by:

$$\log P(\bar{\mathbf{x}}, \bar{q}) = \sum_t [\log P(q_t | q_{t-1}) + \log P(\mathbf{x}_t | q_t)]. \tag{7.1}$$

For ASR, the hidden states $q_t$ and observed outputs $\mathbf{x}_t$ denote phonetic labels and acoustic feature vectors, respectively, and the distributions $P(\mathbf{x}_t | q_t)$ are typically modeled by multivariate Gaussian mixture models (GMMs):

$$P(\mathbf{x}_t | q_t = j) = \sum_{m=1}^{M} \omega_{jm} \mathcal{N}(\mathbf{x}_t; \boldsymbol{\mu}_{jm}, \boldsymbol{\Sigma}_{jm}). \tag{7.2}$$

In Equation (7.2), we have used $\mathcal{N}(\mathbf{x}; \boldsymbol{\mu}, \boldsymbol{\Sigma})$ to denote the Gaussian distribution with mean vector $\boldsymbol{\mu}$ and covariance matrix $\boldsymbol{\Sigma}$, while the constant $M$ denotes the number of mixture components per GMM. The mixture weights $\omega_{jm}$ in Equation (7.2) are constrained to be non-negative and normalized: $\sum_m \omega_{jm} = 1$ for all states $j$.

Let $\boldsymbol{\theta}$ denote all the model parameters including transition probabilities, mixture weights, mean vectors and covariance matrices. The goal of parameter estimation in CD-HMMs is to compute the optimal $\boldsymbol{\theta}^*$ (with respect to a particular measure of optimality), given $N$ pairs of observation and target label sequences $\{\bar{\mathbf{x}}_n, \bar{p}_n\}_{n=1}^{N}$. In what follows, we review well-known frameworks for parameter estimation based on ML, CML and MCE.

### 7.2.1   Maximum Likelihood Estimation

The simplest approach to parameter estimation in CD-HMMs maximizes the joint likelihood of output and label sequences. The corresponding estimator is given by

$$\boldsymbol{\theta}^{\mathrm{ML}} = \arg\max_{\boldsymbol{\theta}} \sum_n \log P(\bar{\mathbf{x}}_n, \bar{p}_n). \tag{7.3}$$

For transition probabilities, ML estimates in this setting are obtained from simple counts (assuming the training corpus provides phonetic label sequences). For GMM parameters, the EM algorithm provides iterative update rules that converge monotonically to local stationary points of the likelihood. The main attraction of the EM algorithm is that no free parameters need to be tuned for its convergence.

## 7.2.2   Conditional Maximum Likelihood

CD-HMMs provide transcriptions of unlabeled speech by inferring the hidden label sequence with the highest posterior probability: $\bar{p} = \arg\max_{\bar{q}} P(\bar{q}|\bar{\mathbf{x}})$. The CML estimator in CD-HMMs directly attempts to maximize the probability that this inference returns the correct transcription. Thus, it optimizes the conditional likelihood:

$$\theta^{\text{CML}} = \arg\max_{\theta} \sum_n \log P(\bar{p}_n|\bar{\mathbf{x}}_n). \tag{7.4}$$

In CML training, the parameters are adjusted to increase the likelihood gap between correct labelings $\bar{p}_n$ and incorrect labelings $\bar{q} \neq \bar{p}_n$. This can be seen more explicitly by rewriting Equation (7.4) as

$$\theta^{\text{CML}} = \arg\max_{\theta} \left[ \log P(\bar{\mathbf{x}}_n, \bar{p}_n) - \log \sum_{\bar{q}} P(\bar{\mathbf{x}}_n, \bar{q}) \right]. \tag{7.5}$$

The CML estimator in Equation (7.4) is closely related to the maximum mutual information (MMI) estimator (Valtchev *et al.* 1997; Woodland and Povey 2002), given by

$$\theta^{\text{MMI}} = \arg\max_{\theta} \sum_n \log \frac{P(\bar{\mathbf{x}}_n, \bar{p}_n)}{P(\bar{\mathbf{x}}_n)P(\bar{p}_n)}. \tag{7.6}$$

Note that Equations (7.4) and (7.6) yield identical estimators in the setting where the (language model) probabilities $P(\bar{p}_n)$ are held fixed.

## 7.2.3   Minimum Classification Error

MCE training is based on minimizing the number of sequence misclassifications. The number of such misclassifications is given by

$$\mathcal{N}_{\text{err}} = \sum_n \text{sign}\left[ -\log P(\bar{\mathbf{x}}_n, \bar{p}_n) + \max_{\bar{q} \neq \bar{p}_n} \log P(\bar{\mathbf{x}}_n, \bar{q}) \right], \tag{7.7}$$

where $\text{sign}[z] = 1$ for $z > 0$ and $\text{sign}[z] = 0$ for $z \leq 0$. To minimize Equation (7.7), the parameters must be adjusted to maintain a likelihood gap between the correct labeling and all competing labelings. Unlike CML training, however, the size of the gap in Equation (7.7) does not matter, as long as it is finite.

The non-differentiability of the sign and max functions in Equation (7.7) makes it difficult to minimize the misclassification error directly. Thus, MCE training (McDermott *et al.* 2006) adopts the surrogate cost function:

$$\mathcal{N}_{\text{err}} \approx \sum_n \sigma\left( -\log P(\bar{\mathbf{x}}_n, \bar{p}_n) + \log\left[ \frac{1}{C} \sum_{\bar{q} \neq \bar{p}_n} e^{\eta \log P(\bar{\mathbf{x}}_n, \bar{q})} \right]^{1/\eta} \right), \tag{7.8}$$

where the sigmoid function $\sigma(z) = (1 + e^{-\alpha z})^{-1}$ replaces the sign function $\text{sign}[z]$, and a softmax function (parameterized by $\eta$) replaces the original max. The parameters $\alpha$ and $\eta$ in this approximation must be set by heuristics. The sum in the second term is taken over the top $C$ competing label sequences.

## 7.3   Large Margin Training

Recently, we proposed a new framework for discriminative training of CD-HMMS based on the idea of margin maximization (Sha and Saul 2007b). Our framework has two salient features: (1) it attempts to separate the accumulated scores of correct versus incorrect label sequences by margins proportional to the number of mislabeled states (Taskar *et al.* 2004); (2) the required optimization is convex, thus avoiding the pitfall of spurious local minima.

### 7.3.1   Discriminant Function

We start by reviewing the discriminant functions in large margin CD-HMMs. These parameterized functions of observations $\bar{\mathbf{x}}$ and states $\bar{q}$ take a form analogous to the log-probability in Equation (7.1). In particular, we define

$$\mathcal{D}(\bar{\mathbf{x}}, \bar{q}) = \sum_t [\lambda(q_{t-1}, q_t) + \rho(\mathbf{x}_t, q_t)] \tag{7.9}$$

in terms of state–state transition scores $\lambda(q_{t-1}, q_t)$ and state–output emission scores $\rho(\mathbf{x}_t, q_t)$. Unlike Equation (7.1), however, Equation (7.9) does not assume that the transition scores $\lambda(q_{t-1}, q_t)$ are derived from the logarithm of normalized probabilities. Likewise, the emission scores $\rho(\mathbf{x}_t, q_t)$ in Equation (7.9) are parameterized by sums of *unnormalized* Gaussian distributions:

$$\rho(\mathbf{x}_t, q_t = j) = \log \sum_m e^{-(\mathbf{x}_t - \boldsymbol{\mu}_{jm})^{\mathsf{T}} \boldsymbol{\Sigma}_{jm}^{-1} (\mathbf{x}_t - \boldsymbol{\mu}_{jm}) - \theta_{jm}}, \tag{7.10}$$

where the non-negative scalar parameter $\theta_{jm} \geq 0$ is entirely independent of $\boldsymbol{\Sigma}_{jm}$ (as opposed to being related to its log-determinant).

To obtain a convex optimization for large margin training, we further reparameterize the emission score in Equation (7.10). In particular, we express each mixture component's parameters $\{\boldsymbol{\mu}_{jm}, \boldsymbol{\Sigma}_{jm}, \theta_{jm}\}$ as elements of the following matrix:

$$\boldsymbol{\Phi}_{jm} = \begin{bmatrix} \boldsymbol{\Sigma}_{jm}^{-1} & -\boldsymbol{\Sigma}_{jm}^{-1} \boldsymbol{\mu}_{jm} \\ -\boldsymbol{\mu}_{jm}^{\mathsf{T}} \boldsymbol{\Sigma}_{jm}^{-1} & \boldsymbol{\mu}_{jm}^{\mathsf{T}} \boldsymbol{\Sigma}_{jm}^{-1} \boldsymbol{\mu}_{jm} + \theta_{jm} \end{bmatrix}. \tag{7.11}$$

Our framework for large margin training optimizes the matrices $\boldsymbol{\Phi}_{jm}$, as opposed to the conventional GMM parameters $\{\boldsymbol{\mu}_{jm}, \boldsymbol{\Sigma}_{jm}, \theta_{jm}\}$. Since the matrix $\boldsymbol{\Sigma}_{jm}$ is positive definite and the scalar $\theta_{jm}$ is non-negative, we also require the matrix $\boldsymbol{\Phi}_{jm}$ to be positive semidefinite (as denoted by the constraint $\boldsymbol{\Phi}_{jm} \succ 0$). With this reparameterization, the emission score in Equation (7.10) can be written as

$$\rho(\mathbf{x}_t, q_t = j) = \log \sum_m e^{-\mathbf{z}_t^{\mathsf{T}} \boldsymbol{\Phi}_{jm} \mathbf{z}_t} \quad \text{where } \mathbf{z}_t = \begin{bmatrix} \mathbf{x}_t \\ 1 \end{bmatrix}. \tag{7.12}$$

Note that this score is convex in the elements of the matrices $\boldsymbol{\Phi}_{jm}$.

### 7.3.2 Margin Constraints and Hamming Distances

For large margin training of CD-HMMs, we seek parameters that separate the discriminant functions for correct and incorrect label sequences. Specifically, for each joint observation-label sequence $(\bar{\mathbf{x}}_n, \bar{p}_n)$ in the training set, we seek parameters such that

$$\mathcal{D}(\bar{\mathbf{x}}_n, \bar{p}_n) - \mathcal{D}(\bar{\mathbf{x}}_n, \bar{q}) \geq \mathcal{H}(\bar{p}_n, \bar{q}), \quad \forall \bar{q} \neq \bar{p}_n, \tag{7.13}$$

where $\mathcal{H}(\bar{p}_n, \bar{q})$ denotes the *Hamming distance* between the two label sequences (Taskar *et al.* 2004). Note how this constraint requires the log-likelihood gap between the target sequence $\bar{p}_n$ and each incorrect decoding $\bar{q}$ to scale in proportion to the number of mislabeled states.

Equation (7.13) actually specifies an exponentially large number of constraints, one for each alternative label sequence $\bar{q}$. We can fold all these constraints into a single constraint by writing

$$-\mathcal{D}(\bar{\mathbf{x}}_n, \bar{p}_n) + \max_{\bar{q} \neq \bar{p}_n} \{\mathcal{H}(\bar{p}_n, \bar{q}) + \mathcal{D}(\bar{\mathbf{x}}_n, \bar{q})\} \leq 0. \tag{7.14}$$

In the same spirit as the MCE derivation for Equation (7.8), we obtain a more tractable (i.e. differentiable) expression by replacing the max function in Equation (7.14) with a 'softmax' upper bound:

$$-\mathcal{D}(\bar{\mathbf{x}}_n, \bar{p}_n) + \log \sum_{\bar{q} \neq \bar{p}_n} e^{\mathcal{H}(\bar{p}_n, \bar{q}) + \mathcal{D}(\bar{\mathbf{x}}_n, \bar{q})} \leq 0. \tag{7.15}$$

Note that the constraint in Equation (7.15) is stricter than the one in Equation (7.14); in particular, Equation (7.15) implies Equation (7.14). The exponential terms in Equation (7.15) can be summed efficiently using a modification of the standard forward–backward procedure.

### 7.3.3 Optimization

In general, it is not possible to find parameters that satisfy the large margin constraint in Equation (7.15) for all training sequences $\{\bar{\mathbf{x}}_n, \bar{p}_n\}_{n=1}^{N}$. For such 'infeasible' scenarios, we aim instead to minimize the total amount by which these constraints are violated. However, as a form of regularization, we also balance the total amount of violation (on the training sequences) against the scale of the GMM parameters. With this regularization, the overall cost function for large margin training takes the form

$$\mathcal{L} = \gamma \sum_{cm} \text{trace}(\mathbf{\Phi}_{cm}) + \sum_{n} \left[ -\mathcal{D}(\bar{\mathbf{x}}_n, \bar{p}_n) + \log \sum_{\bar{q} \neq \bar{p}_n} e^{\mathcal{H}(\bar{p}_n, \bar{q}) + \mathcal{D}(\bar{\mathbf{x}}_n, \bar{q})} \right]_+. \tag{7.16}$$

The first term in the cost function regularizes the scale of the GMM parameters; the scalar $\gamma$ determines the amount of regularization, relative to the second term in the cost function. In the second term, the '+' subscript in Equation (7.16) denotes the hinge function: $[z]_+ = z$ if $z > 0$ and $[z]_+ = 0$ if $z \leq 0$. The minimization of Equation (7.16) is performed subject to the positive semidefinite constraints $\mathbf{\Phi}_{jm} \succ 0$. We can further simplify the minimization by assuming that each emission score $\rho(\mathbf{x}_t, q_t)$ in the first term is dominated by the contribution from a single (pre-specified) Gaussian mixture component. In this case, the overall optimization is convex (see Sha (2007); Sha and Saul (2007b) for further details). The gradients of this cost function with respect to the GMM parameters $\mathbf{\Phi}_{cm}$ and transition parameters $\lambda(q, q')$ can be computed efficiently using dynamic programming, by a variant of the standard forward–backward procedure in HMMs (Sha 2007).

It is worth emphasizing two crucial differences between this optimization and previous ones (Juang and Katagiri 1992; Nádas 1983; Woodland and Povey 2002) for discriminative training of CD-HMMs for ASR. First, due to the reparameterization in Equation (7.11), the discriminant function $\mathcal{D}(\bar{\mathbf{x}}_n, \bar{q}_n)$ and the softmax function are convex in the model parameters. Therefore, the optimization in Equation (7.15) can be cast as a convex optimization, avoiding spurious local minima (Sha 2007). Second, the optimization not only increases the log-likelihood gap between correct and incorrect state sequences, but also drives the gap to grow in proportion to the number of individually incorrect labels (which we believe leads to more robust generalization).

### 7.3.4 Related Work

There have been several related efforts in ASR to incorporate ideas from large margin classification. Many studies have defined margins in terms of log-likelihood gaps between correct and incorrect decodings of speech. Previous approaches have also focused on balancing error rates versus model complexity. However, these approaches have differed from ours in important details. The approach in Jiang *et al.* (2006) and Li *et al.* (2005) was based on maximizing the margin with respect to a *preselected subset* of training examples correctly classified by initial models. The parameters in this approach were re-estimated under constraints that they did not change too much from their initial settings. Later this approach was extended to include all training examples (Jiang and Li 2007), but with the parameter estimation limited only to the mean vectors of GMMs. In Li *et al.* (2007), a large margin training criterion was defined in terms of averaged log-likelihood ratios over *mismatched frames*. This approach was motivated by earlier training criteria for minimum phone error training (Povey and Woodland 2002). Finally, in other work inspired by large margin classification, Yu *et al.* (2006) extended MCE training by introducing and adapting non-zero offsets in the sigmoid approximation for the zero-one loss. By mimicking the effects of margin penalties, these offsets led to improved performance on a large vocabulary task in ASR.

Note that our work differs from all the above approaches by penalizing incorrect decodings in proportion to their Hamming distance from correct ones. In fact, as we discuss in Section 7.4.2, it is possible to incorporate such penalties into more traditional frameworks for CML and MCE training. In very recent work, such an approach has been shown to reduce error rates on several tasks in large vocabulary ASR (Povey *et al.* 2008). It is also possible to formulate margins in terms of string edit distances, as opposed to Hamming distances. Keshet *et al.* (2006) experimented with this approach in the setting of online parameter estimation.

## 7.4   Experimental Results

We used the TIMIT speech corpus (Lamel *et al.* 1986; Lee and Hon 1988; Robinson 1994) to perform experiments in phonetic recognition. We followed standard practices in preparing the training, development and test data. Our signal processing front-end computed 39-dimensional acoustic feature vectors from 13 mel-frequency cepstral coefficients and their first and second temporal derivatives. In total, the training utterances gave rise to roughly 1.2 million frames, all of which were used in training.

We trained CD-HMMs for phonetic recognition in all of the previously described frameworks, including ML, CML, MCE and margin maximization. We also experimented with several variants of these frameworks to explore the effects of different parameterizations, initializations and cost functions.

In all the recognizers, the acoustic feature vectors were labeled by 48 phonetic classes, each represented by one state in a first-order CD-HMM. For each recognizer, we compared the phonetic state sequences obtained by Viterbi decoding with the 'ground-truth' phonetic transcriptions provided by the TIMIT corpus. For the purpose of computing error rates, we followed standard conventions in mapping the 48 phonetic state labels down to 39 broader phone categories.

We computed two different types of phone error rates, one based on Hamming distance, the other based on edit distance. The former was computed simply from the percentage of mismatches at the level of individual frames. The latter was computed by aligning the Viterbi and ground truth transcriptions using dynamic programming (Lee and Hon 1988), then summing the substitution, deletion and insertion error rates from the alignment process. The 'frame based' phone error rate (based on Hamming distances) is more closely tracked by our objective function for large margin training, while the 'string based' phone error rate (based on edit distances) provides a more relevant metric for ASR. We mainly report the latter except in experiments where the frame based error rate provides additional revealing context.

## 7.4.1 Large Margin Training

We trained two different types of large margin recognizers. The large margin recognizers in the first group were 'low-cost' discriminative CD-HMMs whose GMMs were merely trained for frame based classification. In particular, the parameters of these GMMs were estimated by treating different frames of speech as i.i.d. training examples. Because such GMMs can be viewed as a special case of CD-HMMs for 'sequences' containing exactly one observation, large margin training reduces to a much simpler optimization (Sha and Saul 2006) that does not involve Viterbi decoding or dynamic programming. These GMMs were then substituted into first-order CD-HMMs for sequence decoding. The large margin recognizers in the second group were fully trained for sequential classification. In particular, their CD-HMMs were estimated by solving the optimization in Equation (7.15) using multiple mixture components per state and adaptive transition parameters (Sha 2007; Sha and Saul 2007a,b).

Tables 7.1 and 7.2 show the results of these experiments for CD-HMMs with different numbers of Gaussian mixture components per hidden state. For comparison, we also show the performance of baseline recognizers trained by ML estimation using the EM algorithm. For both types of phone error rates (frame based and string based), and across all model sizes, the best performance was consistently obtained by large margin CD-HMMs trained for sequential classification. Moreover, among the two different types of large margin recognizers, utterance-based training generally yielded significant improvement over frame based training.

Table 7.1    Frame-based phone error rates, computed from Hamming distance, of phonetic recognizers from differently trained Continuous Density Hidden Markov Models. Reproduced by permission of © 2007 MIT Press

| Mixture components (per hidden state) | Baseline ML (EM algorithm) | Large margin (frame based) | Large margin (utterance-based) |
|:---:|:---:|:---:|:---:|
| 1 | 45.2% | 37.1% | 29.5% |
| 2 | 44.7% | 36.0% | 29.0% |
| 4 | 42.2% | 34.6% | 28.4% |
| 8 | 40.6% | 33.8% | 27.2% |

ML, Maximum Likelihood; EM, Expectation Maximization.

Table 7.2    String-based phone error rates, computed from edit distance, of phonetic recognizers from differently trained Continuous Density Hidden Markov Models. Reproduced by permission of © 2007 MIT Press

| Mixture components (per hidden state) | Baseline ML (EM algorithm) | Large margin (frame based) | Large margin (utterance-based) |
|:---:|:---:|:---:|:---:|
| 1 | 40.1% | 36.3% | 31.2% |
| 2 | 36.5% | 33.5% | 30.8% |
| 4 | 34.7% | 32.6% | 29.8% |
| 8 | 32.7% | 31.0% | 28.2% |

ML, Maximum Likelihood; EM, Expectation Maximization.

## 7.4.2    Comparison with CML and MCE

Next we compared large margin training with other popular frameworks for discriminative training. Table 7.3 shows the string based phone error rates of different CD-HMMs trained by ML, CML, MCE and margin maximization. As expected, all the discriminatively trained CD-HMMs yield significant improvements over the baseline CD-HMMs trained by ML. On this particular task, the results show that MCE does slightly better than CML, while the largest relative improvements are obtained by large margin training (by a factor of two or more). Using MMI on this task, Kapadia *et al.* (1993) reported larger relative reductions in error rates than we have observed for CML (though not better performance in absolute terms). It is difficult to compare our findings directly with theirs, however, since their ML and MMI recognizers used different front ends and numerical optimizations than those in our work.

## 7.4.3    Other Variants

What factors explain the better performance of CD-HMMs trained by margin maximization? Possible factors include: (1) the relaxation of Gaussian normalization constraints by the parameterization in Equation (7.11), yielding more flexible models, (2) the convexity of the margin-based cost function in Equation (7.16), which ensures that its optimization (unlike those for CML and MCE) does not suffer from spurious local minima, and (3) the closer

Table 7.3    String-based phone error rates, computed from edit distance, of phonetic recognizers from differently trained Continuous Density Hidden Markov Models. Reproduced by permission of © 2007 IEEE

| Mixture components (per hidden state) | Baseline ML (EM algo.) | Discriminative (CML) | Discriminative (MCE) | Discriminative (large margin) |
|---|---|---|---|---|
| 1 | 40.1% | 36.4% | 35.6% | 31.2% |
| 2 | 36.5% | 34.6% | 34.5% | 30.8% |
| 4 | 34.7% | 32.8% | 32.4% | 29.8% |
| 8 | 32.7% | 31.5% | 30.9% | 28.2% |

EM, Expectation Maximization; CML, Conditional Maximum Likelihood; MCE, Minimum Classification Error.

Table 7.4  String-based phone error rates from discriminatively trained Continuous Density Hidden Markov Models with normalized versus unnormalized Gaussian Mixture Models. Reproduced by permission of © 2007 IEEE

| Mixture components (per hidden state) | Normalized (CML) | Unnormalized (CML) | Normalized (MCE) | Unnormalized (MCE) |
|---|---|---|---|---|
| 1 | 36.4% | 36.0% | 35.6% | 36.4% |
| 2 | 34.6% | 36.3% | 34.5% | 35.6% |
| 4 | 32.8% | 33.6% | 32.4% | 32.7% |
| 8 | 31.5% | 31.6% | 30.9% | 32.1% |

CML, Conditional Maximum Likelihood; MCE, Minimum Classification Error.

tracking of phonetic error rates by the margin-based cost function, which penalizes incorrect decodings in direct proportion to their Hamming distance from the target label sequence. To determine which (if any) of these factors played a significant role, we conducted several experiments on the TIMIT corpus with variants of CML and MCE training.

First, we experimented with CML and MCE training procedures that did not enforce GMM normalization constraints. In these experiments, we optimized the usual objective functions for CML and MCE training, but parameterized the CD-HMMs in terms of the discriminant functions for large margin training in Equation (7.9). Table 7.4 compares phone error rates for CML and MCE training with normalized versus unnormalized GMMs. The table shows that lifting the normalization constraints generally leads to slightly worse performance, possibly due to overfitting. It seems that the extra degrees of freedom in unnormalized GMMs help to optimize the objective functions for CML and MCE training in ways that do not correlate with the actual phone error rate.

Next we experimented with different initial conditions for CML and MCE training. Because the optimizations in these frameworks involve highly nonlinear, non-convex objective functions, the results are susceptible to spurious local optima. To examine the severity of this problem, we experimented by re-initializing the CML and MCE training procedures with GMMs that had been discriminatively trained for segment-based phonetic classification

Table 7.5 String-based phone error rates from discriminatively trained Continuous Density Hidden Markov Models with differently initialized Gaussian Mixture Models (GMMs). The baseline GMMs were initialized by Maximum Likelihood estimation. The improved GMMs were initialized by large margin training for segment-based phonetic classification. Reproduced by permission of © 2007 IEEE

| Mixture components (per hidden state) | Baseline (CML) | Improved (CML) | Baseline (MCE) | Improved (MCE) |
|:---:|:---:|:---:|:---:|:---:|
| 1 | 36.4% | 32.6% | 35.6% | 32.9% |
| 2 | 34.6% | 31.7% | 34.5% | 31.3% |
| 4 | 32.8% | 31.2% | 32.4% | 31.1% |
| 8 | 31.5% | 28.9% | 30.9% | 29.0% |

CML, Conditional Maximum Likelihood; MCE, Minimum Classification Error.

(Sha and Saul 2006). These discriminatively trained GMMs provide a much better initialization than those trained by ML estimation. Table 7.5 compares the results from different initializations. For both CML and MCE training, the differences in performance are quite significant: indeed, the improved results approach the performance of CD-HMMs trained by margin maximization. These results highlight a significant drawback of the non-convex optimizations in CML and MCE training – namely that the final results can be quite sensitive to initial conditions.

Finally, we experimented with variants of CML and MCE training that penalize incorrect decodings in proportion to their Hamming distances from the target label sequence. The Hamming distance penalties in large margin training put more emphasis on correcting egregiously mislabeled sequences. Our final experiments were designed to test whether similarly proportional penalties in CML and MCE training would lead to better performance. For CML training in these experiments, we maximized a reweighted version of the conditional likelihood:

$$\theta_{\mathcal{H}}^{\text{CML}} = \arg\max_{\theta} \sum_n \log \frac{P(\bar{\mathbf{x}}_n, \bar{p}_n)}{\sum_{\bar{q}} e^{\mathcal{H}(\bar{p}_n, \bar{q})} P(\bar{\mathbf{x}}_n, \bar{q})}. \tag{7.17}$$

The reweighting in Equation (7.17) penalizes incorrect decodings in proportion to their Hamming distance from the target label sequence, analogous to the cost function of Equation (7.16) for large margin training. For MCE training in these experiments, we applied the same intuition by incorporating the Hamming distance penalty into the surrogate cost function of Equation (7.8):

$$\mathcal{N}_{\text{err}}^{\mathcal{H}} \approx \sum_n \sigma\left(-\log P(\bar{\mathbf{x}}_n, \bar{p}_n) + \log\left[\frac{1}{C} \sum_{\bar{q} \neq \bar{p}_n} e^{\eta[\mathcal{H}(\bar{p}_n, \bar{q}) + \log P(\bar{\mathbf{x}}_n, \bar{q})]}\right]^{1/\eta}\right). \tag{7.18}$$

Note that this approach differs from merely incorporating an adaptive scalar offset into the sigmoid transfer function. Such an offset has been shown to improve performance in large vocabulary ASR (Yu *et al.* 2006). Though a positive scalar offset plays a similar role as a 'margin', it does not explicitly penalize incorrect decodings in proportion to the number of mislabeled states. Table 7.6 shows the results from CML and MCE training with

Table 7.6  String-based phone error rates from discriminatively trained Continuous Density Hidden Markov Models (CD-HMMs) with different cost functions. The CD-HMMs in the third and fifth columns were trained by reweighting the penalties on incorrect decodings to grow in proportion to their Hamming distance from the target label sequence. Reproduced by permission of © 2007 IEEE

| Mixture components (per hidden state) | Baseline (CML) | Reweighted (CML) | Baseline (MCE) | Reweighted (MCE) |
|---|---|---|---|---|
| 1 | 36.4% | 33.6% | 35.6% | 33.3% |
| 2 | 34.6% | 32.8% | 34.5% | 32.3% |
| 4 | 32.8% | 32.8% | 32.4% | 31.5% |
| 8 | 31.5% | 31.0% | 30.9% | 30.8% |

CML, Conditional Maximum Likelihood; MCE, Minimum Classification Error.

normal versus reweighted objective functions. Interestingly, the reweighting led to generally improved performance, but this positive effect diminished for larger models.

The experiments in this section were designed to examine the effects of relaxed normalization constraints, local versus global optima and Hamming distance penalties in CML and MCE training. Overall, though better initializations and reweighted objective functions improved the results from CML and MCE training, none of the variants in this section ultimately matched the performance from large margin training. It is difficult to assess which factors contributed most strongly to this better performance. We suspect that all of them, working together, play an important role in the overall performance of large margin training.

## 7.5  Conclusion

Discriminative learning of sequential models is an active area of research in both ASR (Li *et al.* 2005; Roux and McDermott 2005; Woodland and Povey 2002) and machine learning (Altun *et al.* 2003; Lafferty *et al.* 2001; Taskar *et al.* 2004). In this chapter, we have described a particular framework for large margin training of CD-HMMs which makes contributions to lines of work in both communities. In distinction to previous work in ASR, we have proposed a convex, margin-based cost function that penalizes incorrect decodings in proportion to their Hamming distance from the desired transcription. The use of the Hamming distance in this context is a crucial insight from earlier work (Taskar *et al.* 2004) in the machine learning community, and it differs profoundly from merely penalizing the log-likelihood gap between incorrect and correct transcriptions, as commonly done in ASR. In distinction to previous work in machine learning, we have proposed a framework for sequential classification that naturally integrates with the infrastructure of modern speech recognizers. For real-valued observation sequences, we have shown how to train large margin HMMs via convex optimizations over their parameter space of positive semidefinite matrices. Using the softmax function, we have also proposed a novel way to monitor the exponentially many margin constraints that arise in sequential classification.

On the task of phonetic recognition, we compared our framework for large margin training with two other leading frameworks for discriminative training of CD-HMMs.

In our experiments, CD-HMMs trained by margin maximization achieved significantly better performance than those trained by CML or MCE. Follow-up experiments suggested two possible reasons for this better performance: (1) the convexity of the optimization for large margin training and (2) the penalizing of incorrect decodings in direct proportion to the number of mislabeled states. In future research, we are interested in applying large margin training to large vocabulary ASR, where both CML and MCE training have already demonstrated significant reductions in word error rates (McDermott *et al.* 2006; Valtchev *et al.* 1997; Woodland and Povey 2002).

# References

Altun Y, Tsochantaridis I and Hofmann T 2003 Hidden Markov support vector machines. *Proceedings of the 20th International Conference on Machine Learning (ICML-03)* (eds. T Fawcett and N Mishra), Washington, DC. AAAI Press, pp. 3–10.

Bahl LR, Brown PF, de Souza PV and Mercer RL 1986 Maximum mutual information estimation of hidden Markov model parameters for speech recognition. *Proceedings of the IEEE International Conference on Acoustics, Speech, and Signal Processing*, Tokyo, pp. 49–52.

Gopalakrishnan PS, Kanevsky D, Nádas A and Nahamoo D 1991 An inequality for rational functions with applications to some statistical estimation problems. *IEEE Transactions on Information Theory* **37**(1), 107–113.

Jiang H and Li X 2007 Incorporating training errors for large margin HMMs under semi-definite programming framework. *Proceedings of the IEEE International Conference on Acoustics, Speech, and Signal Processing (ICASSP-07)*, Honolulu, HI, vol. 4, pp. 629–632.

Jiang H, Li X and Liu C 2006 Large margin hidden Markov models for speech recognition. *IEEE Transactions on Audio, Speech and Language Processing* **14**(5), 1584–1595.

Juang BH and Katagiri S 1992 Discriminative learning for minimum error classification. *IEEE Transactions on Signal Processing* **40**(12), 3043–3054.

Kapadia S, Valtchev V and Young S 1993 MMI training for continuous phoneme recognition on the TIMIT database. *Proceedings of the IEEE International Conference on Acoustics, Speech, and Signal Processing (ICASSP-93)*, Minneapolis, MN, vol. 2, pp. 491–494.

Keshet J, Shalev-Shwartz S, Bengio S, Singer Y and Chazan D 2006 Discriminative kernel-based phoneme sequence recognition. *Proceedings of the International Conference on Spoken Language Processing (ICSLP-06)*, Pittsburgh, PA.

Lafferty J, McCallum A and Pereira FCN 2001 Conditional random fields: Probabilistic models for segmenting and labeling sequence data. *Proceedings of the 18th International Conference on Machine Learning (ICML-01)*. Morgan Kaufmann, San Francisco, CA, pp. 282–289.

Lamel LF, Kassel RH and Seneff S 1986 Speech database development: design and analsysis of the acoustic-phonetic corpus. *Proceedings of the DARPA Speech Recognition Workshop* (ed. LS Baumann), pp. 100–109.

Lee KF and Hon HW 1988 Speaker-independent phone recognition using hidden Markov models. *IEEE Transactions on Acoustics, Speech, and Signal Processing* **37**(11), 1641–1648.

Li J, Yuan M and Lee C 2007 Approximate test risk bound minimization through soft margin estimation. *IEEE Transactions on Speech, Audio and Language Processing* **15**(8), 2392–2404.

Li X, Jiang H and Liu C 2005 Large margin HMMs for speech recognition. *Proceedings of the IEEE International Conference on Acoustics, Speech, and Signal Processing (ICASSP-05)*, Philadelphia, PA, pp. 513–516.

McDermott E, Hazen TJ, Roux JL, Nakamura A and Katagiri S 2006 Discriminative training for large vocabulary speech recognition using minimum classification error. *IEEE Transactions on Speech and Audio Processing* **15**(1), 203–223.

Nádas A 1983 A decision-theoretic formulation of a training problem in speech recognition and a comparison of training by unconditional versus conditional maximum likelihood. *IEEE Transactions on Acoustics, Speech and Signal Processing* **31**(4), 814–817.

Povey D, Kanevsky D, Kingsbury B, Ramabhadran B, Saon G and Visweswariah K 2008 Boosted MMI for model and feature-space discriminative training. *Proceedings of the IEEE International Conference on Acoustics, Speech and Signal Processing (ICASSP-08)*, Las Vegas, NV.

Povey D and Woodland P 2002 Minimum phone error and i-smoothing for improved discriminative training. *Proceedings of the IEEE International Conference on Acoustics, Speech, and Signal Processing (ICASSP-02)*, Orlando, FL, pp. 105–108.

Robinson T 1994 An application of recurrent nets to phone probability estimation. *IEEE Transactions on Neural Networks* **5**(2), 298–305.

Roux JL and McDermott E 2005 Optimization methods for discriminative training. *Proceedings of 9th European Conference on Speech Communication and Technology (EuroSpeech-05)*, Lisbon, Portugal, pp. 3341–3344.

Sha F 2007 *Large margin training of acoustic models for speech recognition* University of Pennsylvania, Philadelphia, PA, PhD thesis.

Sha F and Saul LK 2006 Large margin Gaussian mixture modeling for phonetic classification and recognition. *Proceedings of the IEEE International Conference on Acoustics, Speech and Signal Processing (ICASSP-06)*, Toulouse, France, pp. 265–268.

Sha F and Saul LK 2007a Comparison of large margin training to other discriminative methods for phonetic recognition by hidden Markov models. *Proceedings of the IEEE International Conference on Acoustics, Speech and Signal Processing (ICASSP-07)*, Honolulu, HI, pp. 313–316.

Sha F and Saul LK 2007b Large margin hidden Markov models for automatic speech recognition. *Advances in Neural Information Processing Systems 19* (eds. B Schölkopf, J Platt and T Hofmann). MIT Press, Cambridge, MA, pp. 1249–1256.

Taskar B, Guestrin C and Koller D 2004 Max-margin Markov networks. *Advances in Neural Information Processing Systems (NIPS 16)* (eds. S Thrun, L Saul and B Schölkopf). MIT Press, Cambridge, MA, pp. 25–32.

Valtchev V, Odell JJ, Woodland PC and Young SJ 1997 MMIE training of large vocabulary recognition systems. *Speech Communication* **22**, 303–314.

Woodland PC and Povey D 2002 Large scale discriminative training of hidden Markov models for speech recognition. *Computer Speech and Language* **16**, 25–47.

Yu D, Deng L, He X and Acero A 2006 Use of incrementally regulated discriminative margins in MCE training for speech recognition. *Proceedings of the International Conference on Spoken Language Processing (ICSLP-06)*, Pittsburgh, PA, pp. 2418–2421.

# Part III

# Language Modeling

# 8

# A Survey of Discriminative Language Modeling Approaches for Large Vocabulary Continuous Speech Recognition

## Brian Roark

Discriminative training of language models has been the focus of renewed research interest recently, and significant improvements to high accuracy large vocabulary continuous speech recognition have been demonstrated through the use of such models. In this chapter, we review past and present work on this topic, and focus in particular on three key issues: training data, learning algorithms and features. We will show how simple models trained under standard approaches can provide significant accuracy gains with negligible impact on decoding time, and how richer models can provide further accuracy gains. We will argue that while the published results are promising, many practical issues remain under-explored, so that further improvements might be expected as the best practices become more developed.

## 8.1 Introduction

It has long been the case that system improvements in statistical speech recognition have come far more from acoustic modeling innovations than from language modeling innovations – to the extent that some top-notch researchers have quit working on language modeling for speech recognition, sometimes with a parting message of warning to future researchers against expecting system improvements. (See the extended version of Goodman (2001),

*Automatic Speech and Speaker Recognition: Large Margin and Kernel Methods*  Joseph Keshet and Samy Bengio
© 2009 John Wiley & Sons, Ltd

which can be found with the search engine query: 'language modeling' 'all hope abandon.') There are a number of reasons why little progress was observed due to language modeling, including (though certainly not confined to): extreme sparse data; optimization of objectives only very loosely corresponding to the true system objective; and difficulty integrating disparate potential features into a single model. Sparse data is a problem that can be alleviated with increased training data, which some may not consider a particularly interesting solution, though it can require solving difficult problems that arise when scaling up algorithms to handle larger data sets. The continued improvement of performance in speech recognition and machine translation due to language model training on extremely large training sets (Emami *et al.* 2007; Popat *et al.* 2007) illustrates the extent to which sparse data has been impacting language models. Even what have been considered very large corpora in the past do not provide adequate training for multinomial $n$-gram models with millions or even billions of parameters. In Popat *et al.* (2007), for example, the models have hundreds of billions of parameters (corresponding to $n$-gram features of various orders), which are trained on corpora of several trillion tokens; they find a linear gain with each doubling of the training data, with no indication of a plateau.

In this chapter, we will focus on language modeling improvements in large vocabulary continuous speech recognition (LVCSR) that are not due to an increase in the amount of training data, but rather to improvements in the other two dimensions of the problem that were identified in the previous paragraph: optimization of parameters with respect to objectives more closely related to actual system objectives; and the inclusion of a large number of heterogeneous feature sets derived from the output string hypotheses.

Neither of these two considerations is new to research in language modeling for LVCSR. It was suggested in Jelinek (1996) to train an 'acoustic-sensitive' language model whose parameters would be trained to minimize the expected uncertainty of the spoken text given the acoustic signal. Discriminative training methods that were successfully pursued for acoustic modeling, such as MMIE (Bahl *et al.* 1986) or related error-corrective training methods (Bahl *et al.* 1993; Juang *et al.* 1997), spurred initial research into discriminative modeling approaches such as those of Chen *et al.* (2000), Stolcke and Weintraub (1998) and Stolcke *et al.* (2000), with modest utility. In addition, generative maximum entropy models have been explored for combining diverse features within a single model (Khudanpur and Wu 2000; Rosenfeld *et al.* 2001), including $n$-grams, topic labels and syntactic features. Thus the notion of exploiting diverse feature sets is not new to the field; and neither are discriminative approaches to parameter estimation.

Over the past few years, however, there has been a renewed effort in this direction, driven in large part by developments in the machine learning community. The benefits of global conditional modeling for sequence learning (Lafferty *et al.* 2001) and the application of large margin methods to sequence processing through the Perceptron algorithm (Collins 2002; Freund and Schapire 1999) have provided a unifying framework within which many related methods can be pursued and compared. The learning and inference algorithms associated with some of these approaches are efficient enough to be scaled up, so that models making use of very large feature sets can be effectively trained and used for inference. Well-motivated and theoretically grounded regularization techniques have been shown to effectively control for the overtraining that language models have long suffered from. Strong empirical gains have been realized on difficult tasks and continue to be realized up to the present within this paradigm.

This chapter addresses discriminative language modeling approaches in general, not kernel approaches in particular. The approaches that are the focus of this chapter do not represent the only strand of research focused on leveraging competing LVCSR system hypotheses to achieve reductions in error rate. The approaches we describe here are focused on *training* models with a discriminative objective function, not *inference* techniques for optimizing an objective. Hence approaches such as Minimum Bayes-Risk classification (Goel and Byrne 2000) or confusion networks ('sausages') (Mangu *et al.* 2000) will not be covered here, although methods exist for learning their edit distance (Shafran and Byrne 2004).[1] Note that inference using such techniques can also be pursued with models trained as presented here, hence the techniques are complementary. We will also omit any detailed discussion of methods that perform post-processing on LVCSR output to correct specific errors, such as induction of transformational rules on the output (Mangu and Padmanabhan 2001) or methods for proposing alternatives that are not present in the system output and choosing between them (Ringger and Allen 1996; Zhou *et al.* 2006b). Again, such methods could also be applied to the output of systems using models trained with methods presented in this survey, hence they are complementary to what will be discussed here. Also, we have chosen to focus on large vocabulary systems, hence will not discuss the application of similar methods in small vocabulary applications, e.g. that in Kuo *et al.* (2002).

## 8.2 General Framework

Most of the work covered in this chapter falls within the general framework of linear models, which we present here largely following the notation of Collins (2002), much as presented in Roark *et al.* (2007). These approaches assume the presence of several components, as follows.

First, these are supervised approaches, hence requiring the presence of training data, consisting of a set of $m$ input sequences $\bar{x}_i \in \mathcal{X}$ and corresponding reference (ground truth) output sequences $\bar{y}_i^{\text{ref}} \in \mathcal{Y}$, where $\mathcal{X}$ denotes the set of possible inputs and $\mathcal{Y}$ the set of possible outputs. In the case of language modeling for LVCSR, using a vocabulary $\Sigma$, the input set $\mathcal{X}$ consists of acoustic samples, the output set consists of all possible transcriptions, i.e. $\mathcal{Y} = \Sigma^*$, and the reference (ground truth) output sequence for any given utterance is the manual (human) transcription of that utterance.

Next, for each input sequence $\bar{x} \in \mathcal{X}$, the approaches require some mechanism for identifying a set of output candidates, which is often denoted as a function $\mathbf{GEN}(\bar{x}) \subseteq \mathcal{Y}$. For example, $\mathbf{GEN}(\bar{x})$ could be the word lattice output or $n$-best list from a baseline LVCSR system.

Third, there must be a mapping from each input:output pairing $(\bar{x}, \bar{y}) \in \mathcal{X} \times \mathcal{Y}$ to a real valued $d$-dimensional feature vector $\boldsymbol{\psi}(\bar{x}, \bar{y}) \in \mathbb{R}^d$, for some $d$. In the current case, this could be a mapping from a string to the set of $n$-grams of order $\leq k$ that occur in the string. An example feature would then be something like 'feature 1236: number of times the bigram "dog house" occurs in the string'. Each explicit $n$-gram to be included in the model would be a separate feature in the vector.

---

[1] The methods for learning the edit distance used in such techniques are focused on learning a weighted transduction between reference transcripts and LVCSR system outputs. This learned transduction can be seen as a form of discriminative language modeling, since it effectively moves the centroid closer to the reference.

Finally, there must be a real valued parameter vector $\alpha \in \mathbb{R}^d$ of the same dimension $d$ as the feature vector. For example, feature 1236 mentioned above may have a parameter value of 0.2 assigned to it. Methods for determining the parameter values associated with each feature are perhaps the key difference between many of the different approaches that will be presented here, though they are all discriminative rather than generative approaches.

Given these four components, we can define the best scoring (or 1-best) output $\bar{y}_i^{max}$ for a given input $x_i$ as

$$\bar{y}_i^{max} = \underset{\bar{y} \in \mathbf{GEN}(\bar{x}_i)}{\mathrm{argmax}} \; \boldsymbol{\psi}(\bar{x}_i, \bar{y}) \cdot \boldsymbol{\alpha}, \qquad (8.1)$$

where $\boldsymbol{\psi}(\bar{x}_i, \bar{y}) \cdot \boldsymbol{\alpha}$ is the dot product.

In the remainder of this section, we will focus on each of these four components in turn, highlighting some of the key trends and variations in the work from this area.

## 8.2.1 Training Data and the GEN Function

Training data for discriminative language modeling consists of input speech recordings and reference output strings representing the gold standard transcription of the speech. The role of the GEN function is typically taken by a baseline LVCSR system, so that $\mathrm{GEN}(\bar{x})$ for some input speech is an $n$-best list or word lattice. Hence most discriminative language modeling approaches fall within the range of approaches more traditionally known as $n$-best or lattice rescoring. Working from lists and lattices, however, is not a requisite part of the approach, and some recent papers (Kuo *et al.* 2007; Lin and Yvon 2005) have modified the weights directly within the baseline recognition system, rather than adopting a rescoring or reranking paradigm. See Section 8.2.2 for further details on directly updating the baseline recognition system, where it is discussed as a method for encoding features to allow for efficient processing.

The output of LVCSR systems can be thought of as weighted, acyclic finite state automata, with words and weights labeling the arcs. Figure 8.1 shows three kinds of representation of weighted output from an LVCSR system. For all such weighted automata, the semiring must be specified, so that the weights can be interpreted. The automata in Figure 8.1 are in the Real semiring, meaning that the weights along a path are multiplied, and when two paths are combined, their weights are added, which is the appropriate semiring when dealing with probabilities.[2] In the figure, the weights define a conditional probability distribution over three possible strings: 'a b c' with probability 0.3; 'a g c', with probability 0.3; and 'e d c c', with probability 0.4. Although the weights are presented as conditional probabilities, they can be arbitrary.

Not all weighted finite-state automata (WFA) can be determinized, but any acyclic WFA can be determinized (Mohri and Riley 1997). The result of determinization on the $n$-best list automaton in Figure 8.1(a) is the deterministic tree in Figure 8.1(b). As with the original representation, a candidate string is represented by the concatenation of symbols labeling the arcs of any path from the start state at the left to a final state denoted with a double circle.

---

[2]The most common semirings in speech processing are the Tropical and Log semirings, both of which are appropriate for negative log probabilities. Both add weights along the path. The Tropical semiring takes the **min** of two weights when combining paths, and the Log semiring takes the log of the sum of the exponentials: $A \oplus B = -\log(\exp(-A) + \exp(-B))$.

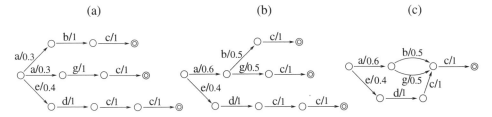

Figure 8.1 Three weighted, acyclic finite-state automata representations of the output from a speech recognizer: (a) $n$-best list; (b) deterministic tree; (c) deterministic word lattice. The automata are in the Real semiring, and all three represent the same weighted distributions over the three strings.

What is key to notice from these figures is that determinization preserves the path weight of every path in the automaton. Recall that weights are combined along a path in the Real semiring by multiplying. Inspection will verify that all three of the paths in the original automaton are present in the deterministic tree representation with the same path weight.

Deterministic WFA can be minimized, i.e. converted to a representation of exactly the same weighted paths, but with a minimal number of states and arcs to represent it (Mohri and Riley 1997). For the current example, this minimized WFA is shown in Figure 8.1(c). These are word lattices, which are simply a more compact representation of a set of distinct weighted hypotheses output from the LVCSR system. Efficient use of this kind of compact representation allows for effective rescoring in scenarios where brute force enumeration of all distinct candidates would be intractable. Effective use of these representations, however, precludes the use of global features, unless they can be decomposed for dynamic programming. Some approaches to discriminative language modeling, e.g. those making use of long-distance dependencies uncovered via context-free parsing, will be forced to operate on undeterminized or unminimized $n$-best representations, which limits the number of candidates that can be tractably scored.

There are a couple of reasons to introduce WFA representations in this context. First, they become important in the next section, when efficient feature mapping is discussed. Also, it should be clear from the discussion that there is no difference between $n$-best lists and word lattices, other than determinization and minimization – they both represent sets of competing (weighted) candidate transcriptions. Even 1-best output of a system is a special case of these acyclic WFA. Hence we can discuss the approaches in general, without having to specify whether the data is in the form of $n$-best lists or lattices. We will thus henceforth refer to them simply as lattices, which may or may not be determinized or minimized. General operations on WFA can be used to specify operations on the lattices, including intersection, composition with a transducer, and finding the minimum cost (or highest scoring) path.

Note that working within a reranking or rescoring paradigm has a few immediate ramifications for the general modeling framework we are presenting. Most critically, it is not guaranteed in this approach that the reference transcription is present in the lattice. Second, if training lattices are produced for utterances that were also used for building the models of the baseline recognizer, this may introduce a mismatch between training conditions and testing conditions. These are two key issues investigated in Roark *et al.* (2004a, 2007), and we discuss them here in turn.

**(a) Reference Transcript**

If the reference transcript for a particular utterance is not in the lattice output in the training data, then standard parameter estimation algorithms will be trying to achieve an objective that cannot be reached. These parameter estimation algorithms will be discussed in Section 8.2.3, but, intuitively, an algorithm (such as the Perceptron algorithm) that penalizes candidates other than the gold standard (reference) will move parameters to penalize every candidate if the gold standard is not in the set of candidates, even the highest accuracy candidate in the whole set.

One quick fix to this problem would be to append the reference transcript to the set of candidates in the training set. The main problem with this is that it creates a great mismatch between the training condition and the testing condition, in that the training examples are being given better candidate sets than will be seen under testing conditions. Of lesser significance is the problem of acquiring features derived from the baseline LVCSR system for the reference transcript, if it was not one of its candidates. For example, the baseline system score is typically a useful feature in the model. Obtaining the baseline system score for the reference may require having some access to the baseline system models, to constrain the decoder to output the reference.

A better solution is to define the gold standard candidate in terms of maximum accuracy or minimum error rate. Then, if the reference is in the candidate set, it will be the gold standard; otherwise, another candidate from the set is chosen as the gold standard, to avoid the problems with parameter estimation. When the reference is present in the lattice, the gold standard is unique. If the reference is not present in the lattice, then there may not be a unique gold standard: more than one candidate may have the minimum error rate. The results in Roark *et al.* (2004a, 2007) empirically validate the utility of using a minimum error rate gold standard versus a reference gold standard in the Perceptron algorithm. In those papers, one gold standard is chosen from among the set of minimum error rate candidates; in Singh-Miller and Collins (2007), all members from the set were collectively used as the gold standard in a loss-sensitive version of the Perceptron algorithm. See Section 8.2.3 for more details about parameter estimation when the reference is not in the lattice.

Note that this is also an issue when training outside of a rescoring or reranking paradigm, as noted in Kuo *et al.* (2007), even with known training data. If the vocabulary is pruned, there will be out-of-vocabulary (OOV) terms in the references which the closed vocabulary LVCSR system cannot recognize. One method for decreasing the OOV rate in any of these training scenarios, however, is to map terms to 'equivalent' items that are in the vocabulary, e.g. legal spelling variations.

**(b) Seen Versus Unseen Data**

The second issue that was mentioned regarding the training data is the potential mismatch between training and testing scenarios when training lattices are produced on previously seen data. This is an issue with any kind of a reranking problem, and similar solutions to the one that will be discussed here have been used for other such problems, e.g. Collins (2000). In fact, the problem is to minimize mismatch, because most problems have a constrained amount of supervised training data (with both input and output sides of the mapping), hence the same data will often be used both for training the baseline models and for training discriminative language models. It is well-known that $n$-gram language models are overparameterized and overtrained; the result of this is that the LVCSR output for seen data will have fewer errors than the output for unseen data, and those errors that do occur may differ from the kinds of error observed for unseen data. The discriminative reranking model will then learn to solve a different problem than the one that will be encountered at test time. To obtain a useful model, the mismatch needs to be reduced by producing training lattices in an unseen condition.

An unseen condition can be achieved through cross-validation, whereby the training data is split into $k$ subsets, and the baseline models for each subset $j$ are trained on the other $k-1$ subsets $(1, \ldots, j-1, j+1, \ldots, k)$. Note that this can still result in a mismatch, because the testing condition will use baseline models trained on all $k$ subsets. The hope is that, if $k$ is large enough, the difference in performance when training on the extra section will be relatively small, particularly compared with the difference in seen versus unseen performance. Again, the results in Roark *et al.* (2004a, 2007) empirically validate the utility of cross-validating the baseline language model training: not cross-validating resulted in models that achieved no gain over the baseline system, whereas cross-validation produced significant word-error rate (WER) reductions of over 1% absolute.

While the presentation here has been on the importance of cross-validating the models used in the baseline systems, previous work has only examined the utility of cross-validating the language models, not the acoustic models. The reasons for this are twofold. First, conventional wisdom is that acoustic models, by virtue of having far fewer parameters than language models, tend not to overtrain to the same degree, so that the seen/unseen difference in performance is relatively small. Second, while baseline $n$-gram models are relatively easy and fast to train (even for large corpora), since they typically use relative frequency estimation and simple smoothing to estimate the parameters, acoustic model training is a far more intensive process, so that training a baseline acoustic model on a large training corpus may take many days or more on multiple processors. Hence the cost of cross-validation is high and the perceived benefit low. Even so, it would be of interest to know exactly how much mismatch exists between seen and unseen data with respect to the acoustic model, and whether this may actually impact the utility of training data for discriminative language modeling.

## 8.2.2  Feature Mapping

The third component of the general framework that we are presenting is the mapping from input/output pairs $(\bar{x}, \bar{y})$ to a feature vector $\psi(\bar{x}, \bar{y}) \in \mathbb{R}^d$. In the current context, there are two scenarios that we have discussed: reranking from lattices; and updating the baseline LVCSR system directly. We will first discuss feature mapping for reranking, where candidates are represented as an acyclic WFA of candidates; then for direct update of the system.

For reranking, one approach to feature mapping is to simply perform feature mapping for each candidate transcript independently. For example, if an existing statistical parser were to be applied to the strings and the result used to extract features, then the particular parser may not be able to handle anything other than a string at a time. This requires explicit enumeration of all candidates, which will limit the number of candidates that can be considered for any given utterance.

A more efficient strategy is to take advantage of shared structure in a word lattice to extract features without having to explicitly enumerate every path in the lattice. This is particularly important in LVCSR, where the number of competing hypotheses can be in the hundreds of thousands or millions, and the feature set in the several millions. $N$-gram features are popular in large part because they have an efficient finite-state structure, and this structure can be exploited for efficient feature mapping that allows for large lattices to be rescored, rather than small $n$-best lists. We will use a toy example to illustrate this. For more details on $n$-gram model structure, see Allauzen *et al.* (2003) and Roark *et al.* (2004a,b, 2007).

For simplicity, assume a vocabulary $\Sigma = \{a,b,c\}$, and all strings consist of one or more symbols from that vocabulary in sequence ($\Sigma^+$). The output symbols will be feature indices, which we will denote $f_k$ for feature with index $k$. For this example, let us assume the following unigram and bigram features: $\{f_1 = a; f_2 = b; f_3 = <s>a; f_4 = <s>b; f_5 = aa; f_6 = ab; f_7 = a</s>; f_8 = ba; f_9 = bb; f_{10} = b </s>\}$. Note that the begin-of-string symbol ($<s>$) and end-of-string symbol ($</s>$) are implicit in the string and not part of the vocabulary. One can see by inspection of this set that there are all unigram and bigram features including the symbols 'a' and 'b', but none including the symbol 'c', hence those $n$-grams will not map to any feature in the set.

The finite-state transducer in Figure 8.2(a) maps from strings in $\{a,b,c\}^+$ to feature indices from this set. The start state (denoted with $>$ to its left) implicitly encodes the begin-of-string symbol, hence the first symbol in the string will be in a bigram beginning with $<s>$. There are two special symbols in the transducer: $\phi$, which represents a failure transition; and $\epsilon$, which represents the empty string. An arc labeled with $\phi$ is traversed if and only if there is no arc leaving the state that is labeled with the next symbol in the input. No input symbols are consumed or output symbols emitted when the $\phi$ transition is traversed. An arc with an $\epsilon$ symbol labeling the input side can always be traversed without consuming an input symbol, while outputting the symbol on the output side. An arc with an $\epsilon$ symbol labeling the output side produces no output while consuming the symbol on the input side. When final states (denoted with double circle) are reached and there is no more input, the transduction is successful.

Let us step through a couple of examples. First the simple string 'acb', which we will compose with transducer $T_1$ from Figure 8.2(a). Beginning at the start state, there is an arc labeled with 'a' on the input side, with $f_3$ on the output side. From the destination state of that arc, there is no arc with 'c' on the input side, so the failure transition (labeled with $\phi$) is traversed. From the destination state of that arc, there is an arc with 'c' on the input side, which is a looping arc with $\epsilon$ (empty string, i.e. no feature) on the output side. The transition from that state with 'b' on the input side has $f_2$ on the output side. The destination state of that arc is not a final state, but there is a transition with an $\epsilon$ on the input side which can be traversed, which gives $f_{10}$ on the output side and reaches a final state. Hence this transduction gives the features: $f_3$, $f_2$ and $f_{10}$. This is not, however, the total feature set, but rather the highest order $n$-grams reached at each word. For example, feature $f_3$ is the bigram '$<s>a$', but there should also be the unigram feature $f_1$ associated with the 'a'. In order to derive the total feature set based on the highest order features, we compose the output sequence to the transducer $T_2$ in Figure 8.2(b). In that transducer, all features map to themselves, but additionally all bigram features ending in an 'a' or 'b' then also map to the appropriate unigram feature. The complete transduction from a string $S$ to the feature set is $S \circ T_1 \circ T_2$, where $\circ$ denotes the composition of finite-state transducers.

Note that these are both deterministic on the input side, i.e. every string in $\{a,b,c\}^+$ has one and only one path through $T_1 \circ T_2$. It should be straightforward for the reader

(a)　　　　　　　　　　　　　　(b)

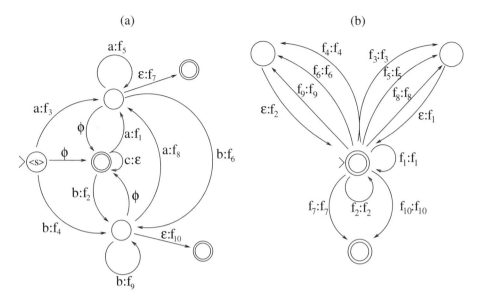

Figure 8.2 String sequence to feature set transducers: (a) from symbols to highest order features; (b) from highest order features to total set of features.

to verify that the input string 'bbacb', when composed with $T_1 \circ T_2$, yields the feature sequence $f_4 f_2 f_9 f_2 f_8 f_1 f_2 f_{10}$. Because this feature mapping uses standard finite-state transducer transduction, word lattices encoded as finite-state automata can be composed with these transducers, to produce output sequences for every path in the lattice corresponding to features in the feature set. Multiple paths that share part of their structure will not need to be mapped independently of each other, as they would be in non-deterministic representations of $n$-best lists.

Note that, at test time, the feature identities are not of interest, but rather the scores of the paths in the lattices, so that the best scoring candidate can be found. There is also a compact finite-state automaton that can score the lattice with $n$-gram parameter weights, without requiring transduction to the feature labels. The automaton in Figure 8.3 is in the Tropical semiring (sum weights along a path), and makes use of roughly the same topology as the transducer in Figure 8.2(a). Note that all arcs corresponding to bigram features must have the sum of both the bigram feature weight and the corresponding unigram feature weight. Parameter weights associated with bigrams containing the end-of-string symbol are encoded in the final weight of the corresponding final state. Similar structure is used when going to $n$-grams of higher order than bigrams. See Roark et al. (2004a,b, 2007) for more information about such finite-state $n$-gram model encoding.

Finite-state transducer encoding of the model is also a central topic in the recent work looking at direct updating of parameters in the baseline LVCSR model rather than reranking (Kuo et al. 2007; Lin and Yvon 2005). These works adopt an approach to efficient LVCSR that involves building a finite-state 'decoding graph' off-line and performing optimizations on this transducer, as described in detail in Mohri et al. (2002). The decoding graph is the

| Idx | Param | Feature |
|-----|-------|---------|
| 1 | $\alpha_1$ | a |
| 2 | $\alpha_2$ | b |
| 3 | $\alpha_3$ | \<s\>a |
| 4 | $\alpha_4$ | \<s\>b |
| 5 | $\alpha_5$ | aa |
| 6 | $\alpha_6$ | ab |
| 7 | $\alpha_7$ | a\</s\> |
| 8 | $\alpha_8$ | ba |
| 9 | $\alpha_9$ | bb |
| 10 | $\alpha_{10}$ | b\</s\> |

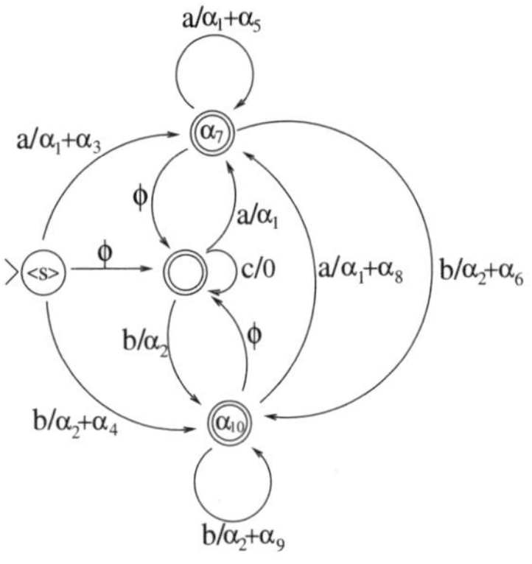

Figure 8.3 Deterministic automaton reading in symbols and assigning parameter weights.

result of combining an automata representation of an $n$-gram language model ($G$) with a pronunciation dictionary represented as a finite-state transducer ($L$) and decision tree models that split the phones in the pronunciations into context-dependent (CD) phone representations ($C$). The composition of these weighted transducers[3] $C \circ L \circ G$ results in a single transducer, which serves to guide decoding. Optimizations to this graph, such as determinization on the input side and weight pushing, can result in a far more efficient search than combining these models on-the-fly during search.

The idea pursued in Lin and Yvon (2005) and extended to full context-dependent LVCSR systems in Kuo *et al.* (2007) is to directly update transition weights within the decoding graph. To do this, they determine the state sequence through the decoding graph of the reference transcription by optimally force aligning the transcription with the speech. Training in Kuo *et al.* (2007) is an on-line procedure discussed in Section 8.2.3 that compares the best scoring state sequence with the state sequence derived from the reference, and updates the transition weights accordingly. Hence each transition weight in the graph is an independent feature within a log linear framework.

This approach to feature mapping brings up some interesting questions. First, what is discriminative language modeling? In this instance, the features include context-dependent phone identities as part of the state definition, and stochastic modeling of such features is typically outside the range of language modeling. While it is not the intent of this chapter to try to philosophically delimit the scope of this topic, it seems clear that the transition weights in these graphs represent discrete symbol sequences rather than the Gaussian densities modeled at the states, hence are certainly modeling the language not the acoustics. However, it highlights the fact that there is little in this general framework to limit the kinds of features that might be used, and that future work may find hybrid feature sets – e.g. for combining syntactic and prosodic features – that will lead to substantial system improvements.

The second question that arises is, what do these features represent? The input stream of the transducer is labeled with context-dependent phones, and the output stream is labeled with words in the vocabulary, so in its simplest form, these features represent

---

[3]Note that an automaton is a special case of a transducer where every symbol maps to itself.

(CD phone × word) sub-sequences. These graphs, however, are optimized to increase the amount of shared structure between words with similar CD phone realizations, hence the transitions themselves may represent CD phones over sets of $n$-grams. These complicated composite features do appear to have strong utility for this task (see Kuo *et al.* (2007) for empirical results), and they capture key distinctions that are clearly lost in most language modeling approaches. The downside of this approach is a loss of generalization from one phone sequence realization of a word sequence to another realization of the same word sequence. One could imagine an extension of this approach that treats arcs as features, but also includes raw $n$-gram features, and updates arcs in the graph with respect to all relevant features. In such a way, the benefits of this approach could be achieved while also gaining generalization to other pronunciations of the same $n$-gram.

## 8.2.3 Parameter Estimation

The final component to the general framework is the parameter estimation algorithm, for choosing a parameter weight vector $\alpha \in \mathbb{R}^d$. Here any technique that has been used for log linear modeling in other domains can be used, with the proviso that the method must scale to very large problems. Language modeling for LVCSR is typically most effective with a large number of candidates (at least in the hundreds) and a very large number of features (in the millions). This means that there have been limited uses of resource intensive methods such as Support Vector Machines (SVMs) (though see Section 8.3.3 for one example using SVMs).

Before getting into specific training objectives, it is probably worth spending a little time to address a potential confusion of terminology. In language modeling, *conditional* models have typically signified the conditional probabilities used via the chain rule to define the joint probability of a string. For example, a standard trigram model is defined as the probability of a word conditioned on the previous two words. In the current context, however, conditional models do not denote the local conditional probabilities that are used to define the joint generative probability, but rather denote the likelihood of the string *given* the input utterance. This sort of conditional model is discriminative.

Given the popularity of Maximum Mutual Information Estimation (MMIE) for acoustic modeling (Bahl *et al.* 1986), which has a conditional likelihood objective in the sense we are using here, direct applications of such techniques have been pursued for language modeling. As in prior work of Stolcke and Weintraub (1998), Chueh *et al.* (2005) examines a log linear model which maximizes an unregularized conditional likelihood objective. Linguistic ($n$-gram) features of the model are combined with a baseline acoustic model score, which can be thought of as just another feature in the model. If $\bar{y}_i^{\text{ref}}$ is the reference output for speech input $\bar{x}_i$, then the conditional log likelihood of the training data of $m$ utterances given a feature mapping $\psi$ and parameter vector $\alpha$ is

$$
\begin{aligned}
\text{LogLike}(\alpha) &= \sum_{i=1}^{m} \log \text{P}(\bar{y}_i^{\text{ref}} \mid \bar{x}_i) \\
&= \sum_{i=1}^{m} \log\left( \frac{\exp(\psi(\bar{x}_i, \bar{y}_i^{\text{ref}}) \cdot \alpha)}{\sum_{\bar{y} \in \text{GEN}(\bar{x}_i)} \exp(\psi(\bar{x}_i, \bar{y}) \cdot \alpha)} \right) \\
&= \sum_{i=1}^{m} \left( \psi(\bar{x}_i, \bar{y}_i^{\text{ref}}) \cdot \alpha - \log \sum_{\bar{y} \in \text{GEN}(\bar{x}_i)} \exp(\psi(\bar{x}_i, \bar{y}) \cdot \alpha) \right). \quad (8.2)
\end{aligned}
$$

This is a convex function, and hence the $\alpha$ that maximizes the log likelihood can be found through standard techniques relying on the gradient (see below).

One key problem with this approach is overtraining, which, as usual with language modeling, is severe and has led to little or no system gain for approaches pursuing it as presented above (Chueh *et al.* 2005; Stolcke and Weintraub 1998). With millions of features, the opportunity for overtraining is much greater for language modeling than for acoustic modeling, hence some method for controlling for overtraining (regularization) needs to be added to the training approach in order to improve generalization to new data. Of course, the reasons why these models provided little or no gain in system performance may also be related to some of the issues in training data preparation presented in Section 8.2.1, but certainly lack of regularization also contributed. Similar lack of system improvement was observed in Paciorek and Rosenfeld (2000) when using Minimum Classification Error (MCE) training (Juang *et al.* 1997) for language model training, as well as in Kuo and Chen (2005) when using Minimum Word Error (MWE) training (Povey and Woodland 2002). Direct application of popular acoustic model training approaches have thus not been successful, and a large reason for this is almost certainly lack of appropriate model regularization.

Adopting regularization techniques from other sequence modeling work making use of global conditional likelihood optimization (Johnson *et al.* 1999; Lafferty *et al.* 2001), the regularized objective in Roark *et al.* (2004b, 2007) becomes

$$\text{RegLogLike}(\alpha) = \sum_{i=1}^{m}\left(\psi(\bar{x}_i, \bar{y}_i^{\text{ref}}) \cdot \alpha - \log \sum_{\bar{y} \in \text{GEN}(\bar{x}_i)} \exp(\psi(\bar{x}_i, \bar{y}) \cdot \alpha)\right) - \frac{\|\alpha\|^2}{2\sigma^2},$$

(8.3)

which is the common zero-mean Gaussian regularizer that imposes a penalty within the objective as the magnitude of the parameter values grow. This is also a convex function, thus guaranteeing convergence to the globally optimal solution. The derivative of this function with respect to any given parameter $\alpha_k$ is

$$\frac{\partial \, \text{RegLogLike}(\alpha)}{\partial \alpha_k} = \sum_{i=1}^{m}\left(\psi_k(\bar{x}_i, \bar{y}_i^{\text{ref}}) - \sum_{\bar{y} \in \text{GEN}(\bar{x}_i)} P(\bar{y} \mid \bar{x}_i)\psi_k(\bar{x}_i, \bar{y})\right) - \frac{\alpha_k}{\sigma^2}.$$

(8.4)

The empirical results in Roark *et al.* (2004b, 2007) validate that the unregularized global conditional likelihood objective in Equation (8.2) did not lead to any system improvements, while the regularized objective in Equation (8.3) does provide statistically significant improvements over the baseline, using an empirically chosen regularization parameter $\sigma$. The gradients of the $n$-gram features in that work were efficiently extracted from large word lattices using general finite-state techniques, which enabled the approach to be scaled up to over 10 million features, using lattices encoding a very large number of candidate transcriptions, for a training corpus consisting of hundreds of hours of speech.

Another technique adopted from the recent machine learning literature is the averaged Perceptron (Collins 2002; Freund and Schapire 1999), which was shown to provide significant word-error rate reductions over a baseline LVCSR system in Roark *et al.* (2004a) and was later compared with global conditional likelihood optimization in Roark *et al.* (2004b, 2007). The reader is referred to the cited papers for a formal presentation of the algorithm. Informally, this is an on-line algorithm that initializes parameters at zero and updates the parameters after every example as follows: using the current model, find the best scoring candidate of the current example ($\bar{y}_i^{\text{max}}$); then update the parameters by penalizing features extracted from $\bar{y}_i^{\text{max}}$ and rewarding features extracted from $\bar{y}_i^{\text{ref}}$, the gold standard. Thus, the parameter update at example ($\bar{x}_i$, $\bar{y}_i^{\text{ref}}$), in its simplest form, is

$$\alpha[i] = \alpha[i-1] + \psi(\bar{x}_i, \bar{y}_i^{\text{ref}}) - \psi\left(\bar{x}_i, \operatorname*{argmax}_{\bar{y} \in \textbf{GEN}(\bar{x}_i)} \psi(\bar{x}_i, \bar{y}) \cdot \alpha[i-1]\right), \qquad (8.5)$$

where $\alpha[i-1]$ is the parameter vector from the previous time step.[4] Of course, if $\bar{y}_i^{\text{ref}} = \operatorname{argmax}_{\bar{y} \in \textbf{GEN}(\bar{x}_i)} \psi(\bar{x}_i, \bar{y}) \cdot \alpha[i-1]$, then the parameters are unchanged.

Like the conditional likelihood optimization, this approach is prone to overtraining, hence requires some regularization. The most common technique used for these models is *averaging* the parameters over the models at every timestep of the algorithm. The averaged parameter $\text{avg}_i(\alpha_k)$ at time $i$ is defined as

$$\text{avg}_i(\alpha_k) = \frac{1}{i} \sum_{j=1}^{i} \alpha_k[j]. \qquad (8.6)$$

The raw parameter values are used while training the model, but this averaged parameter value is then used for evaluating the model on held aside or test data. Typically, some held-aside data is evaluated after every iteration over the training data, to determine when to stop training.

One benefit of Perceptron training versus regularized conditional likelihood optimization is that the Perceptron training typically arrives at a more parsimonious solution, i.e. the number of features with non-zero parameter weights are fewer, at least with commonly used regularizers. The reason for this is clear – parameters can only be updated in the Perceptron algorithm if the feature is extracted from either the gold standard candidate $\bar{y}_i^{\text{ref}}$ or the best scoring candidate $\bar{y}_i^{\text{max}}$, hence many features never move away from their initial values of zero. In the comparisons between Perceptron and conditional likelihood optimization in Roark *et al.* (2004b, 2007), the conditional likelihood estimation resulted in larger gains over the baseline than the Perceptron estimation, but the Perceptron was useful as an initial step prior to the conditional likelihood optimization, both for initializing parameter weights for fast convergence and for selecting features to include in the model.

Recent work has looked at improving on Perceptron language model training by introducing loss-sensitivity to the on-line parameter update in Equation (8.5). In the approach of Singh-Miller and Collins (2007), the contribution of an example to the change in the parameter values is based on a loss value, which in the cited paper is taken as the number of errors beyond the minimum error rate hypotheses in the $n$-best lists. Features extracted from highly errorful hypotheses will be more heavily penalized. In addition, this approach provides a well-motivated way of dealing with multiple candidates with the same minimum error rate in the set of candidates. The previously discussed (Section 8.2.2) work in Kuo *et al.* (2007) also uses an on-line parameter estimation approach related to the Perceptron algorithm, where the magnitude of the parameter value update is determined via a sigmoid function.

While the above-mentioned parameter estimation techniques make use of well behaved (convex) functions, the objectives are not precisely what is typically evaluated, which is

---

[4] The algorithm may perform multiple passes over the training data, so the index $i$ may indicate the number of updates, rather than the absolute example number in the corpus.

word-error rate (WER): the number of insertions, deletions or substitutions per 100 words of reference text. The global conditional likelihood approach is maximizing the (regularized) conditional likelihood of the gold standard; and the Perceptron is optimizing the string error rate rather than the WER. MWE training (Povey and Woodland 2002) optimizes a continuous function more closely related to error rate than either of these two objective functions, so perhaps a regularized version of that would be worth investigating. Others have tried methods of optimizing the WER function directly, which cannot be done with any guarantees of convergence or finding a global optimum. For example, minimum sample risk training (Gao *et al.* 2005) performs a greedy search of parameterizations by performing a line search along each dimension in sequence. Although no guarantees are given for finding an optimal solution, reported empirical results (Gao *et al.* 2005) are competitive with other methods such as Boosting and the averaged Perceptron. Similarly unguaranteed parameter search techniques were used in Banerjee *et al.* (2003) to update bigram parameter weights for word-error reduction.

We have presented a basic, general characterization of how these sorts of problems can be approached, and have pointed to specific papers pursuing specific versions of this general approach. Some of these have achieved relatively large improvements (more than 1% absolute) over baseline LVCSR systems, using just $n$-gram features. In the following section, we will survey some extensions to the basic approach, in three key directions: novel features; novel objective functions; and within the context of domain adaptation.

# 8.3    Further Developments

## 8.3.1    Novel Features

Up to now, we have discussed models that rely on $n$-grams as features alongside the baseline model score, or treat arcs within a decoding graph as features, e.g. Kuo *et al.* (2007). In this section, we point to three relatively recent papers that look to expand upon these simple feature sets to achieve additional discrimination. Note that this sort of enterprise has been notoriously difficult in generative language modeling for LVCSR, where WER reductions due to features other than $n$-grams have been hard to come by.

Syntactic language models have been a topic of some interest over the past decade, with some notable successes in achieving WER reductions, e.g. Charniak (2001), Chelba and Jelinek (2000), Roark (2001) and Wang *et al.* (2003). In these works, generative context-free (Charniak 2001; Chelba and Jelinek 2000; Roark 2001) or finite-state (Wang *et al.* 2003) models are used to define a distribution over strings that can be used as a language model. Within a discriminative language modeling framework, however, Collins *et al.* (2005) used a generative context-free parsing model not to derive a score, but rather to derive a syntactic annotation which was then used within the feature mapping $\psi$ for discriminative modeling with the Perceptron algorithm. Example features included part-of-speech (POS) and shallow parse sub-sequence features; context-free rule instances in the parses; and bilexical head-to-head dependencies. This stands in marked contrast to the generative models, where the generative model scores were used for modeling and the structural annotations were an unused byproduct. The WER reductions in Collins *et al.* (2005) were statistically significant (though modest) versus just using $n$-gram features, and most of the reduction was achieved with POS-tag sequence features.

A similar approach was taken for incorporating morphological features into language models in Czech (Shafran and Hall 2006). Unlike generative morphology-based language models, which have been used profitably in, for example, Arabic speech recognition (Vergyri *et al.* 2004), the use of the kinds of sub-word features that morphological analysis affords does not necessitate complicated smoothing techniques. Rather, the annotations from morphological analyzers can be used within the feature mapping to provide additional features in the model. This is particularly important in highly inflected or agglutinative languages, and promises substantial improvements in such languages, as demonstrated in Shafran and Hall (2006).

One key issue in discriminative syntactic and morphological language modeling approaches is feature selection. Features can be derived from rich structural annotations in many ways, which can lead to an astronomical number of potential features in the model. Features were selected in Shafran and Hall (2006) via a $\chi^2$ statistic derived from the co-occurrence of features and the classes of interest, in this case 'correct' and 'incorrect' candidates. A large $\chi^2$ statistic indicates a high correlation between a feature and either the set of correct candidates or the set of incorrect candidates, hence indicating that the feature would have high discriminative utility.

Trigger features (Lau *et al.* 1993) are a way of capturing the bursty and topical nature of language – if a word is observed, then there is typically a high likelihood of that word occurring again or related words occurring. A self-trigger is a feature that exploits an instance of a word or term to increase the likelihood of observing another, or in the case of discriminative models, to reward candidates that have the triggered term. In Singh-Miller and Collins (2007), self-triggers of various sorts were used within a loss-sensitive Perceptron modeling framework to reduce WER significantly over using just $n$-gram features. To do so, they define a history $h$ of previous utterances in the conversation, and derive trigger features based on unigrams and bigrams that either a) occur multiple times in the candidate transcription; or b) occur in the candidate transcription and in transcriptions in the history $h$. In addition, they create backoff bins of unigrams and bigrams based on TF-IDF statistics, and derive another set of features based on these bins, as a way of tying the parameters for words that are similarly distributed. The combination of raw and binned trigger features, along with standard $n$-gram features, resulted in the best performing system.

Real gains are being achieved by expanding the feature set beyond simple $n$-gram features, and the discriminative language modeling paradigm that we have been presenting is a natural framework for work on novel methods for feature extraction and selection.

## 8.3.2 Novel Objectives

While most research in LVCSR over the years has focused on WER as an objective, it is increasingly becoming the case that speech recognition is a sub-process within a larger application such as call-routing, spoken language understanding or spoken document retrieval. Within such a scenario, discriminative language modeling approaches can be applied with a downstream objective function in training rather than an objective function based on the accuracy of the speech recognition itself. This sort of optimization can occur within the context of generative language model estimation, as with the spoken language understanding applications in Goel (2004) and Goel *et al.* (2005), in which bigram probabilities were iteratively re-estimated to improve utterance annotation (Goel 2004) or call-routing (Goel *et al.* 2005) performance. Discriminative language modeling methods, however, allow for great flexibility of feature mapping along with direct optimization of such an objective.

In Saraclar and Roark (2005, 2006), utterances within a call-routing application were annotated with class labels (97 classes), and discriminative language model estimation was done using either the Perceptron algorithm (Saraclar and Roark 2005) or regularized global conditional likelihood maximization (Saraclar and Roark 2006) for various objective functions: a WER objective; an utterance class error rate (CER) objective; and a joint WER/CER objective. Features included $n$-grams, class labels, and combination class label and $n$-gram. Optimization of the joint objective led to a very slight (not statistically significant) increase of CER versus training with an exclusively CER objective, but no increase in WER was observed versus training with an exclusively WER objective. That is, the trade-off inherent in a joint optimization came at the expense of CER rather than WER, which is unsurprising given that each utterance consists of many words but receives just one class label. Inclusion of the utterance class in the model achieved modest (though statistically significant) improvements in WER versus just $n$-gram features, hence this can also be seen as a method for increasing the feature set.

Given the growing interest in machine translation (MT) of speech, one might expect future work in pushing MT objective functions back into LVCSR model training, and discriminative language modeling of the sort described here is one way to achieve this. For example, the presence or absence of phrases from a phrase table within a phrase-based MT system may impact the final quality of translation, hence optimizing the LVCSR system to do a better job of recognizing such phrases, perhaps at the expense of overall WER, which may result in an overall system improvement. Similar approaches could be pushed in speech information retrieval or other areas of spoken language understanding. Other than the papers discussed here, we are unaware of any published results in this direction.

### 8.3.3   Domain Adaptation

All of the work discussed up to now has been within the context of in-domain trained systems, but language model adaptation to novel domains is an area of great interest where they have also been shown to have utility. An adaptation scenario does change a couple of key aspects of the training that impact the best practices. First, to the extent that the new domain is not part of the baseline training data (which is the whole point of domain adaptation) then the issue of cross validating the training set to get training data is moot. However, there remains the issue of the reference transcription not being in the lattice output of the recognizer. In fact, the oracle accuracy of the lattice, i.e. the best possible accuracy achievable from among the candidates, will likely be low for the novel domain, which can reduce the efficacy of these discriminative approaches, relative to generative alternatives such as Maximum a Posteriori (MAP) adaptation (Bacchiani *et al.* 2006; Gauvain and Lee 1994).

In Bacchiani *et al.* (2004), a large vocabulary general voicemail transcription system was adapted for customer service support voicemail recognition. The baseline out-of-domain training of roughly 100 hours was augmented with 17 hours of in-domain training data. In a straight head-to-head comparison, MAP adaptation of the generative language model, based on counts derived from the in-domain data, led to significantly larger improvements in transcription accuracy over the baseline (nearly 8% absolute) than using the Perceptron algorithm (5% absolute). Note that the vocabulary was kept constant in this application, so the MAP adaptation did not benefit from additional in-domain vocabulary. The hypothesis was that the re-ranking nature of the Perceptron was hurting it, since the MAP adapted generative model was used in the first pass, hence could find good candidate transcriptions that were not in the word lattices produced by the baseline out-of-domain trained recognizer.

To remedy this, a hybrid of MAP adaptation followed by Perceptron reranking was pursued in Bacchiani *et al.* (2004). Of course, once the in-domain data was included in the baseline LVCSR training data, appropriate cross-validation techniques had to be used to produce training lattices for the discriminative language model. The result of this approach was an additional 0.7% absolute improvement versus using MAP adaptation alone.

In Zhou *et al.* (2006), several discriminative training algorithms were compared for domain adaptation in Chinese LVCSR, including the Perceptron algorithm, boosting (Schapire *et al.* 1998), ranking SVMs (Joachims 2002) and minimum sample risk (Gao *et al.* 2005, see Section 8.2.3). They found the Perceptron algorithm provided the best domain adaptation (by more than 1% absolute), which they attributed to the large size of the feature set chosen by the Perceptron algorithm relative to the other methods investigated. In addition, they found that the ranking SVM took two orders of magnitude more training time than the other methods.

What remain unexplored are semi-supervised adaptation techniques for this problem – making use of in-domain text without recordings or recordings without transcriptions to adapt language models using discriminative parameter estimation techniques. The flexibility of generative adaptation approaches, such as MAP estimation, permits the opportunistic exploitation of available resources for model training. Achieving such flexibility within discriminative language modeling approaches is an important topic that deserves further attention.

## 8.4 Summary and Discussion

In this chapter, we have presented a general discriminative modeling framework and its application to language modeling for LVCSR. Nearly all of the work on discriminative language modeling fits this general characterization, and it is useful for defining the dimensions within which approaches differ. We have made an effort to be comprehensive in the survey, within the scope declared at the outset of the chapter. Overall the number of papers that have been written on this topic are relatively few, which can be attributed to a number of factors.

First, there are significant barriers to entry in researching this topic. Unlike other application areas, such as machine translation, achieving state-of-the-art performance in LVCSR is not accessible to researchers outside of a small number of research labs and centers. The number of techniques that must be combined within a full-blown LVCSR system to achieve this performance – including a range of acoustic modeling techniques in multiple stages, first-pass large vocabulary decoding, speaker adaptation and lattice rescoring techniques – are such that a researcher must expect a major time investment to even produce the training data needed to explore discriminative language modeling. What should be a priority within the speech community is to lower these barriers by making a large amount of training and evaluation word lattices available for research purposes to the community at large. That would almost certainly spur researchers in machine learning and NLP to begin applying their work in this domain.

It used to be the case that perplexity reduction was the primary mode for evaluating language model improvements. Given the poor correlation between reductions in perplexity and WER reduction, this is no longer widely accepted as a language modeling objective. It did have the virtue, however, of allowing evaluation on text corpora, which meant that research on language modeling could be conducted without access to LVCSR systems. For the discriminative models presented here, this is no longer the case: LVCSR systems are required for training data, and WER is the objective – perplexity cannot be calculated from these models. One step towards increasing research on language modeling would be to make word lattices from competitive LVCSR systems available to the community at large.

Even if such data were made available, however, this remains a very large and difficult problem. Scaling up machine learning algorithms to handle problem spaces with many millions of dimensions and training sets with millions or billions of examples (or more) is a challenge. There is a reason for discussing finite-state methods in this survey, and that is the scale of the problem. Coupled with the widespread perception that language modeling for LVCSR is hopeless, these challenges present another kind of barrier to entry.

This is a shame, because there has been progress in language modeling in this direction, and there remain a large number of very interesting and relatively unexplored directions in which to take it, such as the topics in the following short list:

- semi-supervised techniques for leveraging large text corpora alongside the kind of training data we have been discussing;

- further research on richer feature sets, such as syntactic features or hybrid acoustic/language features;

- feature selection and/or dimensionality reduction;

- scaling up compute intensive parameter estimation techniques.

This survey suggests that it is not unreasonable to foster the hope that real LVCSR system improvements can and will be achieved along these lines over the next decade, to the point where resulting techniques will become part of the collective best practices in the field.

# References

Allauzen C, Mohri M and Roark B 2003 Generalized algorithms for constructing language models. *Proceedings of the 41st Annual Meeting of the Association for Computational Linguistics (ACL)*, pp. 40–47.

Bacchiani M, Riley M, Roark B and Sproat R 2006 MAP adaptation of stochastic grammars. *Computer Speech and Language* **20**(1), 41–68.

Bacchiani M, Roark B and Saraclar M 2004 Language model adaptation with MAP estimation and the perceptron algorithm. *Proceedings of the Human Language Technology Conference of the North American Chapter of the Association for Computational Linguistics (HLT-NAACL)*, pp. 21–24.

Bahl L, Brown P, de Souza P and Mercer R 1986 Maximum mutual information estimation of hidden Markov model parameters for speech recognition. *Proceedings of the IEEE International Conference on Acoustics, Speech, and Signal Processing (ICASSP)*, pp. 49–52.

Bahl L, Brown P, de Souza P and Mercer R 1993 Estimating hidden Markov model parameters so as to maximize speech recognition accuracy. *IEEE Transactions on Speech and Audio Processing* **1**(1), 77–83.

Banerjee S, Mostow J, Beck J and Tam W 2003 Improving language models by learning from speech recognition errors in a reading tutor that listens. *Proceedings of the 2nd International Conference on Applied Artificial Intelligence*, Fort Panhala, Kolhapur, India.

Charniak E 2001 Immediate-head parsing for language models. *Proceedings of the 39th Annual Meeting of the Association for Computational Linguistics (ACL)*.

Chelba C and Jelinek F 2000 Structured language modeling. *Computer Speech and Language* **14**(4), 283–332.

Chen Z, Lee KF and Li MJ 2000 Discriminative training on language model. *Proceedings of the International Conference on Spoken Language Processing (ICSLP)*.

Chueh C, Chien T and Chien J 2005 Discriminative maximum entropy language model for speech recognition. *Proceedings of the European Conference on Speech Communication and Technology (InterSpeech)*.

Collins M 2002 Discriminative training methods for hidden Markov models: Theory and experiments with perceptron algorithms. *Proceedings of the Conference on Empirical Methods in Natural Language Processing (EMNLP)*, pp. 1–8.

Collins M, Saraclar M and Roark B 2005 Discriminative syntactic language modeling for speech recognition. *Proceedings of the 43rd Annual Meeting of the Association for Computational Linguistics (ACL)*, pp. 507–514.

Collins MJ 2000 Discriminative reranking for natural language parsing. *Proceedings of the 17th International Conference on Machine Learning*.

Emami A, Papineni K and Sorensen J 2007 Large-scale distributed language modeling. *Proceedings of the IEEE International Conference on Acoustics, Speech, and Signal Processing (ICASSP)*.

Freund Y and Schapire R 1999 Large margin classification using the perceptron algorithm. *Machine Learning* **3**(37), 277–296.

Gao J, Yu H, Yuan W and Xu P 2005 Minimum sample risk methods for language modeling. *Proceedings of the Conference on Human Language Technology Conference and Empirical Methods in Natural Language Processing (HLT-EMNLP)*, pp. 209–216.

Gauvain JL and Lee CH 1994 Maximum a posteriori estimation for multivariate Gaussian mixture observations of Markov chains. *IEEE Transactions on Speech and Audio Processing* **2**(2), 291–298.

Goel V 2004 Conditional maximum likelihood estimation for improving annotation performance of n-gram models incorporating stochastic finite state grammars. *Proceedings of the International Conference on Spoken Language Processing (ICSLP)*.

Goel V and Byrne W 2000 Minimum Bayes-risk automatic speech recognition. *Computer Speech and Language* **14**(2), 115–135.

Goel V, Kuo H, Deligne S and Wu C 2005 Language model estimation for optimizing end-to-end performance of a natural language call routing system. *Proceedings of the IEEE International Conference on Acoustics, Speech, and Signal Processing (ICASSP)*, pp. I/565–568.

Goodman J 2001 A bit of progress in language modeling. *Computer Speech and Language* **15**(4), 403–434.

Jelinek F 1996 Acoustic sensitive language modeling. Center for Language and Speech Processing, Johns Hopkins University, Baltimore, MD. *Technical Report*.

Joachims T 2002 Optimizing search engines using clickthrough data. *Proceedings of the ACM Conference on Knowledge Discovery and Data Mining (KDD)*.

Johnson M, Geman S, Canon S, Chi Z and Riezler S 1999 Estimators for stochastic "unification-based" grammars. *Proceedings of the 37th Annual Meeting of the Association for Computational Linguistics (ACL)*, pp. 535–541.

Juang B, Chou W and Lee C 1997 Minimum classification error rate methods for speech recognition. *IEEE Transactions on Speech and Audio Processing* **5**(3), 257–265.

Khudanpur S and Wu J 2000 Maximum entropy techniques for exploiting syntactic, semantic and collocational dependencies in language modeling. *Computer Speech and Language* **14**(4), 355–372.

Kuo H, Kingsbury B and Zweig G 2007 Discriminative training of decoding graphs for large vocabulary continuous speech recognition. *Proceedings of the IEEE International Conference on Acoustics, Speech, and Signal Processing (ICASSP)*, pp. IV/45–48.

Kuo HKJ, Fosler-Lussier E, Jiang H and Lee CH 2002 Discriminative training of language models for speech recognition. *Proceedings of the IEEE International Conference on Acoustics, Speech, and Signal Processing (ICASSP)*, Orlando, FL.

Kuo J and Chen B 2005 Minimum word error based discriminative training of language models. *Proceedings of the European Conference on Speech Communication and Technology (InterSpeech)*.

Lafferty J, McCallum A and Pereira F 2001 Conditional random fields: Probabilistic models for segmenting and labeling sequence data. *Proceedings of the 18th International Conference on Machine Learning*, pp. 282–289.

Lau R, Rosenfeld R and Roukos S 1993 Trigger-based language models: a maximum entropy approach. *Proceedings of the IEEE International Conference on Acoustics, Speech, and Signal Processing (ICASSP)*, pp. 45–48.

Lin S and Yvon F 2005 Discriminative training of finite-state decoding graphs. *Proceedings of the European Conference on Speech Communication and Technology (InterSpeech)*.

Mangu L, Brill E and Stolcke A 2000 Finding consensus in speech recognition: word error minimization and other application of confusion networks. *Computer Speech and Language* **14**(4), 373–400.

Mangu L and Padmanabhan M 2001 Error corrective mechanisms for speech recognition. *Proceedings of the IEEE International Conference on Acoustics, Speech, and Signal Processing (ICASSP)*.

Mohri M and Riley M 1997 Weighted determinization and minimization for large vocabulary speech recognition. *Proceedings of the European Conference on Speech Communication and Technology (InterSpeech)*.

Mohri M, Pereira FCN and Riley M 2002 Weighted finite-state transducers in speech recognition. *Computer Speech and Language* **16**(1), 69–88.

Paciorek C and Rosenfeld R 2000 Minimum classification error training in exponential language models. *Proceedings of the NIST Speech Transcription Workshop*.

Popat TBA, Xu P, Och F and Dean J 2007 Large language models in machine translation. *Proceedings of the Joint Conference on Empirical Methods in Natural Language Processing and Computational Natural Language Learning (EMNLP-CoNLL)*, pp. 858–867.

Povey D and Woodland P 2002 Minimum phone error and I-smoothing for improved discriminative training. *Proceedings of the International Conference on Spoken Language Processing (ICSLP)*.

Ringger EK and Allen JF 1996 Error corrections via a post-processor for continuous speech recognition. *Proceedings of the IEEE International Conference on Acoustics, Speech, and Signal Processing (ICASSP)*.

Roark B 2001 Probabilistic top-down parsing and language modeling. *Computational Linguistics* **27**(2), 249–276.

Roark B, Saraclar M and Collins M 2004a Corrective language modeling for large vocabulary ASR with the perceptron algorithm. *Proceedings of the IEEE International Conference on Acoustics, Speech, and Signal Processing (ICASSP)*, pp. I/749–752.

Roark B, Saraclar M, Collins M and Johnson M 2004b Discriminative language modeling with conditional random fields and the perceptron algorithm. *Proceedings of the 42nd Annual Meeting of the Association for Computational Linguistics (ACL)*, pp. 47–54.

Roark B, Saraclar M and Collins M 2007 Discriminative $n$-gram language modeling. *Computer Speech and Language* **21**(2), 373–392.

Rosenfeld R, Chen S and Zhu X 2001 Whole-sentence exponential language models: a vehicle for linguistic-statistical integration. *Computer Speech and Language* **15**(1), 55–73.

Saraclar M and Roark B 2005 Joint discriminative language modeling and utterance classification. *Proceedings of the IEEE International Conference on Acoustics, Speech, and Signal Processing (ICASSP)*, pp. I/561–564.

Saraclar M and Roark B 2006 Utterance classification with discriminative language modeling. *Speech Communication* **48**(3-4), 276–287.

Schapire RE, Freund Y, Bartlett P and Lee WS 1998 Boosting the margin: A new explanation for the effectiveness of voting methods. *The Annals of Statistics* **26**(5), 1651–1686.

Shafran I and Byrne W 2004 Task-specific minimum Bayes-risk decoding using learned edit distance. *Proceedings of the International Conference on Spoken Language Processing (ICSLP)*.

Shafran I and Hall K 2006 Corrective models for speech recognition of inflected languages. *Proceedings of the Conference on Empirical Methods in Natural Language Processing (EMNLP)*, pp. 390–398.

Singh-Miller N and Collins M 2007 Trigger-based language modeling using a loss-sensitive perceptron algorithm. *Proceedings of the IEEE International Conference on Acoustics, Speech, and Signal Processing (ICASSP)*, pp. IV/25–28.

Stolcke A, Bratt H, Butzberger J, Franco H, Gadde VRR, Plauche M, Richey C, Shriberg E, Sonmez K, Weng F and Zheng J 2000 The SRI March 2000 Hub-5 conversational speech transcription system. *Proceedings of the NIST Speech Transcription Workshop*.

Stolcke A and Weintraub M 1998 Discriminitive language modeling. *Proceedings of the 9th Hub-5 Conversational Speech Recognition Workshop*.

Vergyri D, Kirchoff K, Duh K and Stolcke A 2004 Morphology-based language modeling for Arabic speech recognition. *Proceedings of the International Conference on Spoken Language Processing (ICSLP)*.

Wang W, Harper MP and Stolcke A 2003 The robustness of an almost-parsing language model given errorful training data. *Proceedings of the IEEE International Conference on Acoustics, Speech, and Signal Processing (ICASSP)*.

Zhou Z, Gao J, Soong F and Meng H 2006 A comparative study of discriminative methods for reranking LVCSR *n*-best hypotheses in domain adaptation and generalization. *Proceedings of the IEEE International Conference on Acoustics, Speech, and Signal Processing (ICASSP)*, pp. I/141–144.

Zhou Z, Meng H and Lo W 2006 A multi-pass error detection and correction framework for Mandarin LVCSR. *Proceedings of the International Conference on Spoken Language Processing (ICSLP)*.

# 9

# Large Margin Methods for Part-of-Speech Tagging

## Yasemin Altun

Part-of-Speech (POS) tagging, an important component of speech recognition systems, is a sequence labeling problem which involves inferring a state sequence from an observation sequence, where the state sequence encodes a labeling, annotation or segmentation of an observation sequence. In this chapter we give an overview of discriminative methods developed for this problem. Special emphasis is put on large margin methods by generalizing multiclass Support Vector Machines (SVMs) and AdaBoost to the case of label sequences. Experimental evaluation on POS tagging demonstrates the advantages of these models over classical approaches like Hidden Markov Models (HMMs) and their competitiveness with methods like Conditional Random Fields (CRFs).

## 9.1 Introduction

A language model is one of the two important components of speech recognition systems. By assigning probabilities to sequences of words, the language model acts as setting a prior distribution on what utterances can be produced. The acoustic model, which is the second important component of speech recognition systems, defines a conditional probable distribution of observing an acoustic signal given that a particular sequence of words is uttered. The predominant approach in speech recognition, namely *source-channel model*, combines the distributions given by these two system in order to maximize the joint probability of the acoustic signal and the word sequence and in turn to minimize the error rate.

Until the last decade, standard word-based $n$-gram models (i.e. $n$-gram models that are trained generatively) were the most common tools for language models. One of the important

*Automatic Speech and Speaker Recognition: Large Margin and Kernel Methods*   Joseph Keshet and Samy Bengio
© 2009 John Wiley & Sons, Ltd

improvements over these models came from the idea of class-based $n$-gram models (Brown *et al.* 1992). These models outperform the word-based $n$-gram models when the training data is small and sparse (Niesler and Woodland 1996b). Moreover, combining these models with word-based models on large datasets improves the perplexity, which is the standard evaluation criteria in speech recognition (Niesler and Woodland 1996a).

In class-based models, the classes can be defined either syntactically or semantically with various refinement levels. In this chapter, we consider syntactic classes, in particular POS tags. POS tagging is the process of marking up the words in a natural language text with their corresponding grammatical type (part of speech) e.g. noun, verb, adjective, adverb, based on the word itself as well as its context such as the adjacent words and the related words in a phrase, sentence or paragraph. Given a sequence of words, we are interested in finding the best POS tag sequence. Since neighboring POS tags give very informative statistics, we model this task as a *sequence labeling* problem, where the goal is to learn a model that uses both the input and the output context to make accurate predictions. In this chapter, we consider bigrams of POS tags as output context; however, generalization to higher order models is trivial. It is worthwhile to note that one can use the methods described in this chapter to learn other class-based models, as long as they respect a sequence model.

An important design choice is how to train the class model, i.e. generatively (using standard log-likelihood techniques) or discriminatively (using generalization of state of the art discriminative learning methods). The generative approach models the joint distribution of observation and label sequences and predicts a label sequence of an observation sequence using the conditional probability obtained by the Bayes rule. This model is the well-known HMMs. Efficient inference and learning in this setting often requires making questionable conditional independence assumptions. More precisely, it is assumed that all dependencies on past and future observations are mediated through neighboring labels which is clearly violated in many applications (especially where long distance dependencies are important). This is particularly true for POS tagging. The discriminative approaches overcome this problem by either modeling the conditional distribution of the response variables given the observation or by directly learning a discriminative function over the observation-label sequence pairs such that the correct prediction achieves a higher score than other label sequences. This approach provides more flexibility in modeling additional dependencies such as direct dependencies of the $t$th label on past or future observations.

In this chapter, we formulate the sequence labeling problem and outline discriminative learning methods for sequence labeling. In particular, we focus on large margin approaches. We present the generalization of two of the most competitive large margin methods for classification, namely AdaBoost and SVMs, to the problem of sequence labeling. These methods combine the advantages of the discriminative methods described with the efficiency of dynamic programming. We then apply our methods to POS tagging. Experimental evaluation demonstrates the advantages of our models over classical approaches like HMMs and their competitiveness with methods like CRFs.

## 9.2    Modeling Sequence Labeling

We are interested in the inference of a sequence of labels $\bar{y} = (y_1, \ldots, y_t, \ldots, y_T) \in \mathcal{Y}$ for a sequence of observations $\bar{x} = (x_1, \ldots, x_t, \ldots, x_T) \in \mathcal{X}$ where $T$ denotes the length of

Table 9.1 POS tags used in the Penn TreeBank Project

| Label | Description | Label | Description | Label | Description |
|-------|-------------|-------|-------------|-------|-------------|
| CC | Conjunction | CD | Cardinal | DT | Determiner |
| EX | Existential | FW | Foreign word | IN | Preposition |
| JJ | Adjective (Adj) | JJR | Adj comparative | JJS | Adj superlative |
| LS | List marker | MD | Modal | NN | Noun, singular |
| NNS | Noun, plural | NNP | Proper noun, singular | NNPS | Proper noun, plural |
| PDT | Predeterminer | POS | Possessive ending | PRP | Personal pronoun |
| PRP | Possessive pronoun | RB | Adverb (Adv) | RBR | Adv comparative |
| RBS | Adv superlative | RP | Particle | SYM | Symbol |
| TO | To | UH | Interjection | VB | Verb |
| VBD | Verb, Past | VBG | Verb, Pres participle | VBN | Verb, Past participle |
| VBP | Verb, Present | VBZ | Verb, 3P Sing Pres | WDT | Wh-determiner |
| WP | Wh-pronoun | WP | Possessive wh-pronoun | WRB | Wh-adverb |

the sequence. Let $y_t \in \Sigma$, then $\mathcal{Y} = \Sigma^+$ is the set of arbitrary length sequences generated from $\Sigma$. In POS tagging, $x_t$ denotes the $t$th word, $y_t$ denotes the grammatical type (POS tag) of $x_t$ and $\Sigma$ denotes the set of all grammatical types, for example as given in Table 9.1.

The goal is to learn a mapping $f : \mathcal{X} \to \mathcal{Y}$ where $\mathcal{X}$ and $\mathcal{Y}$ are observation and label sequence spaces. Hence, our learning problem is a generalization of supervised classification where values of the response variables (labels) are predicted not only with respect to the observations but also with respect to values of other response variables. The approach we pursue is to learn a *discriminant function* $F : \mathcal{X} \times \mathcal{Y} \to \Re$ where $F(\bar{x}, \bar{y})$ can be interpreted as measuring the compatibility of $\bar{x}$ and $\bar{y}$. Each discriminant function $F$ induces a mapping $f$,

$$f(\bar{x}; \mathbf{w}) = \arg \max_{\bar{y} \in \mathcal{Y}} F(\bar{x}, \bar{y}; \mathbf{w}), \tag{9.1}$$

where $\mathbf{w}$ denotes a parameter vector and ties are arbitrarily broken. We restrict the space of $F$ to linear functions over some feature representation $\boldsymbol{\phi}$, which is defined on the joint input–output space. $\boldsymbol{\phi}$ is chosen such that it models the sequential structure of $\bar{y}$ and its dependency on $\bar{x}$. As in HMMs, we assume the model to be stationary. This results in the sharing of $\mathbf{w}$ parameters across components in the sequence

$$F(\bar{x}, \bar{y}; \mathbf{w}) = \langle \mathbf{w}, \boldsymbol{\phi}(\bar{x}, \bar{y}) \rangle = \sum_t \langle \mathbf{w}, \boldsymbol{\phi}(\bar{x}, \bar{y}; t) \rangle.$$

Given this definition of $F$, we need to define the features $\boldsymbol{\phi}$ to complete the design of the architecture. Our main design goal in defining $\boldsymbol{\phi}$ is to make sure that $f$ can be computed from $F$ efficiently, i.e. using a Viterbi-like decoding algorithm.

### 9.2.1 Feature Representation

We extract two types of features from an observation-label sequence pair $(\bar{x}, \bar{y})$. The first type of features captures inter-label dependencies

$$\phi^1_{s\sigma\sigma'}(\bar{x}, \bar{y}; t) = \mathbb{1}_{\{y_s = \sigma\}} \mathbb{1}_{\{y_t = \sigma'\}},$$

where $\sigma, \sigma' \in \Sigma$ denote labels for single observations and $\mathbb{1}_{\{\cdot\}}$ denotes the indicator function. Here, $\phi^1_{s\sigma\sigma'}(\bar{x}, \bar{y}; t)$ corresponds to a feature $\phi^1_k(\bar{x}, \bar{y}; t)$ where $k$ is indexed by $(s, \sigma, \sigma')$, the label $\sigma$ at position $s$ and the label $\sigma'$. These features simply indicate the statistics of how often a particular combination of labels occurs at neighboring sites. We restrict the inter-label dependencies to neighboring labels (e.g. $s \in \{t + 1, t + 2\}$ for third order Markov features) to achieve our design goal, i.e. to ensure the availability of an efficient dynamic programming algorithm.

The second type of features captures the dependency between a label and the observation sequence. Let $\varphi$ be the set of observation attributes relevant for the particular application. For example, in POS tagging, the information that a word ends with '-ing' can be very informative and therefore should be an attribute of the observation. We describe the attributes for POS tagging problems in detail in Section 9.5.1. The observation-label dependency features combine the attributes with individual labels, $\sigma \in \Sigma$. These features indicate the statistics of how often an attribute occurs with a particular label conjunctively:

$$\phi^2_{sr\sigma}(\bar{x}, \bar{y}; t) = \mathbb{1}_{\{y_t = \sigma\}} \varphi_r(x_s).$$

Again $\phi^2_{sr\sigma}(\bar{x}, \bar{y}; t)$ corresponds to a feature $\phi^2_k(\bar{x}, \bar{y}; t)$ where $k$ is indexed by $(s, r, \sigma)$, the attribute $r$ of the observation at position $s$ and the label $\sigma$. For example, in POS tagging, if $\varphi_r(x)$ denotes whether the word $x$ is ending with '-ing', $\sigma$ is 'VBG' (Verb, Present participle) and $s = t - 1$, then the feature $\phi^2_{sr\sigma}(\bar{x}, \bar{y}; t)$ denotes whether the $t$th label in the label sequence $\bar{y}$ is a 'VBG' and the previous word $((t - 1)$th word) ends with '-ing'.

Since in the discriminative framework we *condition on* the observation sequence (as opposed to *generating* it as in the generative framework), extracting features from arbitrarily past or future observations does not increase the complexity of the model. For this reason, there are no restrictions on the relationship of $s$ and $t$ for observation-label features. The features for which $s \neq t$ are usually called *overlapping* features (or *sliding window* features), since the same input attribute $\varphi_k(x_s)$ is encoded in the model multiple times within different contexts. These features are used widely in discriminative sequence learning (Collins 2002; Lafferty *et al.* 2001) and have been shown to improve classification accuracy in different applications.

Finally, our feature map is the concatenation of the inter-label and observation-label dependency features (where $\oplus$ denotes concatenation),

$$\phi(\bar{x}, \bar{y}; t) = \phi^1(\bar{x}, \bar{y}; t) \oplus \phi^2(\bar{x}, \bar{y}; t).$$

Various generalizations of this feature representation are possible, such as extracting different input features dependent on the relative distance $|ts|$ in the chain. We use this representation as a prototypical example. In Section 9.4, we investigate how to extend this via kernel functions.

## 9.2.2 Empirical Risk Minimization

We investigate a supervised learning setting, where observation-label sequence pairs $(\bar{x}, \bar{y})$ are generated from a fixed but unknown distribution $P$ over $\mathcal{X} \times \mathcal{Y}$ and the goal is to find the discriminant function $F$ such that the *risk*, i.e. the cost of making an incorrect prediction, is minimized. Since $P$ is unknown, we use a sample of input–output pairs

$S = \{(\bar{x}^1, \bar{y}^1), \dots, (\bar{x}^n, \bar{y}^n)\}$, that are assumed to be generated i.i.d. according to $P$ and we minimize the empirical risk with respect to $S$.

The standard zero-one risk typically used in classification is not appropriate for the sequence labeling problem, since zero-one risk cannot capture the degrees of incorrect prediction of sequences. In particular, in POS tagging the evaluation measure is the Hamming distance, which measures the number of incorrect components of the predicted sequence. The empirical Hamming risk is given by

$$\mathcal{R}^{\mathrm{hm}}(\mathbf{w}; S) = \sum_{i=1}^{n} \sum_{t=1}^{T} \mathbb{1}_{\{y_t^i \neq f(\bar{x}^i; \mathbf{w})_t\}},$$

where $f(\bar{x}; \mathbf{w})_t$ denotes the $t$th component of the predicted label sequence.

Another risk measure that can be used for label sequences is the rank function (Freund 1998; Schapire 1999), which measures the fraction of incorrect label sequences that are ranked higher than the correct one. The empirical rank risk is given by

$$\mathcal{R}^{\mathrm{rk}}(\mathbf{w}; S) = \sum_{i=1}^{n} \sum_{\bar{y} \neq \bar{y}^i} \mathbb{1}_{\{F(\bar{x}^i, \bar{y}; \mathbf{w}) \geq F(\bar{x}^i, \bar{y}^i; \mathbf{w})\}}.$$

## 9.2.3 Conditional Random Fields and Sequence Perceptron

Due to the nonconvexity of the risk functions, it is common to optimize convex surrogate functions that are upper bounds of the risk functions. Logarithmic, exponential and hinge loss functions are the most common surrogate functions of the zero-one risk. Among these functions, the logarithmic loss has been proposed for sequence labeling, where the goal is to minimize the negative conditional log-likelihood of the data,

$$\mathcal{R}^{\log}(\mathbf{w}; S) = - \sum_{i=1}^{n} \log p(\bar{y}^i | \bar{x}^i; \mathbf{w}), \quad \text{where}$$

$$p(\bar{y} | \bar{x}; \mathbf{w}) = \frac{e^{F(\bar{x}, \bar{y}; \mathbf{w})}}{\sum_{\bar{y}' \in \mathcal{Y}} e^{F(\bar{x}, \bar{y}'; \mathbf{w})}}. \tag{9.2}$$

This gives the optimization problem of the widely used CRFs (Lafferty *et al.* 2001). $\mathcal{R}^{\log}$ is a convex function and is generally optimized using gradient methods. The gradient of $\mathcal{R}^{\log}(\mathbf{w}; S)$ is given by

$$\nabla_{\mathbf{w}} \mathcal{R}^{\log} = \sum_{i} \mathbf{E}_{\bar{y} \sim p(.|\bar{x})} \left[ \sum_{t} \boldsymbol{\phi}(\bar{x}, \bar{y}; t) | \bar{x} = \bar{x}^i \right] - \sum_{t} \boldsymbol{\phi}(\bar{x}^i, \bar{y}^i; t). \tag{9.3}$$

The expectation of the feature functions can be computed using the Forward–Backward algorithm. When it is not regularized, $\mathcal{R}^{\log}$ is prone to overfitting, especially with noisy data. For this reason, it is common to regularize this objective function by adding a Gaussian prior (a term proportional to the squared $L_2$ norm of the parameter $\mathbf{w}$) to avoid overfitting the training data (Chen and Rosenfeld 1999; Johnson *et al.* 1999).

Another discriminative learning method for sequences is Hidden Markov Perceptron (Collins 2002), which generalizes the Perceptron algorithm for sequence labeling.

---

Initialize $\mathbf{w} = \mathbf{0}$.

**Repeat**:
    **for** all training patterns $\bar{x}^i$
        compute $\bar{y}' = \arg\max_{\bar{y}} F(\bar{x}^i, \bar{y}; \mathbf{w})$
        **if** $\bar{y}^i \neq \bar{y}'$
            $\mathbf{w} \leftarrow \mathbf{w} + \phi(\bar{x}^i, \bar{y}^i) - \phi(\bar{x}^i, \bar{y}')$
**Until** convergence or a maximum number of iterations reached.

---

Figure 9.1  Hidden Markov Perceptron algorithm.

This approach is outlined in Figure 9.1. Investigating the parameter update of Figure 9.1, we can view Hidden Markov Perceptron as an online approximation for CRFs, where the expectation term in Equation (9.3) is approximated with the contribution from the most likely label. This approximation can be tight if the distribution $p(\bar{y}|\bar{x})$ is peaked. Then, the contribution of the most likely label can dominate the expectation values.

In this chapter, we focus on the exponential loss and the hinge loss as surrogate functions of the empirical risk. This leads to the generalizations of two well-known large margin methods, Boosting and SVMs, to label sequence learning problems. These methods, either explicitly or implicitly, maximize the margin of the data, where the margin for an observation sequence $\bar{x}$ is defined as

$$\gamma(\bar{x}, \bar{y}; \mathbf{w}) = F(\bar{x}, \bar{y}; \mathbf{w}) - \max_{\bar{y}' \neq \bar{y}} F(\bar{x}, \bar{y}'; \mathbf{w}). \tag{9.4}$$

This is a simple generalization of the multiclass separation margin (Crammer and Singer 2000) and it measures the differences of the compatibility scores of $(\bar{x}, \bar{y})$ and $(\bar{x}, \bar{y}')$ where $\bar{y}'$ is the most competitive labeling of $\bar{x}$ that is not $\bar{y}$. Intuitively, when this value is positive for the correct labeling of an observation sequence $\bar{x}$, the prediction function Equation (9.1) makes a correct prediction for $\bar{x}$. The goal of large margin classifiers is to maximize the margin in all training instances in the sample, which in turn may lead to correct predictions not only on the sample but also on unobserved observations.

## 9.3  Sequence Boosting

AdaBoost (Freund and Schapire 1995) is one of the most powerful learning ideas introduced to the machine learning community in the last decade. We present a generalization of AdaBoost to label sequence learning. This method, which we refer to as Sequence AdaBoost, was first proposed in Altun *et al.* (2003a). It employs the advantages of large margin methods as well as its implicit feature selection property. This aspect is quite important for efficient inference in high dimensional feature spaces and for cascaded systems, for which speech recognition is a canonical example. Sequence AdaBoost also makes use of the availability of a dynamic programming (DP) algorithm for efficient learning and inference due to the Markov chain dependency structure between labels.

## 9.3.1 Objective Function

The key idea of Boosting is to obtain a high accuracy classifier by combining weak learners that are trained on different distributions of the training examples. In Sequence AdaBoost, we consider the feature functions $\phi$ defined in Section 9.2.1 as weak learners and learn their combination coefficients $\mathbf{w}$ by training over the sample $S$ with distributions $D$. $D$ assigns a weight to each training instance $\bar{x}^i$ and its possible labelings $\bar{y} \in \mathcal{Y}^{T_i}$, where $T_i$ is the length of $\bar{x}^i$.

In particular, motivated by Schapire and Singer (1999), we propose optimizing the exponential risk function on the sequential data weighted according to $D$,

$$\mathcal{R}^{\exp}(\mathbf{w};\, S) \equiv \sum_{i=1}^{n} \sum_{\bar{y} \neq \bar{y}^i} D(i,\, \bar{y}) e^{F(\bar{x}^i, \bar{y};\mathbf{w}) - F(\bar{x}^i, \bar{y}^i;\mathbf{w})}.$$

Intuitively, by minimizing this function, we increase the compatibility of the correct observation-label sequence pairs, while decreasing the compatibility of all the other incorrect pairs. Due to the exponential function, $\mathcal{R}^{\exp}$ is dominated by the term(s) corresponding to $(x^i, \bar{y})$ with largest relative compatibility score for $\bar{y} \neq \bar{y}^i$ ($\max_{\bar{y} \neq \bar{y}^i} F(\bar{x}^i, \bar{y};\, \mathbf{w}) - F(\bar{x}^i, \bar{y}^i;\, \mathbf{w}) = -\gamma(\bar{x}^i, \bar{y}^i;\, \mathbf{w})$). Hence, $\mathcal{R}^{\exp}$ performs an implicit margin maximization where the margin is defined in Equation (9.4). Moreover, it is easy to show that $\mathcal{R}^{\exp}$ is a convex surrogate of the rank risk $\mathcal{R}^{\text{rk}}$.

**Proposition 9.3.1** *The exponential risk is an upper bound on the rank risk,* $\mathcal{R}^{\text{rk}} \leq \mathcal{R}^{\exp}$.

*Proof.* (i) If $F(\bar{x}^i, \bar{y}^i;\, \mathbf{w}) > F(\bar{x}^i, \bar{y};\, \mathbf{w})$ then $\mathbb{1}_{\{F(\bar{x}^i, \bar{y};\mathbf{w}) \geq F(\bar{x}^i, \bar{y}^i;\mathbf{w})\}} = 0 \leq e^z$ for any $z$.
(ii) Otherwise, $\mathbb{1}_{\{F(\bar{x}^i, \bar{y};\mathbf{w}) \geq F(\bar{x}^i, \bar{y}^i;\mathbf{w})\}} = 1 = e^0 \leq e^{F(\bar{x}^i, \bar{y};\mathbf{w}) - F(\bar{x}^i, \bar{y}^i;\mathbf{w})}$. Performing a weighted sum over all instances and label sequences $\bar{y}$ completes the proof. □

Interestingly, when $D$ is uniform for all $\bar{y}$ of each training instance $\bar{x}^i$, this loss function is simply the weighted inverse conditional probability of the correct label sequence as pointed out by Lafferty *et al.* (2001),

$$\mathcal{R}^{\exp}(\mathbf{w};\, S) = \sum_i \left[ \frac{1}{p(\bar{y}^i | \bar{x}^i;\, \mathbf{w})} - 1 \right].$$

where $p$ is defined in Equation (9.2). Introducing the distribution $D$ allows the loss function to focus on difficult training instances.

## 9.3.2 Optimization Method

Clearly, $\mathcal{R}^{\exp}$ is a convex function and can be optimized using standard gradient based methods. Instead, we present a Boosting algorithm that generalizes the AdaBoost.MR algorithm for multiclass classification (Schapire and Singer 2000) to label sequence learning. This is a sparse greedy approximate algorithm that optimizes an upper bound on $\mathcal{R}^{\exp}$ with sequential updates. In general, Sequence AdaBoost leads to a sparse feature representation due to these sequential updates.

As standard AdaBoost.MR, Sequence AdaBoost forms a linear combination of weak learners, an ensemble, by selecting the best performing weak learner and its optimal weight parameter to minimize $\mathcal{R}^{\text{exp}}$ at each round. It also maintains a weight distribution over observation-label sequence pairs and updates this distribution after every round of Boosting such that training instances $\bar{x}^i$ with small margin and their most competitive incorrect labelings are weighted higher. This allows the weak learners in the next round to focus on instances that are difficult to predict correctly.

## (a) Weight Distribution of Examples

We define a sequence of distributions $(D_1, \ldots, D_r, \ldots, D_R)$ where $D_r$ denotes the distribution over $(\bar{x}^i, \bar{y})$ pairs in the $r$th round of Boosting and $R$ is the total number of rounds. $D_r(i, \bar{y})$ is given by the recursive formula of $D_{r-1}$, the feature selected in the $r$th round (denoted by $\phi_{k(r)}$) and its optimal weight update (denoted by $\triangle w_{k(r)}$)

$$D_{r+1}(i, \bar{y}) \equiv \frac{D_r(i, \bar{y})}{Z_r} e^{\triangle w_{k(r)}(\sum_t \phi_{k(r)}(\bar{x}^i, \bar{y}; t) - \phi_{k(r)}(\bar{x}^i, \bar{y}^i; t))}, \tag{9.5}$$

where $Z_r$ is the normalization constant. We set $D_0(i, \bar{y})$ to be uniform over all $\bar{y} \neq \bar{y}^i$ and $D_0(i, \bar{y}^i) = 0$ for all $i$. We denote $\mathbf{w}_r$ as the sum of all weight updates from round 1 to $r$. One can relate $\mathcal{R}^{\text{exp}}$ to the weight distribution.

**Proposition 9.3.2** $\mathcal{R}^{\text{exp}}$ *is the product of the normalization constants,* $\mathcal{R}^{\text{exp}}(\mathbf{w}_R; S) = \prod_{r=1}^{R} Z_r$.

*Proof.* It can be shown that $D_r(i, \bar{y}) = \exp[F(\bar{x}^i, \bar{y}; \mathbf{w}_r) - F(\bar{x}^i, \bar{y}^i; \mathbf{w}_r)]/Z_r$. Then, we see that $\mathcal{R}^{\text{exp}}(\mathbf{w}_R; S) = Z_R$. The recursive definition of the distribution Equation (9.5) reveals that $Z_R = \prod_r Z_r$. $\square$

It can be shown that $D_r(i, \bar{y})$ can be decomposed into a relative weight for each training instance, $D_r(i)$, and a relative weight of each sequence, $\pi_{ri}(\bar{y})$, where

$$D_r(i, \bar{y}) = D_r(i)\pi_{ri}(\bar{y}),$$

$$D_r(i) \equiv \frac{[p(\bar{y}^i|\bar{x}^i; \mathbf{w}_r)^{-1} - 1]}{\sum_j [p(\bar{y}^j|\bar{x}^j; \mathbf{w}_r)^{-1} - 1]}, \quad \forall i \text{ and}$$

$$\pi_{ri}(\bar{y}) \equiv \frac{p(\bar{y}|\bar{x}^i; \mathbf{w}_r)}{1 - p(\bar{y}^i|\bar{x}^i; \mathbf{w}_r)}, \quad \forall \bar{y} \neq \bar{y}^i.$$

The relative weight of each training instance $D_r(i)$ enforces the optimization problem to focus on difficult instances $\bar{x}^i$ in the sense that the probability of the correct labeling $p(\bar{y}^i|\bar{x}^i; \mathbf{w}_r)$ is low with respect to the current parameter values $\mathbf{w}_r$. $D_r(i) \to 0$ as we approach the perfect prediction case of $p(\bar{y}^i|\bar{x}^i; \mathbf{w}_r) \to 1$. The relative weight of each sequence $\pi_{ri}(\bar{y})$ is simply the re-normalized conditional probability $p(\bar{y}|\bar{x}^i; \mathbf{w}_r)$. For a given $\bar{x}^i$, it enforces the optimization problem to focus on the most competitive labelings with respect to $\mathbf{w}_r$.

## (b) Selecting Optimal Feature and its Weight Update

Proposition 9.3.2 motivates a Boosting algorithm which minimizes $Z_r$ with respect to a weak learner $\phi_{r(k)}$ and its optimal weight update $\triangle w_{r(k)}$ at each round. In order to simplify

---

Initialize $D_0(i, \bar{y}) = \dfrac{1}{\sum_j (|\mathcal{Y}^{T_j}| - 1)}$ for $\bar{y} \neq \bar{y}^i$, $D_0(i, \bar{y}^i) = 0$.

Initialize $\mathbf{w} = 0$.

**Loop:** for $r = 1, \ldots, R$

    Perform Forward–Backward algorithm over $S$ to compute $s_k$, $\forall k$ via Equation (9.8).

    Select $\phi_k$ that minimizes the upper bound in Equation (9.7).

    Compute optimal increment $\triangle w_k$ using Equation (9.9).

    Update $D_{r+1}$ using Equation (9.5).

---

Figure 9.2  Sequence AdaBoost.

the notation, we define $u_k(\bar{x}^i, \bar{y}) \equiv \sum_t \phi_k(\bar{x}^i, \bar{y}; t) - \phi_k(\bar{x}^i, \bar{y}^i; t)$. Moreover, we drop the round index $r$ focusing on one round of Boosting. Given these definitions, we can rewrite $Z$ as

$$Z(\triangle w) \equiv \sum_i D(i) \sum_{\bar{y} \neq \bar{y}^i} \pi_i(\bar{y}) \exp[\triangle w\, u(\bar{x}^i, \bar{y})]. \tag{9.6}$$

Minimizing $Z$ exactly on large datasets and/or a large number of features is not tractable, since this requires programming run with accumulators (Lafferty *et al.* 2001) for every feature. To overcome this problem, we propose minimizing an upper bound on $Z$ that is linear in $\phi$. Defining $U_k$ and $L_k$ as the maximum and minimum values of $u_k$ for all $\bar{x}^i$ and all $\bar{y}$, one can write an upper bound of $Z$ as

$$Z(\triangle w) \leq s e^{\triangle wL} + (1 - s)e^{\triangle wU}, \text{ where} \tag{9.7}$$

$$s \equiv \sum_i D(i) \sum_{\bar{y} \neq \bar{y}^i} \pi_i(\bar{y}) \frac{U - u(\bar{x}^i, \bar{y})}{U - L}. \tag{9.8}$$

We refer the interested reader to Altun *et al.* (2003a) for the derivation of this bound. Note that $0 \leq s \leq 1$ and it is large when $\sum_t \phi_k(\bar{x}^i, \bar{y}^i; t) > \sum_t \phi_k(\bar{x}^i, \bar{y}; t)$. Then $s$ can be interpreted as a measure of correlation of $\phi$ and the correct label sequences with respect to the distribution $D(i, \bar{y})$. We call $s$ the *pseudo-accuracy* of feature $\phi$.

The advantage of Equation (9.7) over $Z$ is that one can compute $s$ for all features simultaneously by running the Forward–Backward algorithm once, since $s$ is simply an expectation of the sufficient statistics. The optimal weight update $\triangle w$ that minimizes Equation (9.7) is given by

$$\triangle w = \frac{1}{U - L} \log\left(\frac{-Ls}{U(1 - s)}\right). \tag{9.9}$$

Figure 9.2 describes the iterative procedure of Sequence AdaBoost. In each iteration, the Forward–Backward algorithm is performed over the training data. This provides $s_k$ values of all features. The optimal updates are computed with respect to Equation (9.9). These values are plugged into Equation (9.7) to find the feature with minimum $Z(\triangle w)$. This feature is added to the ensemble and the distribution is updated according to Equation (9.5).

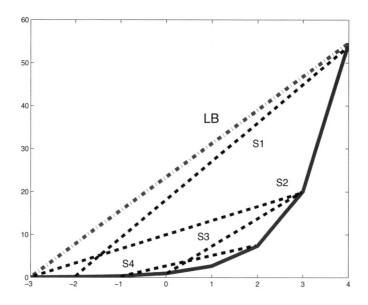

Figure 9.3  An example of the tight and loose bounds on the normalization constant $Z$.

### (c) A Tighter Approximation

In sequence learning, $U_k$ and $-L_k$ can be as large as the maximum length of the training sequences if the feature $\phi_k$ occurs frequently. Then, the bound in Equation (9.7) can be very loose, especially when there exists a very long sequence in the training set. The bound can be refined with piece-wise linear functions defined with respect to each training instance. Defining $U_k^i$ and $L_k^i$ as the maximum and minimum values of $u_k$ for all $\bar{y}$ for a fixed observation sequence $\bar{x}^i$, we give a tighter bound on $Z$ as

$$Z(\triangle w) \leq \sum_i D(i)(s^i e^{\triangle w L^i} + (1 - s^i)e^{\triangle w U^i}), \quad \text{where} \tag{9.10}$$

$$s^i \equiv \sum_{\bar{y} \neq \bar{y}^i} \pi_i(\bar{y}) \frac{U^i - u(\bar{x}^i, \bar{y})}{U^i - L^i}. \tag{9.11}$$

Figure 9.3 illustrates the difference between the two bounds on an example of four observation-label sequence pairs of equal length. The line labeled 'LB' is the loose bound given by Equation (9.7). The gap between $Z$ Equation (9.6) and the loose bound Equation (9.7) is four times the area between LB and the exponential function (solid line). The gap between $Z$ and the tight bound Equation (9.10) is the sum of the area between each line $S^*$ and the exponential curve, which is much smaller than the gap of the loose bound.

Unfortunately, there is no analytical solution of $\triangle w$ that minimizes Equation (9.10). However, since it is convex in $\triangle w$, it can be minimized with a simple line search. The algorithm optimizing this tighter bound is given by changing lines 4, 5 and 6 in Figure 9.2 accordingly.

**(d) Algorithmic Details**

Both of the upper bounds may lead to conservative estimates of the optimal step sizes. One might use more elaborate techniques to find the optimal $\triangle w_{k(r)}$, once $\phi_{k(r)}$ has been selected. For example, $Z_r$ can be optimized exactly for the feature $\phi_{k(r)}$ that minimizes the upper bound on $Z_r$. This involves performing a dynamic program with accumulators on the sequences in which $\phi_{k(r)}$ can be observed.

As observed in Collins (2000), the computation can be reduced drastically by taking sparsity into account. To achieve this, we perform some bookkeeping before Sequence AdaBoost. In particular, for each training instance $\bar{x}^i$ we keep a list of features that can be observed in $(\bar{x}^i, \bar{y})$ pairs for any $\bar{y}$. Moreover, for each feature $\phi_k$ we keep a list of the training sequences $\bar{x}^i$ in which $\phi_k$ can be observed. During Boosting, we store the pseudo-accuracy $s_k$ of all features. Then, at each round we only need to update $s_k$ of features $\phi_k$ that may co-occur with $\phi_{k(r-1)}$. This can be achieved efficiently by restricting the dynamic program to only the training instances in which $\phi_{k(r-1)}$ can be observed using the lists described above.

# 9.4 Hidden Markov Support Vector Machines

In this section, we present the second large margin method for sequence which is a generalization of SVMs to label sequence learning first presented in Altun *et al.* (2003b). This method is named Hidden Markov Support Vector Machines (HM-SVMs), since it combines the advantages of HMMs and SVMs. It is a discriminative sequence learning method with the power of maximum (soft) margin criteria and the strength of learning with nonlinear feature mappings defined jointly on the observation-label space by using kernel functions, two genuine properties of SVMs. It also inherits DP techniques from HMMs to exploit the sequence structure of labels.

## 9.4.1 Objective Function

In HM-SVMs, we define an optimization problem that uses the margin $\gamma$ where the margin is defined in Equation (9.4). Note that $\gamma(\bar{x}, \bar{y}; \mathbf{w}) > 0$ implies that the correct label sequence receives the highest score. In the maximum margin setting, the goal is to find a hyperplane that not only assigns maximum score to the correct label sequence of each observation sequence, but also separates them with some margin, which is measured by $\gamma$. Then, we propose maximizing the minimum of the margins,

$$\mathbf{w}^* = \arg\max_{\mathbf{w}} \ \min_i \ \gamma(\bar{x}^i, \bar{y}^i; \mathbf{w}). \tag{9.12}$$

As in standard SVMs, the margin can be made arbitrarily large by scaling $\mathbf{w}$, if a minimal margin of $\gamma > 0$ can be achieved. We overcome this multiple solution problem by fixing the margin to be equal to 1 and minimizing the squared norm of $\mathbf{w}$, $\|\mathbf{w}\|^2$ subject to the margin constraints. In order to allow margin violations for non-separable cases, we add slack variables $\xi(i)$ for every training sequence:

$$\text{SVM}_1 : \min_{\mathbf{w}, \xi} \frac{1}{2} \|\mathbf{w}\|^2 + C \sum_{i=1}^{n} \xi(i), \quad \text{s.t. } \xi(i) \geq 0, \quad \forall i$$

$$F(\bar{x}^i, \bar{y}^i; \mathbf{w}) - F(\bar{x}^i, \bar{y}; \mathbf{w}) \geq 1 - \xi(i), \quad \forall i, \bar{y} \neq \bar{y}^i,$$

where the optimal solution of the slack variables can be found with respect to the $\mathbf{w}$ parameters, $\xi(i; \mathbf{w}) = \max\{0, 1 - \gamma(\bar{x}^i, \bar{y}^i; \mathbf{w})\}$.

**Proposition 9.4.1** *The hinge risk $\mathcal{R}^{\text{svm}}(\mathbf{w}) = (1/n) \sum_{i=1}^{n} \xi(i; \mathbf{w})$ is an upper bound on 0-1 risk.*

*Proof.* (i) If $\xi(i; \mathbf{w}) < 1$, then one gets $F(\bar{x}^i, \bar{y}^i; \mathbf{w}) - \max_{\bar{y} \neq \bar{y}^i} F(\bar{x}^i, \bar{y}; \mathbf{w}) = \gamma(\bar{x}^i, \bar{y}^i)$ $> 0$, which means the data point is correctly classified and $\mathbb{1}_{\{f(\bar{x}^i; \mathbf{w}) \neq \bar{y}^i\}} = 0 \leq \xi(i; \mathbf{w})$.
(ii) If $\xi(i; \mathbf{w}) \geq 1$, then the bound holds automatically, since $\mathbb{1}_{\{f(\bar{x}^i; \mathbf{w}) \neq \bar{y}^i\}} \leq 1 \leq \xi(i; \mathbf{w})$.
Summing over all $i$ completes the proof.                                                                □

One can generalize SVM$_1$ by imposing the margin to be more than some cost function, such as Hamming distance (Taskar *et al.* 2003; Tsochantaridis *et al.* 2005). Then, the margin constraints in SVM$_1$ are stated as

$$F(\bar{x}^i, \bar{y}^i; \mathbf{w}) - F(\bar{x}^i, \bar{y}; \mathbf{w}) \geq \triangle(\bar{y}, \bar{y}^i) - \xi(i), \quad \forall i, \bar{y} \neq \bar{y}^i, \qquad (9.13)$$

where $\triangle$ denotes the Hamming distance. By incorporating the cost function into the optimization problem, one can hope to achieve better performance than optimizing a surrogate function of the risk. This formulation provides an upper bound on the Hamming loss.

**Proposition 9.4.2** *The risk $\mathcal{R}^{\text{hs}}(\mathbf{w}) = (1/n) \sum_{i=1}^{n} \xi^{\text{hm}}(i; \mathbf{w})$ is an upper bound on $\mathcal{R}^{\text{hm}}$, where $\xi^{\text{hm}}$ is the optimal solution of SVM$_1$ with Equation (9.13) constraints.*

*Proof.* $\xi^{\text{hm}}(i; \mathbf{w}) = \max\{0, \triangle(\bar{y}, \bar{y}^i) - \gamma(\bar{x}^i, \bar{y}^i; \mathbf{w})\}$ is guaranteed to upper bound $\triangle(\bar{y}, \bar{y}^i)$ for $\bar{y}$ such that $\gamma(\bar{x}^i, \bar{y}^i; \mathbf{w}) \leq 0$. Summing over all $i$ completes the proof.       □

As an alternative to SVM$_1$, one can introduce one slack variable for every training instance and every sequence $\bar{y}$ leading to a similar QP with slack variables $\xi(i, \bar{y}; \mathbf{w}) = \max\{0, 1 - [F(\bar{x}^i, \bar{y}^i; \mathbf{w}) - F(\bar{x}^i, \bar{y}; \mathbf{w})]\}$. This provides an upper bound on the rank loss, $\mathcal{R}^{\text{rk}}$.

**Proposition 9.4.3** $(1/n) \sum_{i=1}^{n} \sum_{\bar{y} \neq \bar{y}^i} \xi(i, \bar{y}; \mathbf{w}) \geq \mathcal{R}^{\text{rk}}(\mathbf{w}, \mathbf{w})$.

*Proof.* (i) If $\xi(i, \bar{y}; \mathbf{w}) < 1$, then $F(\bar{x}^i, \bar{y}^i; \mathbf{w}) > F(\bar{x}^i, \bar{y}; \mathbf{w})$, which implies that $\bar{y}$ is ranked lower than $\bar{y}^i$, in which case $\xi(i, \bar{y}; \mathbf{w}) \geq 0$ establishes the bound.
(ii) If $\xi(i, \bar{y}; \mathbf{w}) \geq 1$, then the bound holds trivially, since the contribution of every pair $(\bar{x}^i, \bar{y})$ to $\mathcal{R}^{\text{rk}}$ can be at most 1.                                           □

## 9.4.2 Optimization Method

In this chapter, we focus on **SVM**$_1$, as we expect the number of active inequalities in **SVM**$_1$ to be much smaller compared with the latter formulation since **SVM**$_1$ only penalizes the *largest* margin violation for each example. The dual QP of SVM$_1$ is given by solving the Lagrangian with respect to the Lagrange multipliers $\alpha_{(i,\bar{y})}$ for every margin inequality.

$$\textbf{DSVM}_1 : \max_\alpha \frac{1}{2} \sum_{i,\bar{y}\neq\bar{y}_i} \alpha_{(i,\bar{y})} \tag{9.14a}$$

$$-\frac{1}{2} \sum_{i,\bar{y}\neq\bar{y}_i} \sum_{j,\bar{y}'\neq\bar{y}_j} \alpha_{(i,\bar{y})}\alpha_{(j,\bar{y}')}\langle \delta\phi(i,\bar{y}), \delta\phi(j,\bar{y}')\rangle$$

$$\text{s.t. } \alpha_{(i,\bar{y})} \geq 0, \sum_{\bar{y}\neq\bar{y}_i} \alpha_{(i,\bar{y})} \leq C, \quad \forall i,\bar{y}, \tag{9.14b}$$

where $\delta\phi(i,\bar{y}) = \phi(i,\bar{y}_i) - \phi(i,\bar{y})$. Note that the optimality equation of **w** is given by

$$\mathbf{w}^* = \sum_j \sum_{\bar{y}} \alpha_{(j,\bar{y})}\delta\phi(j,\bar{y}).$$

Then, both the optimization problem **DSVM**$_1$ and the discriminative function $F(\bar{x},\bar{y})$ are computed with respect to inner products of two observation-label sequence pairs. As a simple consequence of the linearity of the inner products, one can replace this with some kernel functions. The only restriction on the types of kernel function to consider here is to compute the arg max operator in the prediction function Equation (9.1) efficiently. We discuss this issue later in Section 9.4.3

**DSVM**$_1$ is a quadratic program parameterized with Lagrange parameters $\alpha$, whose number scales exponentially with the length of the sequences. However, we expect that only a very small fraction of these parameters (corresponding to *support sequences*) will be active at the solution for two reasons. First, the hinge loss leads to sparse solutions as in standard SVMs. Second, and more importantly, many of the parameters are closely related because of the sequence structure, i.e. large amount of overlap of the information represented by each parameter.

The interactions between these parameters are limited to parameters of the same training instances, thus the parameters of the observation sequence $\bar{x}^i$, $\alpha_{(i,.)}$ are independent of the parameters of other observation sequences, $\alpha_{(j,.)}$. Our optimization method exploits this dependency structure of the parameters and the anticipated sparseness of the solution to achieve computational efficiency.

## 9.4.3 Algorithm

We propose using a row selection procedure to incrementally add inequalities to the problem. We maintain an active set of label sequences, $S^i$, for every instance, which are initially $\{\bar{y}^i\}$, the correct label sequences. We call these sets *working sets* and define the optimization problem only in terms of the Lagrange parameters corresponding to the working set of a particular observation sequence. We incrementally update the working sets by adding Lagrange parameter(s) (corresponding to observation-label sequence pair(s)) to the

---

Initialize $S^i \leftarrow \{\bar{y}^i\}, \alpha_{(i,.)} = \mathbf{0}, \forall i$
**Repeat**:
   **for** $i = 1, \ldots, n$
      Compute $\bar{y}' = \arg \max_{\bar{y} \neq \bar{y}^i} F(\bar{x}^i, \bar{y}; \alpha)$
      **if** $F(\bar{x}^i, \bar{y}^i; \alpha) - F(\bar{x}^i, \bar{y}'; \alpha) < 1 - \xi_i$
         $S^i \leftarrow S^i \cup \{\bar{y}'\}$
         Optimize $\mathbf{DSVM}_1$ over $\alpha_{(i,\bar{y})}, \forall \bar{y} \in S^i$
         Remove $\bar{y} \in S^i$ with $\alpha_{(i,\bar{y})} < \epsilon$
**Until**: no margin constraint is violated

---

Figure 9.4  Row selection optimization for HM-SVM.

optimization problem. This is done by iterating over training sequences and finding the label sequence $\bar{y}$ that achieves a best score with respect to the current classifier $F$ other than the correct one. Such a sequence is found by performing a two-best Viterbi decoding (Schwarz and Chow 1990). The satisfaction of the margin constraint by this label sequence implies that all the other label sequences satisfy their margin constraints as well. If $\bar{y}$ violates the margin constraint, we add it into the working set of $\bar{x}^i$ and optimize the quadratic program with respect to the Lagrange parameters in the working set of $\bar{x}^i$, while keeping the remaining variables fixed. Thus, we add at most one negative pseudo-example to the working set at each iteration. This procedure can be viewed as a version of a cyclic coordinate ascent. The algorithm is described in Figure 9.4.

It is possible to show that this algorithm terminates in a polynomial number of steps with respect to the length of sequences, if the constraints are satisfied up to some precision (Tsochantaridis *et al.* 2005). Thus, even though there is an exponential number of constraints, the optimal solution can be obtained by evaluating only a very small percentage of the total constraints, only a polynomial number of constraints.

**Algorithmic Details**

One can replace the inner products of two observation-label sequence pairs in $\mathbf{DSVM}_1$ and in the discriminative function $F$ with a kernel function. Let us recall the feature representation presented in Section 9.2.1:

$$\phi(\bar{x}, \bar{y}) = \sum_t \phi(\bar{x}, \bar{y}; t) = \sum_t \phi^1(\bar{x}, \bar{y}; t) \oplus \phi^2(\bar{x}, \bar{y}; t).$$

Then, the inner product can be written as

$$\langle \phi(\bar{x}, \bar{y}), \phi(\bar{x}', \bar{y}') \rangle = \sum_{s,t} \mathbb{1}_{\{y_{s-1}=y'_{t-1} \wedge y_s=y'_t\}} + \sum_{s,t} \mathbb{1}_{\{y_s=y'_t\}} k(x_s, x'_t) \qquad (9.15)$$

$$\text{where} \quad k(x_s, x'_t) = \langle \varphi(x_s), \varphi(x'_t) \rangle. \qquad (9.16)$$

Hence the similarity between two sequences depends on the number of common two-label fragments as well as the inner product between the feature representation of patterns with a common label. Here, one can replace the inner product on the input attributes with any kernel function. Although one can represent the indicator functions of labels as inner products, it is not beneficial to replace them with nonlinear kernels as this can introduce long range dependencies across labels and therefore it may render the DP algorithm intractable.

Given this decomposition, it is easy to see that the discriminative function $F$ also decomposes into two terms, $F(\bar{x}, \bar{y}) = F_1(\bar{x}, \bar{y}) + F_2(\bar{x}, \bar{y})$, where

$$F_1(\bar{x}, \bar{y}) = \sum_{\sigma, \tau} \delta(\sigma, \tau) \sum_s \mathbb{1}_{\{y_{s-1}=\sigma \wedge y_s=\tau\}}, \tag{9.17a}$$

$$\delta(\sigma, \tau) = \sum_{i, \bar{y}'} \alpha_{(i, \bar{y}')} \sum_{t: y'_{t-1} \neq \sigma \vee y'_t \neq \tau} \mathbb{1}_{\{y'_{t-1}=\sigma \wedge y'_t=\tau\}} \tag{9.17b}$$

and where

$$F_2(\bar{x}, \bar{y}) = \sum_{s, \sigma} \mathbb{1}_{\{y_s=\sigma\}} \sum_{i, t} \beta(i, t, \sigma) k(x_s, x_t^i), \tag{9.18a}$$

$$\beta(i, t, \sigma) = \begin{cases} 0 & \text{if } y_t^i = \sigma \\ \sum_{\bar{y}} \mathbb{1}_{\{y_t=\sigma\}} \alpha_{(i, \bar{y})} & \text{otherwise} \end{cases} \tag{9.18b}$$

Thus, we only need to keep track of the number of times an individual label pair $(\sigma, \tau)$ was predicted incorrectly and the number of times a particular observation $x_s^i$ was incorrectly classified. This representation leads to an efficient computation as it is independent of the number of incorrect sequences $\bar{y}'$.

In order to perform the Viterbi decoding, we have to compute the transition cost matrix and the observation cost matrix $H^i$ for the $i$th sequence. The transition matrix is given by $\delta$ and the observation matrix is given by

$$H_{s\sigma}^i = \sum_j \sum_t \beta(j, t, \sigma) k(x_s^i, x_t^j) \tag{9.19}$$

Given these two matrices, Viterbi decoding amounts to finding the values that maximize the potential function at each position in the sequence.

## 9.5 Experiments

### 9.5.1 Data and Features for Part-of-Speech Tagging

We experimentally evaluated Sequence AdaBoost and HM-SVMs on POS tagging. We used the Penn TreeBank corpus for the POS tagging experiments. This corpus consists of approximately 7 million words of POS tagged *Wall Street Journal* articles. We used the standard experiment setup for Penn TreeBank: sections 2–21 training, section 24 development, section 23 testing. The observation attributes are the standard features used in computational linguistics, such as word identity, sentence initial, last ($n$)-letters of the word, contains dot/hyphen/digit, etc. Overlapping features with respect to these attributes are extracted from a window of size 3.

Table 9.2  Accuracy of POS tagging on Penn TreeBank

| POS | $\mathcal{R}^{\log}$ | $\mathcal{R}^{\exp}$ | Boost |
|-----|------|------|-------|
| $S1$ | 94.91 | 94.57 | 89.42 |
| $S2$ | 95.68 | 95.25 | 94.91 |

## 9.5.2   Results of Sequence AdaBoost

We ran experiments comparing optimization of $\mathcal{R}^{\log}$ and $\mathcal{R}^{\exp}$ using the BFGS (Broyden–Fletcher–Goldferb–Shanno) method, a quasi-Newton convex optimization method (Press *et al.* 1992). We also performed experiments with different formulations of Sequence AdaBoost. Note that the BFGS method performs parallel updates, thus uses all the features available to the model, whereas Sequence AdaBoost performs sequential updates leading to sparse solutions. The number of Boosting rounds $R$ is selected by cross-validation. We used the tight bound Equation (9.10) to select the features and optimized $Z$ exactly to find the optimal weight update for the selected feature, unless stated otherwise.

We investigated the feature sparseness properties of Sequence AdaBoost by defining incremental sets of features. The first set $S1$ consists of only HMM features, i.e. inter-label dependency features (POS tag pairs) as well as word identity features (thus we do not use the other attributes defined above in $S1$). $S2$ also includes features generated from all attributes as well as the inter-label features. These results are summarized in Table 9.2. Sequence AdaBoost performs substantially worse than the BFGS optimization of $\mathcal{R}^{\exp}$ when only HMM features are used, since there is not much information in the features other than the identity of the word to be labeled. Consequently, Sequence AdaBoost needs to include almost all weak learners in the ensemble and cannot exploit feature sparseness. When there are more detailed features such as spelling features, the Boosting algorithm is competitive with the BFGS method, but has the advantage of generating sparser models. For example, in POS tagging, the feature *'The current word ends with -ing and the current tag is VBG.'* replaces many word features. The BFGS method uses all of the available features, whereas Boosting uses only about 10% of the features. Thus, the sparse nature of the problem is necessary for the success of Sequence AdaBoost.

We investigated different methods of optimizing $Z$ in Sequence AdaBoost, namely using the loose and tight bounds, Equation (9.7) and Equation (9.10) respectively, and optimizing $Z$ exactly when the optimal feature is chosen with respect to the tighter bound. Even with the tighter bound in the boosting formulation, the same features are selected many times, because of the conservative estimate of the step size for parameter updates. We observed a speeding up of the convergence of the boosting algorithm when $Z$ is optimized exactly. In order to evaluate this, we collected the features selected in the first 100 rounds and recorded the number of times each feature is selected. We found that the tight bound substantially improves the loose bound. For example, a feature selected 30 times in the first 100 rounds of Boosting using the loose bound optimization is selected only five times by the tight bound optimization. The features that are selected most frequently by the loose bound are the features that are likely to be observed at any position in the sentence, which renders Equation (9.7) very loose. Also, the step sizes achieved by the exact optimization are more

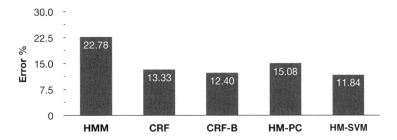

Figure 9.5  Test error of POS task over a window of size 3 using five fold cross validation.

accurate than those of the loose bound. The same feature is selected only once with this optimization (as opposed to five times with the tight bound optimization).

Overall, we observed that the performance of Sequence AdaBoost is comparable to CRFs. The advantage of the Sequence AdaBoost over CRFs is its implicit feature selection property, which results in very sparse models, using about 10% of the feature space. However, it is important that the application naturally has a sparse representation. The tighter bound leads to more accurate step sizes and therefore is preferable over the loose bound. The selection between the exact optimization versus optimizing the tight bound is a tradeoff between the number of Boosting iterations and the size of the training data. When the training data is not very large, one may choose to optimize $Z$ directly in order to reduce the Boosting iterations.

### 9.5.3   Results of Hidden Markov Support Vector Machines

We extracted a corpus consisting of 300 sentences from the Penn TreeBank corpus for the POS tagging experiments. We compared the performance of HMMs and CRFs (optimizing $\mathcal{R}^{\log}$) with HM-SVM as well as an HM-Perceptron, according to their test errors in five fold cross validation. We used second degree polynomial kernel for both the HM-Perceptron and the HM-SVM. $C$ in $SVM_1$ is set to be 1. Although in a generative model like an HMM overlapping features violate the model, we observed that HMMs using the overlapping features described above outperformed the ordinary HMMs. For this reason, we only report the results of HMMs with overlapping features. In Figure 9.5, 'CRF' refers to optimizing $\mathcal{R}^{\log}$, 'CRF-B' refers to optimizing regularized $\mathcal{R}^{\log}$ as described in Section 9.2.3, 'HM-PC' refers to HM-Perceptron.

Figure 9.5 summarizes the experimental results obtained on this task. The results demonstrate the competitiveness of HM-SVMs. As expected, CRFs perform better than the HM-Perceptron algorithm (HM-PC), since CRFs use the derivative of the log-loss function at every step, whereas the Perceptron algorithm uses only an approximation of it. HM-SVMs achieve the best results, which validates our approach of explicitly maximizing a soft margin criterion.

In order to examine the nature of the extracted support sequences, we investigated a subset of the support sequences $\bar{y}$. We observed that most of the support sequences differ in only a few positions from the correct label sequence, resulting in sparse solutions, on average 41 support sequences whereas the size of $\mathcal{Y}$ is exponential in the length of sentences.

It should also be noted that there are no support sequences for many of the training examples, i.e. $\alpha_i(\bar{y}_i) = 0$, since these examples already fulfill the margin constraints.

## 9.6   Discussion

In this chapter, we formulated the sequence labeling problem and outlined discriminative learning methods for sequence labeling with a special focus on large margin approaches.

Sequence Boosting, a generalization of Boosting to label sequence learning, implicitly minimizes the ensemble margin defined over sequences. Taking advantage of the convexity of the exponential function, we defined an efficient algorithm that chooses the best weak learner at each iteration by using the Forward–Backward algorithm. Its performance in accuracy is competitive with the state-of-the-art sequence model of recent years, CRFs. As in standard AdaBoost, Sequence Boosting induces sparse solutions and therefore is preferable over CRFs or other sequence methods in situations where efficiency during inference is crucial. This is of special importance in speech recognition.

HM-SVMs, a generalization of SVMs to label sequence learning problem, inherit the maximum-margin principle and the kernel-centric approach of SVMs. We presented an algorithm that makes use of the sparseness properties of hinge loss and the structure of the parameters. Experimental results show that the algorithm outperforms CRFs in terms of accuracy and is computationally feasible.

We used discriminative sequence labeling methods to predict POS tagging. POS tagging is a particular class-based model used in the language model of speech recognizers. Training these models for other class definitions is of interest. An interesting direction to pursue is the use of these methods to train multiple class-based models jointly in order to improve the accuracy of each of these models and to use them in combination for the language model of speech recognition.

## References

Altun Y, Hofmann T and Johnson M 2003 Discriminative learning for label sequences via boosting *Neural Information Processing Systems 15*.

Altun Y, Tsochantaridis I and Hofmann T 2003 Hidden Markov support vector machines. *Proceedings of the 20th International Conference on Machine Learning* (eds. T Fawcett and N Mishra). AAAI Press, pp. 4–11.

Brown PF, Pietra VJD, deSouza PV, Lai JC and Mercer RL 1992 Class-based n-gram models of natural language. *Computational Linguistics* **18**(4), 467–479.

Chen S and Rosenfeld R 1999 A Gaussian prior for smoothing maximum entropy models. *Technical Report CMUCS-99-108*, Carnegie Mellon University.

Collins M 2000 Discriminative reranking for natural language parsing. *Proceedings of the 17th International Conference on Machine Learning*. Morgan Kaufmann, San Francisco, CA, pp. 175–182.

Collins M 2002 Discriminative training methods for hidden Markov models. *Proceedings of the 2002 Conference on Empirical Methods in Natural Language Processing* (eds. J Hajic and Y Matsumoto). Association for Computational Linguistics, pp. 1–8.

Crammer K and Singer Y 2000 On the learnability and design of output codes for multiclass problems. *Proceedings of the Annual Conference on Computational Learning Theory* (eds. N Cesa-Bianchi and S Goldman). Morgan Kaufmann, San Francisco, CA, pp. 35–46.

Freund Y 1998 Self bounding learning algorithms. *Proceedings of the Annual Conference on Computational Learning Theory*, Madison, Wisconsin. ACM, pp. 247–258.

Freund Y and Schapire RE 1995 A decision-theoretic generalization of on-line learning and an application to boosting. *Proceedings of the European Conference on Computational Learning Theory*. Springer, pp. 23–37.

Johnson M, Geman S, Canon S, Chi Z and Riezler S 1999 Estimators for stochastic unification-based grammars. *Proceedings of the 37th Annual Meeting of the Association for Computational Linguistics*, Maryland. Association for Computational Linguistics, pp. 535–541.

Lafferty JD, McCallum A and Pereira F 2001 Conditional random fields: Probabilistic modeling for segmenting and labeling sequence data. *Proceedings of the 18th International Conference on Machine Learning* (eds. C Brodley and AP Danyluk). Morgan Kaufmann, pp. 282–289.

Niesler T and Woodland P 1996a Combination of word-based and category-based language models. *Proceedings of the International Conference on Spoken Language Processing*, Philadelphia, PA, vol. 1, pp. 220–223.

Niesler T and Woodland P 1996b A variable-length category-based n-gram language model. *Proceedings of the IEEE International Conference on Acoustics, Speech, and Signal Processing*, Atlanta, GA, pp. 164–167.

Press WH, Teukolsky SA, Vetterling WT and Flannery BP 1992 *Numerical Recipes in C: The Art of Scientific Computing*, (2nd edn). Cambridge University Press, Cambridge.

Schapire RE 1999 Drifting games. *Proceedings of the 12th Annual Conference on Computional Learning Theory*. ACM Press, New York, NY, pp. 114–124.

Schapire RE and Singer Y 1999 Improved boosting algorithms using confidence-rated predictions. *Machine Learning* 37(3), 297–336.

Schapire RE and Singer Y 2000 Boostexter: A boosting-based system for text categorization. *Machine Learning* 39(2/3), 135–168.

Schwarz R and Chow Y 1990 The n-best algorithm: An efficient and exact procedure for finding the n most likely hypotheses. *Proceedings of the IEEE International Conference on Acoustics, Speech and Signal Processing*. IEEE, pp. 81–84.

Taskar B, Guestrin C and Koller D 2003 Max-margin Markov networks *Neural Information Processing Systems* (eds. S Thrun, L Saul and B Schölkopf). MIT Press, Cambridge, MA, pp. 25–32.

Tsochantaridis I, Joachims T, Hofmann T and Altun Y 2005 Large margin methods for structured and interdependent output variables. *Journal of Machine Learning Research* 6, 1453–1484.

# 10

# A Proposal for a Kernel Based Algorithm for Large Vocabulary Continuous Speech Recognition

## Joseph Keshet

We present a proposal for a kernel based model for a large vocabulary continuous speech recognizer (LVCSR). Continuous speech recognition is described as a problem of finding the best phoneme sequence and its best time span, where the phonemes are generated from all permissible word sequences. A non-probabilistic score is assigned to every phoneme sequence and time span sequence, according to a kernel based acoustic model and a kernel based language model. The acoustic model is described in terms of segments, where each segment corresponds to a whole phoneme, and it generalizes segment models for the non-probabilistic setup. The language model is based on the discriminative language model recently proposed by Roark *et al.* (2007). We devise a loss function based on the word error rate and present a large margin training procedure for the kernel models, which aims at minimizing this loss function. Finally, we discuss the practical issues of the implementation of a kernel based continuous speech recognition model by presenting an efficient iterative algorithm and considering the decoding process. We conclude the chapter with a brief discussion on the model limitations and future work. This chapter does not report any experimental results.

## 10.1 Introduction

Automatic speech recognition (ASR) is the process of converting a speech signal to a sequence of words, by means of an algorithm implemented as a computer program.

*Automatic Speech and Speaker Recognition: Large Margin and Kernel Methods*   Joseph Keshet and Samy Bengio
© 2009 John Wiley & Sons, Ltd

While ASR does work today, and it is commercially available, it is extremely sensitive to noise, speaker and environmental variability. The current state-of-the-art automatic speech recognizers are based on generative models that capture some temporal dependencies such as Hidden Markov Models (HMMs). While HMMs have been immensely important in the development of large-scale speech processing applications and in particular speech recognition, their performance is far from that of a human listener. This chapter presents a proposal for a different approach to speech recognition which is based not on the HMM but on recent advances in large margin and kernel methods.

Despite their popularity, traditional HMM-based approaches have several known drawbacks such as conditional independence of observations given the state sequence, duration model that is implicitly given by a geometric distribution, and the restriction of using acoustic features which are imposed by frame based processing (Ostendorf 1996). Other drawbacks are the convergence of the training algorithm (expectation-maximization EM) to a local maximum, a training objective which is not aimed at optimizing the evaluation objective (e.g. word error rate) and the fact that the likelihood is dominated by the observation probabilities, often leaving the transition probabilities unused (Ostendorf 1996; Young 1996). A detailed discussion about the shortcomings of HMMs can be found in Chapter 1.

Since HMMs were presented for speech recognition in the 1970s, two major efforts have been made to tackle their shortcomings. The first effort was proposed by Ostendorf and her colleagues (Ostendorf 1996; Ostendorf *et al.* 1996) and was termed *segment models*. In summary, segment models can be thought of as a higher dimensional version of HMMs, where Markov states generate random sequences rather than a single random vector observation. The basic segment model includes an explicit segment-level duration distribution and a family of length-dependent joint distributions, both addressing the conditional independence assumption, the weak duration model and the feature extraction imposed by frame based observations. Although segment models are beautiful a generalization of HMMs, they were not found to be very effective for speech recognition in practice.

More recently, a second effort has been proposed toward discriminative training of HMMs as an alternative to likelihood maximization (see Bahl *et al.* (1986); Fu and Juang (2007); Juang *et al.* (1997); Povey (2003) and the many references therein). Discriminative training of HMMs aims at both maximizing the probability of correct transcription given an acoustic sequence, and minimizing the probability of incorrect transcription given an acoustic sequence, thus bringing the training objective closer to the evaluation objective. Generally speaking, discriminative training methods outperform the maximum likelihood training of HMMs. Nevertheless, the basic underlying model is always generative, and therefore all discriminative training of HMMs suffers from the same shortcomings as the traditional HMMs.

The model proposed in this work tries to combine the segmental models and discriminative training. Our proposed model generalizes segment models for the non-probabilistic setup and hence addresses the same shortcomings of HMMs already addressed by the segment models. In addition, it addresses the minimization of the word error rate and the convergence of the training algorithm to a global minimum rather than to a local one. Moreover, our model is a large margin model which has been found to be robust to noise and can be easily transformed into a nonlinear model using Mercer kernels.

The remainder of the chapter is organized as follows. In Section 10.2 we review the segment models and HMMs for continuous speech recognition. In Section 10.3, we introduce

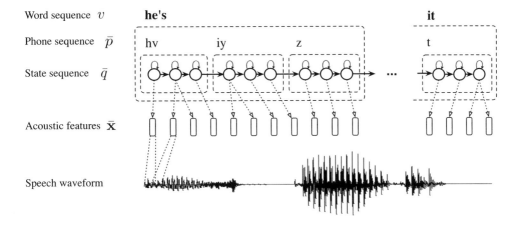

Figure 10.1 The basic notation of a Hidden Markov Model-based speech recognizer.

the kernel based model for large vocabulary continuous speech recognition and show its relationship to the segment models. Then, in Section 10.4 we present the large margin approach for learning the model parameters as a quadratic optimization problem. The implementation details of our method are presented in Section 10.5, where we give an iterative algorithm to approximate the large margin approach, propose a feature function set for the acoustic and language models and discuss the implementation of the decoder and its complexity. We conclude the chapter in Section 10.6.

## 10.2   Segment Models and Hidden Markov Models

The problem of speech recognition involves finding a sequence of words, $\bar{v} = (v_1, \ldots, v_N)$, given a sequence of $d$-dimensional acoustic feature vectors, $\bar{x} = (x_1, x_2, \ldots, x_T)$, where $x_t \in \mathcal{X}$, and $\mathcal{X} \subset \mathbb{R}^d$ is the domain of the acoustic vectors. Each word $v_i \in \mathcal{V}$ belongs to a fixed and known vocabulary $\mathcal{V}$. Typically, each feature vector covers a period of 10 ms and there are approximately $T = 300$ acoustic vectors in a 10 word utterance. Our basic notation is depicted in Figure 10.1.

In the segment model or the traditional HMM speech recognition systems, the problem of speech recognition is formulated as a statistical inference problem of finding the sequence of words $\bar{v}$ that is most likely given the acoustic signal $\bar{x}$ by the *Maximum a Posteriori* (MAP) rule as follows:

$$\bar{v}' = \arg \max_{\bar{v}} P(\bar{v}|\bar{x}) = \arg \max_{\bar{v}} \frac{p(\bar{x}|\bar{v}) P(\bar{v})}{p(\bar{x})}, \tag{10.1}$$

where we used Bayes' rule to decompose the posterior probability in Equation (10.1). The term $p(\bar{x}|\bar{v})$ is the likelihood of observing the acoustic vector sequence $\bar{x}$ given a specified word sequence $\bar{v}$ and it is known as the *acoustic model*. The term $P(\bar{v})$ is the probability of observing a word sequence $\bar{v}$ and it is known as the *language model*. The term $p(\bar{x})$ can be disregarded, since it is constant under the max operation.

The HMM decoding process starts with a postulated word sequence $\bar{v}$. Each word $v_i$ is converted into a sequence of phones $\bar{p}$, where $\bar{p} \in \mathcal{P}^*$ and $\mathcal{P}$ is the set of all phone symbols. The conversion between a word to its corresponding phone sequence is done using a pronunciation lexicon *lex*, where *lex* is a function from the vocabulary to the domain of the phone sequences, $lex : \mathcal{V} \to \mathcal{P}^*$. Recall that a whole utterance is a concatenation of $N$ words, hence its phonetic transcription is the concatenation of the pronunciation of each word:

$$\bar{p} = (lex(v_1), \ldots, lex(v_N)).$$

The phone sequence is then converted to a state sequence, $\bar{q} \in \mathcal{Q}^*$, where $\mathcal{Q}$ is the set of all HMM states. Usually every phone is represented as a sequence of three to five HMM states connected right-to-left with or without jump connections. The probability of the postulated sequence $p(\bar{x}|\bar{p})$ is computed by concatenating the models of the contextual phones composing the sequence. That is,

$$p(\bar{x}|\bar{p}) = \sum_{\bar{q}} p(\bar{x}, \bar{q}|\bar{p}) = \sum_{\bar{q}} p(\bar{x}|\bar{q}, \bar{p}) \cdot P(\bar{q}|\bar{p}) \tag{10.2}$$

$$= \sum_{\bar{q}} \prod_{t=1}^{T} p(\mathbf{x}_t|q_t) \cdot \mathbb{1}_{\{\bar{q}, \bar{p}\}} \prod_{t=1}^{T} P(q_t|q_{t-1}), \tag{10.3}$$

where $\mathbb{1}_{\{\bar{q}, \bar{p}\}}$ is an indicator function that equals 1 if the state sequence $\bar{q}$ is permissible by the phone sequence $\bar{p}$ and zero otherwise. The Viterbi procedure in HMMs involves finding the most likely state sequence,

$$\bar{q}' = \arg \max_{\bar{q}} p(\bar{x}|\bar{q}) P(\bar{q})$$

In the segment model (Ostendorf 1996; Ostendorf *et al.* 1996) each phone $p_l$ in $\bar{p}$ is modeled as a segment, including several frames. Let $\bar{s}$ be the timing (alignment) sequence corresponding to a phone sequence $\bar{p}$. Each $s_l \in \mathbb{N}$ is the start-time of phone $p_l$ in frame units, that is, $s_l$ is the start-time of segment $l$. The probability of the postulated sequence $p(\bar{x}|\bar{p})$ using the segment model is computed as follows:

$$p(\bar{x}|\bar{p}) = \sum_{\bar{s}} p(\bar{x}, \bar{s}|\bar{p}) = \sum_{\bar{s}} p(\bar{x}|\bar{s}, \bar{p}) \cdot P(\bar{s}|\bar{p}) \tag{10.4}$$

$$= \sum_{\bar{s}} \prod_{l=1}^{L} p(\bar{x}_{s_l}^{s_{l+1}-1}|s_l, p_l) \cdot \prod_{l=1}^{L} P(s_l|s_{l-1}, p_l, p_{l-1}), \tag{10.5}$$

where $\bar{x}_{s_l}^{s_{l+1}-1} = (\mathbf{x}_{s_l}, \ldots, \mathbf{x}_{s_{l+1}-1})$ is the sub-sequence of feature vectors constituting the segment, and $L$ is the number of phones (segments), $L = |\bar{p}|$. Segment-based recognition involves finding

$$\bar{p}' = \arg \max_{\bar{p}, L} \left[ \max_{\bar{s}} p(\bar{x}|\bar{s}, \bar{p}) P(\bar{s}|\bar{p}) P(\bar{p}) \right]. \tag{10.6}$$

The kernel based model is closely related to the segment model and can be considered as a generalization of it. In the next section we present the kernel based model and show that it generalizes the segment model.

## 10.3   Kernel Based Model

Recall that the problem of speech recognition can be stated as being the problem of finding the most likely sequence of words $\bar{v}'$ given the acoustic feature vectors $\bar{x}$. In its logarithmic form it can be written as

$$\bar{v}' = \arg \max_{\bar{v}} \log p(\bar{x}|\bar{v}) + \log P(\bar{v}),$$

where $p(\bar{x}|\bar{v})$ is the probability of the acoustic features given the word sequence known as the acoustic model, and $P(\bar{v})$ is the probability of the sequence of words known as the language model.

The discriminative kernel based construction for speech recognition is based on a predefined vector feature function $\boldsymbol{\phi} : \mathcal{X}^{\star} \times \mathcal{V}^{\star} \to \mathcal{H}$, where $\mathcal{H}$ is a reproducing kernel Hilbert space (RKHS). Thus, the input of this function is an acoustic representation, $\bar{x}$, together with a candidate word sequence $\bar{v}$. The feature function returns a vector in $\mathcal{H}$, where, intuitively, each element of the vector represents the confidence that the suggested word sequence $\bar{v}$ is said in the speech signal $\bar{x}$.

The discriminative kernel based speech recognizer is not formulated with probabilities, but rather as a dot product of the vector feature function $\boldsymbol{\phi}$ and a weight vector $\mathbf{w} \in \mathcal{H}$,

$$\bar{v}' = \arg \max_{\bar{v}} \mathbf{w} \cdot \boldsymbol{\phi}(\bar{x}, \bar{v}),$$

where $\mathbf{w} \in \mathcal{H}$ is a weight vector that should be learned. In the same spirit of generative HMM-based speech recognizer, this function can be written as a sum of the acoustic model and the language model:

$$\bar{v}' = \arg \max_{\bar{v}} \mathbf{w}^{\text{acoustic}} \cdot \boldsymbol{\phi}^{\text{acoustic}}(\bar{x}, \bar{v}) + \mathbf{w}^{\text{language}} \cdot \boldsymbol{\phi}^{\text{language}}(\bar{v}), \qquad (10.7)$$

where both $\mathbf{w}^{\text{acoustic}}$ and $\mathbf{w}^{\text{language}}$ are sub-vectors of the vector $\mathbf{w}$, i.e. $\mathbf{w} = (\mathbf{w}^{\text{acoustic}}, \mathbf{w}^{\text{language}})$, and similarly $\boldsymbol{\phi}(\bar{x}, \bar{v}) = (\boldsymbol{\phi}^{\text{acoustic}}(\bar{x}, \bar{v}), \boldsymbol{\phi}^{\text{language}}(\bar{v}))$. The first term $\mathbf{w}^{\text{acoustic}} \cdot \boldsymbol{\phi}^{\text{acoustic}}(\bar{x}, \bar{v})$ in the equation is the *acoustic model*, and it assigns a confidence for every candidate word sequence $\bar{v}$ and a given acoustic signal $\bar{x}$. Since we do not know how to state the feature functions of the acoustic modeling in terms of word sequences explicitly, we state them in terms of the phoneme[1] sequences and phoneme timing sequences. We define the timing (alignment) sequence $\bar{s}$ corresponding to a phoneme sequence $\bar{p}$ as the sequence of start times, where we denote by $s_l \in \mathbb{N}$ the start time of phoneme $p_l \in \mathcal{P}$. Since we do not know the start time sequence while decoding, we search for the best timing sequence, hence

$$\bar{v}' = \arg \max_{\bar{v}, \bar{p} = lex(\bar{v})} \left[ \max_{\bar{s}} \mathbf{w}^{\text{acoustic}} \cdot \boldsymbol{\phi}^{\text{acoustic}}(\bar{x}, \bar{p}, \bar{s}) + \mathbf{w}^{\text{language}} \cdot \boldsymbol{\phi}^{\text{language}}(\bar{v}) \right], \qquad (10.8)$$

where we use a lexicon *lex* mapping to generate phonemes from words, $\bar{p} = (lex(v_1), \ldots, lex(v_N))$, and we overload the definition of the acoustic modeling feature function as $\boldsymbol{\phi} : \mathcal{X}^{\star} \times \mathcal{P}^{\star} \times \mathbb{N}^{\star} \to \mathcal{H}$. The description of the concrete form of the feature function is deferred to Section 10.5.2.

---

[1]Note that in the kernel based model we use *phonemes* rather than *phones*.

The second term $\mathbf{w}^{\text{language}} \cdot \boldsymbol{\phi}^{\text{language}}(\bar{v})$ in Equation (10.7) is the *language model*. Traditionally, the language model assigns a probability to a sequence of words by means of a probability distribution. In the discriminative setting, the language model gives confidence to a string of words in a natural language. Chapter 8 described discriminative language modeling for a large vocabulary speech recognition task which would be suitable for the above setting. Another way to build a discriminative language model is by using prediction suffix trees (PSTs) (Dekel *et al.* 2004; Ron *et al.* 1996) as demonstrated in Pereira *et al.* (1995).

Comparing the logarithmic form of the segment model Equation (10.6) with the kernel based model Equation (10.8), it can be seen that the kernel based model generalizes the segment model. Specifically, we can decompose the weight vector $\mathbf{w}^{\text{acoustic}}$ into two sub-vectors, $\mathbf{w}^{\text{acoustic}} = (\mathbf{w}_1^{\text{acoustic}}, \mathbf{w}_2^{\text{acoustic}})$, and similarly decompose the vector feature function as $\boldsymbol{\phi}^{\text{acoustic}}(\bar{\mathbf{x}}, \bar{p}, \bar{s}) = (\boldsymbol{\phi}_1^{\text{acoustic}}(\bar{\mathbf{x}}, \bar{p}, \bar{s}), \boldsymbol{\phi}_2^{\text{acoustic}}(\bar{p}, \bar{s}))$. We get

$$\bar{v}' = \arg \max_{\bar{v}, \bar{p} = lex(\bar{v})} \left[ \max_{\bar{s}} \mathbf{w}_1^{\text{acoustic}} \cdot \boldsymbol{\phi}^{\text{acoustic}}(\bar{\mathbf{x}}, \bar{p}, \bar{s}) + \mathbf{w}_2^{\text{acoustic}} \cdot \boldsymbol{\phi}^{\text{acoustic}}(\bar{p}, \bar{s}) \right.$$
$$\left. + \mathbf{w}^{\text{language}} \cdot \boldsymbol{\phi}^{\text{language}}(\bar{v}) \right]. \tag{10.9}$$

Setting each element of the vector $\boldsymbol{\phi}$ to be an indicator function for every probability event in the segment model and each element in $\mathbf{w}$ to be the log of the probability estimation of the corresponding indicator in $\boldsymbol{\phi}$, we get the segment model.

There are several advantages of the kernel based model over the segment model. First as we show in Section 10.5.1 the kernel based model estimated the weight vector as having a small word error rate, and the process is guaranteed to converge to a global minimum. Moreover, we can prove that this estimation of the weight vector leads to a word error rate on unseen speech utterance (and not only on the training set). Last, the kernel based model can be easily transformed into a nonlinear model. As we see in the next sections the algorithm for estimating $\mathbf{w}$ solely depends on the dot product between feature functions $\boldsymbol{\phi}$. Wherever such a dot product is used, it is replaced with the kernel function, which expresses a nonlinear dot product operation.

## 10.4 Large Margin Training

The ultimate goal in speech recognition is usually to minimize the word error rate, that is, the Levenshtein distance (edit distance) between the predicted sequence of words and the correct one. In this section we present an algorithm for learning the weight vector $\mathbf{w}$, which aims at minimizing the word error rate. Throughout this chapter we denote by $\gamma(\bar{v}, \bar{v}')$ the Levenshtein distance between the predicted word sequence $\bar{v}'$ and the true word sequence $\bar{v}$.

We now describe a large margin approach for learning the weight vectors $\mathbf{w}^{\text{acoustic}}$ and $\mathbf{w}^{\text{language}}$ which defines the continuous speech recognizer in Equation (10.8), from a training set $\mathcal{T}$ of examples. Each example in the training set $\mathcal{T}$ is composed of a speech utterance $\bar{\mathbf{x}}$ and its corresponding word sequence $\bar{v}$. Overall we have $m$ examples, that is, $\mathcal{T} = \{(\bar{\mathbf{x}}_1, \bar{v}_1), \ldots, (\bar{\mathbf{x}}_m, \bar{v}_m)\}$. We assume that we have a pronunciation lexicon $lex$ which maps every word to a phoneme sequence. We also assume that given a speech utterance $\bar{\mathbf{x}}$ and its corresponding phonetic transcription $\bar{p}$, we have access to the correct time alignment sequence $\bar{s}$ between them. This assumption is actually not restrictive since such an alignment can be inferred relying on a forced alignment algorithm described in Chapter 4.

Recall that we would like to train the acoustic model and the language model discriminatively so as to minimize the Levenshtein distance between the predicted word sequence and the correct one. Similar to the Support Vector Machine (SVM) algorithm for binary classification (Cortes and Vapnik 1995; Vapnik 1998), our approach for choosing the weight vector $\mathbf{w}^{acoustic}$ and $\mathbf{w}^{language}$ is based on the idea of large-margin separation. Theoretically, our approach can be described as a two-step procedure: first, we construct the feature functions $\boldsymbol{\phi}^{acoustic}(\bar{\mathbf{x}}_i, \bar{p}_i, \bar{s}_i)$ and $\boldsymbol{\phi}^{language}(\bar{v}_i)$ based on each instance $(\bar{\mathbf{x}}_i, \bar{v}_i, \bar{p}_i, \bar{s}_i)$, its phonetic transcription $\bar{p}_i$ and its timing sequence $\bar{s}_i$. We also construct the feature functions $\boldsymbol{\phi}^{acoustic}(\bar{\mathbf{x}}_i, \bar{p}, \bar{s})$ and $\boldsymbol{\phi}^{language}(\bar{v})$ based on $\bar{\mathbf{x}}_i$ and every possible word sequence $\bar{v}$, where $\bar{p}$ is the phoneme transcription corresponding to the word sequence $\bar{v}$. Second, we find the weight vectors $\mathbf{w}^{acoustic}$ and $\mathbf{w}^{language}$, such that the projection of feature functions onto $\mathbf{w}^{acoustic}$ and $\mathbf{w}^{language}$ ranks the feature functions constructed for the correct word sequence above the feature functions constructed for any other word sequence. Ideally, we would like the following constraint to hold:

$$\mathbf{w}^{acoustic} \cdot \boldsymbol{\phi}^{acoustic}(\bar{\mathbf{x}}_i, \bar{p}_i, \bar{s}_i) + \mathbf{w}^{language} \cdot \boldsymbol{\phi}^{language}(\bar{v}_i)$$

$$- \max_{\bar{s}} \mathbf{w}^{acoustic} \cdot \boldsymbol{\phi}^{acoustic}(\bar{\mathbf{x}}_i, \bar{p}, \bar{s}) - \mathbf{w}^{language} \cdot \boldsymbol{\phi}^{language}(\bar{v})$$

$$\geq \gamma(\bar{v}_i, \bar{v}) \quad \forall \bar{v} \neq \bar{v}_i, \ \bar{p}_i = lex(\bar{v}_i). \tag{10.10}$$

That is, $\mathbf{w} = (\mathbf{w}^{acoustic}, \mathbf{w}^{language})$ should rank the correct word sequence above any other word sequence by at least the Levenshtein distance between them. We refer to the difference on the left-hand side of Equation (10.10) as the *margin* of $\mathbf{w}$ with respect to the best alignment. Note that if the prediction of $\mathbf{w}$ is incorrect then the margin is negative. Naturally, if there exists $\mathbf{w}$ satisfying all the constraints in Equation (10.10), the margin requirements are also satisfied by multiplying $\mathbf{w}$ by a large scalar. The SVM algorithm solves this problem by selecting the weights $\mathbf{w}$ minimizing $\frac{1}{2}\|\mathbf{w}\|^2 = \frac{1}{2}\|\mathbf{w}^{acoustic}\|^2 + \frac{1}{2}\|\mathbf{w}^{language}\|^2$ subject to the constraints given in Equation (10.10), as it can be shown that the solution with the smallest norm is likely to achieve better generalization (Vapnik 1998).

In practice, it might be the case that the constraints given in Equation (10.10) cannot be satisfied. To overcome this obstacle, we follow the soft SVM approach (Cortes and Vapnik 1995; Vapnik 1998) and define the following hinge loss function:

$$\ell(\mathbf{w}; (\bar{\mathbf{x}}_i, \bar{v}_i, \bar{p}_i, \bar{s}_i))$$

$$= \max_{\bar{v}, \bar{p}=lex(\bar{v})} \Big[ \gamma(\bar{v}_i, \bar{v}) - \mathbf{w}^{acoustic} \cdot \boldsymbol{\phi}^{acoustic}(\bar{\mathbf{x}}_i, \bar{p}_i, \bar{s}_i) - \mathbf{w}^{language} \cdot \boldsymbol{\phi}^{language}(\bar{v}_i)$$

$$+ \max_{\bar{s}} \mathbf{w}^{acoustic} \cdot \boldsymbol{\phi}^{acoustic}(\bar{\mathbf{x}}_i, \bar{p}, \bar{s}) + \mathbf{w}^{language} \cdot \boldsymbol{\phi}^{language}(\bar{v}) \Big]_+, \tag{10.11}$$

where $[a]_+ = \max\{0, a\}$. The hinge loss measures the maximal violation for any of the constraints given in Equation (10.10). The soft SVM approach for our problem is to choose the vector $\mathbf{w}$ which minimizes the following optimization problem:

$$\min_{\mathbf{w}} \frac{1}{2}\|\mathbf{w}^{acoustic}\|^2 + \frac{1}{2}\|\mathbf{w}^{language}\|^2 + C \sum_{i=1}^{m} \ell(\mathbf{w}; (\bar{\mathbf{x}}_i, \bar{v}_i, \bar{p}_i, \bar{s}_i)), \tag{10.12}$$

where the parameter $C$ serves as a complexity–accuracy trade-off parameter: a low value of $C$ favors a simple model, while a large value of $C$ favors a model which solves all training constraints (see Cristianini and Shawe-Taylor (2000)).

## 10.5 Implementation Details

In this section we describe the implementation details of the proposed system. Solving the optimization problem given in Equation (10.12) is expensive since it involves a maximization for each training example. Most of the solvers for this problem, like sequential minimal optimization (SMO) (Platt 1998), iterate over the whole dataset several times until convergence. In the next section, we propose a slightly different method, which visits each example only once and is based on our previous work (Crammer *et al.* 2006). This algorithm is a variant of the algorithm presented in Chapter 4, in Chapter 5 and in Chapter 11.

### 10.5.1 Iterative Algorithm

We now describe a simple iterative algorithm for learning the weight vectors $\mathbf{w}^{\text{acoustic}}$ and $\mathbf{w}^{\text{language}}$. The algorithm receives as input a training set $\mathcal{T} = \{(\bar{\mathbf{x}}_1, \bar{v}_1, \bar{p}_1, \bar{s}_1), \ldots, (\bar{\mathbf{x}}_m, \bar{v}_m, \bar{p}_m, \bar{s}_m)\}$ of examples. Each example is constituted of a spoken acoustic signal $\bar{\mathbf{x}}_i$, its corresponding word sequence $\bar{v}_i$, phoneme sequence $\bar{p}_i$ and alignment sequence $\bar{s}_i$. At each iteration the algorithm updates $\mathbf{w}^{\text{acoustic}}$ and $\mathbf{w}^{\text{language}}$ according to the $i$th example in $\mathcal{T}$ as we now describe. Denote by $\mathbf{w}^{\text{acoustic}}_{i-1}$ and $\mathbf{w}^{\text{language}}_{i-1}$ the value of the weight vectors before the $i$th iteration. Let $\bar{v}'_i$, $\bar{p}'_i$ and $\bar{s}'_i$ be the predicted word sequence, phoneme sequence and timing sequence for the $i$th example, respectively, according to $\mathbf{w}^{\text{acoustic}}_{i-1}$ and $\mathbf{w}^{\text{language}}_{i-1}$,

$$(\bar{v}'_i, \bar{s}'_i) = \arg\max_{\bar{v}, \bar{p}=lex(\bar{v})} \left[ \max_{\bar{s}} \mathbf{w}^{\text{acoustic}}_{i-1} \cdot \boldsymbol{\phi}^{\text{acoustic}}(\bar{\mathbf{x}}_i, \bar{p}, \bar{s}) + \mathbf{w}^{\text{language}}_{i-1} \cdot \boldsymbol{\phi}^{\text{language}}(\bar{v}) \right].$$
$$(10.13)$$

Denote by $\gamma(\bar{v}_i, \bar{v}'_i)$ the Levenshtein distance between the correct word sequence $\bar{v}_i$ and the predicted word sequence $\bar{v}'_i$. Also denote by $\Delta\boldsymbol{\phi}_i = (\Delta\boldsymbol{\phi}^{\text{acoustic}}_i, \Delta\boldsymbol{\phi}^{\text{language}}_i)$ the difference between the feature function of the correct word sequence minus the feature function of the predicted word sequence as

$$\Delta\boldsymbol{\phi}^{\text{acoustic}}_i = \boldsymbol{\phi}^{\text{acoustic}}(\bar{\mathbf{x}}_i, \bar{p}_i, \bar{s}_i) - \boldsymbol{\phi}^{\text{acoustic}}(\bar{\mathbf{x}}_i, \bar{p}', \bar{s}'),$$

and

$$\Delta\boldsymbol{\phi}^{\text{language}}_i = \boldsymbol{\phi}^{\text{language}}(\bar{v}_i) - \boldsymbol{\phi}^{\text{language}}(\bar{v}').$$

We set the next weight vectors $\mathbf{w}^{\text{acoustic}}_i$ and $\mathbf{w}^{\text{language}}_i$ to be the minimizer of the following optimization problem:

$$\min_{\mathbf{w}\in\mathcal{H}, \xi\geq 0} \tfrac{1}{2}\|\mathbf{w}^{\text{acoustic}} - \mathbf{w}^{\text{acoustic}}_{i-1}\|^2 + \tfrac{1}{2}\|\mathbf{w}^{\text{language}} - \mathbf{w}^{\text{language}}_{i-1}\|^2 + C\xi \qquad (10.14)$$

$$\text{s.t.} \quad \mathbf{w}^{\text{acoustic}} \cdot \Delta\boldsymbol{\phi}^{\text{acoustic}}_i + \mathbf{w}^{\text{language}} \cdot \Delta\boldsymbol{\phi}^{\text{language}}_i \geq \gamma(\bar{v}_i, \bar{v}'_i) - \xi,$$

where $C$ serves as a complexity–accuracy trade-off parameter (see Crammer *et al.* (2006)) and $\xi$ is a non-negative slack variable, which indicates the loss of the $i$th example. Intuitively, we would like to minimize the loss of the current example, i.e. the slack variable $\xi$, while keeping the weight vector $\mathbf{w}$ as close as possible to the previous weight vector $\mathbf{w}_{i-1}$. The constraint makes the projection of the feature functions of the correct word sequence onto $\mathbf{w}$ higher than the projection of the feature functions of any other word sequence onto $\mathbf{w}$ by at least $\gamma(\bar{v}_i, \bar{v}_i')$. It can be shown (see Crammer *et al.* (2006)) that the solution to the above optimization problem is

$$\mathbf{w}_i^{\text{acoustic}} = \mathbf{w}_{i-1}^{\text{acoustic}} + \alpha_i \Delta\boldsymbol{\phi}_i^{\text{acoustic}} \tag{10.15}$$

$$\mathbf{w}_i^{\text{language}} = \mathbf{w}_{i-1}^{\text{language}} + \alpha_i \Delta\boldsymbol{\phi}_i^{\text{language}}.$$

The value of the scalar $\alpha_i$ is based on the difference $\Delta\boldsymbol{\phi}_i$, the previous weight vector $\mathbf{w}_{i-1}$ and a parameter $C$. Formally

$$\alpha_i = \min\left\{ C, \frac{[\gamma(\bar{v}_i, \bar{v}_i') - \mathbf{w}_{i-1}^{\text{acoustic}} \cdot \Delta\boldsymbol{\phi}_i^{\text{acoustic}} - \mathbf{w}_{i-1}^{\text{language}} \cdot \Delta\boldsymbol{\phi}_i^{\text{language}}]}{\|\Delta\boldsymbol{\phi}_i^{\text{acoustic}}\|^2 + \|\Delta\boldsymbol{\phi}_i^{\text{language}}\|^2} \right\}. \tag{10.16}$$

A pseudo-code of our algorithm is given in Figure 10.2.

We note in passing that the current state-of-the-art language models are trained on datasets which are many orders of magnitude bigger than those used for acoustic models. Hence, if we use the proposed approach to train the acoustic and language models jointly we will used a small dataset and hence may not get state-of-the-art results. A possible solution to this problem would be to train a discriminative language model on a big corpus, and use it as the initial condition of the iterative algorithm. The iterative algorithm will train the acoustic model and will refine the language model on a smaller corpus. We denote by $\mathbf{w}_0^{\text{acoustic}}$ and $\mathbf{w}_0^{\text{language}}$ the initial acoustic and language models, respectively. As discussed here, $\mathbf{w}_0^{\text{acoustic}} = \mathbf{0}$ and $\mathbf{w}_0^{\text{language}}$ is the pre-trained language model.

Last we would like to note that under some mild technical conditions, it can be shown that the cumulative word error rate of an iterative procedure , $\sum_{i=1}^m \gamma(\bar{v}_i, \bar{v}_i')$, is likely to be small. Moreover, it can be shown (Cesa-Bianchi *et al.* 2004) that if the cumulative word error rate of the iterative procedure is small, there exists among the vectors $\{\mathbf{w}_1, \ldots, \mathbf{w}_m\}$ at least one weight vector which attains small averaged word error rate on unseen examples as well. The analysis follows the same lines as the analysis of the phoneme recognition algorithm presented in Chapter 5.

## 10.5.2 Recognition Feature Functions

Our construction is based on a set of acoustic modeling feature functions, $\{\phi_j^{\text{acoustic}}\}_{j=1}^n$, which maps an acoustic–phonetic representation of a speech utterance as well as a suggested timing sequence into a vector-space. For this task we can utilize the same set of features successfully used in the task of forced alignment (Chapter 4) and keyword spotting (Chapter 11). Our model is also based on a set of language modeling feature functions, $\{\phi_k^{\text{language}}\}_{k=1}^\ell$. The set of features that can be used for language modeling is described in Chapter 9 and in Roark *et al.* (2007).

We can extend the family of linear word sequence recognizers given to nonlinear recognition functions. This extension is based on Mercer kernels often used in SVM algorithms (Vapnik 1998). Recall that the update rule of the algorithm is $\mathbf{w}_i = \mathbf{w}_{i-1} + \alpha_i \Delta\boldsymbol{\phi}_i$

**Input**: Training set $\mathcal{T}=\{(\bar{\mathbf{x}}_i,\,\bar{v}_i,\,\bar{p}_i,\,\bar{s}_i)\}_{i=1}^{m}$; valid. set $\mathcal{T}_{\text{valid}}=\{(\bar{\mathbf{x}}_i,\,\bar{v}_i,\,\bar{p}_i,\,\bar{s}_i)\}_{i=1}^{m_{\text{valid}}}$;
parameter $C$.

**Initialize**: $\mathbf{w}_0 = (\mathbf{w}_0^{\text{acoustic}},\,\mathbf{w}_0^{\text{language}})$

**Loop** for $i = 1, \ldots, m$

**Predict:**

$$(\bar{v}',\,\bar{p}',\,\bar{s}') = \arg\max_{\bar{v},\,\bar{p}=lex(\bar{v})}\left[\max_{\bar{s}}\ \mathbf{w}_{i-1}^{\text{acoustic}} \cdot \boldsymbol{\phi}^{\text{acoustic}}(\bar{\mathbf{x}}_i,\,\bar{p},\,\bar{s}) + \mathbf{w}_{i-1}^{\text{language}} \cdot \boldsymbol{\phi}^{\text{language}}(\bar{v})\right]$$

**Set:**

$$\Delta\boldsymbol{\phi}_i^{\text{acoustic}} = \boldsymbol{\phi}^{\text{acoustic}}(\bar{\mathbf{x}}_i,\,\bar{p}_i,\,\bar{s}_i) - \boldsymbol{\phi}^{\text{acoustic}}(\bar{\mathbf{x}}_i,\,\bar{p}',\,\bar{s}')$$

$$\Delta\boldsymbol{\phi}_i^{\text{language}} = \boldsymbol{\phi}^{\text{language}}(\bar{v}_i) - \boldsymbol{\phi}^{\text{language}}(\bar{v}').$$

**Set:**

$$\ell(\mathbf{w}_{i-1};\,(\bar{\mathbf{x}}_i,\,\bar{v}_i,\,\bar{p}_i,\,\bar{s}_i))$$
$$= \max_{\bar{v},\,\bar{p}=lex(\bar{v})}\left[\gamma(\bar{v}_i,\,\bar{v}') - \mathbf{w}_{i-1}^{\text{acoustic}}\cdot\Delta\boldsymbol{\phi}_i^{\text{acoustic}} - \mathbf{w}_{i-1}^{\text{language}}\cdot\Delta\boldsymbol{\phi}_i^{\text{language}}\right]_+$$

**If** $\ell(\mathbf{w}_{i-1};\,(\bar{\mathbf{x}}_i,\,\bar{v}_i,\,\bar{p}_i,\,\bar{s}_i)) > 0$

**Set:**

$$\alpha_i = \min\left\{C,\ \frac{\ell(\mathbf{w}_{i-1};\,(\bar{\mathbf{x}}_i,\,\bar{v}_i,\,\bar{p}_i,\,\bar{s}_i))}{\|\Delta\boldsymbol{\phi}_i^{\text{acoustic}}\|^2 + \|\Delta\boldsymbol{\phi}_i^{\text{language}}\|^2}\right\}$$

**Update:**

$$\mathbf{w}_i^{\text{acoustic}} = \mathbf{w}_{i-1}^{\text{acoustic}} + \alpha_i\,\Delta\boldsymbol{\phi}_i^{\text{acoustic}}$$

$$\mathbf{w}_i^{\text{language}} = \mathbf{w}_{i-1}^{\text{language}} + \alpha_i\,\Delta\boldsymbol{\phi}_i^{\text{language}}.$$

**Output:** The weight vector $\mathbf{w}^*$ which achieves best performance on a validation set $\mathcal{T}_{\text{valid}}$:

$$\mathbf{w}^* = \arg\min_{\mathbf{w}\in\{\mathbf{w}_1,\ldots,\mathbf{w}_m\}}\sum_{j=1}^{m_{\text{valid}}}\gamma\left(\bar{v}_j,\,f(\bar{\mathbf{x}}_j)\right)$$

Figure 10.2  An iterative algorithm for phoneme recognition.

and that the initial weight vector is $\mathbf{w}_0 = \mathbf{0}$. Thus, $\mathbf{w}_i$ can be rewritten as, $\mathbf{w}_i = \sum_{j=1}^{i}\alpha_j\Delta\boldsymbol{\phi}_j$ and $f$ can be rewritten as

$$f(\bar{\mathbf{x}}) = \arg\max_{\bar{p}}\ \max_{\bar{s}}\ \sum_{j=1}^{i}\alpha_j(\Delta\boldsymbol{\phi}_j \cdot \boldsymbol{\phi}(\bar{\mathbf{x}},\,\bar{v})). \qquad (10.17)$$

By substituting the definition of $\Delta\boldsymbol{\phi}_j$ and replacing the inner-product in Equation (10.17) with a general kernel operator $K(\cdot,\,\cdot)$ that satisfies Mercer's conditions (Vapnik 1998), we obtain a nonlinear phoneme recognition function:

$$f(\bar{\mathbf{x}}) = \arg\max_{\bar{p}}\ \max_{\bar{s}}\ \sum_{j=1}^{i}\alpha_j(K(\bar{\mathbf{x}}_j,\,\bar{v}_j;\,\bar{\mathbf{x}},\,\bar{v}) - K(\bar{\mathbf{x}}_j,\,\bar{v}_j';\,\bar{\mathbf{x}},\,\bar{v})). \qquad (10.18)$$

It is easy to verify that the definition of $\alpha_i$ given in Equation (10.16) can also be rewritten using the kernel operator.

### 10.5.3    The Decoder

Assuming we know the optimal weight vectors $\mathbf{w}^{\text{acoustic}}$ and $\mathbf{w}^{\text{language}}$, the maximization over all word sequences in Equation (10.8) is a search problem (known also as the *inference problem*). A direct search for the maximizer is not feasible since the number of possible word sequences (and hence the number of possible phoneme and timing sequences) is exponential in the size of the vocabulary, $|\mathcal{V}|$. In this section we provide details about performing the search with the kernel based model. Basically, the search algorithm in the kernel based model is similar to that used for the HMMs and segment models, using dynamic programming to find the most likely phoneme sequence.

First we derive the domain of all word sequences, which is the feasible region of our search problems. Let us denote the grammar network $\mathcal{G}$ as a directed graph of all possible word sequences (sentences) in the given language. The network $\mathcal{G}$ is also known as a finite-state transducer (FST) (Mohri *et al.* 2007). Also denote the lexicon network $\mathcal{L}$ as a set of graphs, where each graph in the set is the phonetic network, generated from the mapping *lex*, corresponding to every word in the vocabulary $\mathcal{V}$. The network $\mathcal{L}$ is also an FST. The FST, generated from the composition of the grammar graph $\mathcal{G}$ and the set of the lexicon network $\mathcal{L}$, namely $\mathcal{L} \circ \mathcal{G}$, is the mapping from phoneme sequences to word sequences. As an optimization step the resulting FST is determined and minimized. We denote the final FST by $\mathcal{N}$, that is, $\mathcal{N} = \min(\det(\mathcal{L} \circ \mathcal{G}))$. The best word sequence is found by efficiently searching the FST network $\mathcal{N}$ for a sequence of phonemes (and a sequence of words) which is optimally aligned to the acoustic signal and hence maximizes Equation (10.8). A detailed description of the recursive algorithm is given in Section 4.6, where the search is done over the FST network $\mathcal{N}$ rather than the known phoneme sequence in the problem of forced alignment. Similarly to the complexity of the efficient search presented in Section 4.6, the complexity of the search over the FST here is $O(Q_{\mathcal{N}}||\mathcal{P}||\bar{\mathbf{x}}|^3)$, where $|Q_{\mathcal{N}}|$ is the number of states in the FST $\mathcal{N}$. Again, in practice, we can use the assumption that the maximal length of an event is bounded. This assumption reduces the complexity of the algorithm to be $O(|Q_{\mathcal{N}}||\mathcal{P}||\bar{\mathbf{x}}|L^2)$.

### 10.5.4    Complexity

To conclude this section we discuss the global complexity of our proposed method. In the training phase, our algorithm performs $m$ iterations, one iteration per training example. At each iteration the algorithm evaluates the recognition function once, updates the recognition function, if needed, and evaluates the new recognition function on a validation set of size $m_{\text{valid}}$. Each evaluation of the recognition function takes an order of $O(|Q_{\mathcal{N}}||\mathcal{P}||\bar{\mathbf{x}}|L^2)$ operations. Therefore the total complexity of our method becomes $O(mm_{\text{valid}}|Q_{\mathcal{N}}||\mathcal{P}||\bar{\mathbf{x}}|L^2)$. Finally, we compare the complexity of our method with the complexity of other algorithms which directly solve the SVM optimization problem given in Equation (10.14). The algorithm given in Taskar *et al.* (2003) is based on the SMO algorithm for solving SVM problems. While there is no direct complexity analysis for this algorithm, in practice it usually requires at least $m^2$ iterations, which results in a total complexity of the order $O(m^2 |Q_{\mathcal{N}}||\mathcal{P}||\bar{\mathbf{x}}|L^2)$. The complexity of the algorithm presented in Tsochantaridis *et al.* (2004) depends on the choice of several parameters. For reasonable choice of these parameters the total complexity is also of the order $O(m^2|Q_{\mathcal{N}}||\mathcal{P}||\bar{\mathbf{x}}|L^2)$.

Note that if the model is initiated with an already trained language model, the complexity of its training should be added to the total complexity.

## 10.6  Discussion

This chapter proposes a framework for a large-margin kernel based continuous speech recognizer. This framework is based on adaptation of known algorithms and techniques to this specific task, while minimizing the word error rate directly, by jointly training both the acoustic and the language models.

Most (if not all) of the successful speech recognizers are based on triphones. It is claimed that contextual effects cause large variations in the way that different phonemes are uttered (Young 1996). In order to address this issue, it is common to refine the set of phone labels and use a distinct label for every phone and every pair of left and right contexts. This distinct label is called a triphone. This refinement of the phone label set actually increases the size of the output labels exponentially by 3: with 40 phonemes, there are $40^3 = 64\,000$ possible triphones. Some of them cannot occur due to linguistic concerns, leaving around $45\,000$ possible triphones. Others can rarely be found in the training data, resulting in a poor estimation of the model parameters. While in generative models like HMM there is an efficient way to train such a huge set of labels (tied-mixture system), this, however, is still an open problem for kernel based models.

Another important issue is the efficiency of the kernel based algorithm when working with nonlinear kernels. It was found out that the speech recognition task uses a huge amount of supports (see Chapter 5), which dramatically affects the use of such models in real-life speech applications. This acute point is also left for future work.

## Acknowledgments

We are in debt to John Dines for his important remarks regarding the manuscript. We would like to thank Philip Garner for his clarifications and simplifications of the HMM-based decoder.

## References

Bahl L, Brown P, de Souza P and Mercer R 1986 Maximum mutual information estimation of hidden Markov model parameters for speech recognition. *Proceedings of the IEEE International Conference on Audio, Speech and Signal Processing (ICASSP)*, pp. 49–52.

Cesa-Bianchi N, Conconi A and Gentile C 2004 On the generalization ability of on-line learning algorithms. *IEEE Transactions on Information Theory* **50**(9), 2050–2057.

Cortes C and Vapnik V 1995 Support-vector networks. *Machine Learning* **20**(3), 273–297.

Crammer K, Dekel O, Keshet J, Shalev-Shwartz S and Singer Y 2006 Online passive aggressive algorithms. *Journal of Machine Learning Research* **7**, 551–585.

Cristianini N and Shawe-Taylor J 2000 *An Introduction to Support Vector Machines*. Cambridge University Press.

Dekel O, Shalev-Shwartz S and Singer Y 2004 The power of selective memory: Self-bounded learning of prediction suffix trees. *Advances in Neural Information Processing Systems 17*.

Fu Q and Juang BH 2007 Automatic speech recognition based on weighted minimum classification error (W-MCE) training method. *Proceedings of the IEEE Automatic Speech Recognition & Understanding Workshop*, Kyoto, Japan, December 9–13, pp. 278–283.

Juang BH, Chou W and Lee CH 1997 Minimum classification error rate methods for speech recognition. *IEEE Transactions on Speech and Audio Processing* **5**(3), 257–265.

Mohri M, Pereira F and Riley M 2007 Speech recognition with weighted finite-state transducers, in *Springer Handbook of Speech Processing* (eds. J Benesty, M Sondhi and Y Huang). Springer.

Ostendorf M 1996 From HMM's to segment models: Stochastic modeling for CSR, in *Automatic Speech and Speaker Recognition – Advanced Topics* (eds. CH Lee, F Soong and K Plaiwal). Kluwer Academic Publishers.

Ostendorf M, Digalakis V and Kimball O 1996 From HMM's to segment models: A unified view of stochastic modeling for speech recognition. *IEEE Transactions on Speech and Audio Processing* **4**, 360–378.

Pereira F, Singer Y and Tishby N 1995 Beyond word N-grams. *Proceedings of the Third Workshop on Very Large Corpora* (eds. D Yarowsky and K Church). MIT Press, Cambridge, MA.

Platt JC 1998 Fast training of Support Vector Machines using sequential minimal optimization, in *Advances in Kernel Methods – Support Vector Learning* (eds. B Schölkopf, C Burges and A Smola). MIT Press.

Povey D 2003 Discriminative Training for Large Vocabulary Speech Recognition. Engineering Dept., Cambridge University, PhD thesis.

Roark B, Saraclar M and Collins M 2007 Discriminative *n*-gram language modeling. *Computer Speech and Language* **21**, 373–392.

Ron D, Singer Y and Tishby N 1996 The power of amnesia: learning probabilistic automata with variable memory length. *Machine Learning* **25**(2), 117–150.

Taskar B, Guestrin C and Koller D 2003 Max-margin Markov networks. *Advances in Neural Information Processing Systems 17*.

Tsochantaridis I, Hofmann T, Joachims T and Altun Y 2004 Support vector machine learning for interdependent and structured output spaces. *Proceedings of the 21st International Conference on Machine Learning*, Banff, Alberta, Canada, July 4–8.

Vapnik VN 1998 *Statistical Learning Theory*. John Wiley & Sons.

Young S 1996 A review of large-vocabulary continuous speech recognition. *IEEE Signal Processing Mag.* pp. 45–57.

# Part IV

# Applications

# 11

# Discriminative Keyword Spotting

## David Grangier, Joseph Keshet and Samy Bengio

This chapter introduces a discriminative method for detecting and spotting keywords in spoken utterances. Given a word represented as a sequence of phonemes and a spoken utterance, the keyword spotter predicts the best time span of the phoneme sequence in the spoken utterance along with a confidence. If the prediction confidence is above certain level the keyword is declared to be spoken in the utterance within the predicted time span, otherwise the keyword is declared as not spoken. The problem of keyword spotting training is formulated as a discriminative task where the model parameters are chosen so the utterance in which the keyword is spoken would have higher confidence than any other spoken utterance in which the keyword is not spoken. It is shown theoretically and empirically that the proposed training method resulted with a high area under the Receiver Operating Characteristic (ROC) curve, the most common measure to evaluate keyword spotters. We present an iterative algorithm to train the keyword spotter efficiently. The proposed approach contrasts with standard spotting strategies based on Hidden Markov Models (HMMs), for which the training procedure does not maximize a loss directly related to the spotting performance. Several experiments performed on TIMIT and WSJ corpora show the advantage of our approach over HMM-based alternatives.

## 11.1  Introduction

Keyword spotting aims at detecting any given keyword in spoken utterances. This task is important in numerous applications, such as voice mail retrieval, voice command detection and spoken term detection and retrieval. Previous work has focused mainly on several variants of HMMs to address this intrinsically sequential problem. While the HMM-based

*Automatic Speech and Speaker Recognition: Large Margin and Kernel Methods*    Joseph Keshet and Samy Bengio

approaches constitute the state-of-the-art, they suffer from several known limitations. Most of these limitations are not specific to the keyword spotting problem, and are common to other tasks such as speech recognition, as pointed out in Chapter 1. For instance, the predominance of the emission probabilities in the likelihood, which tends to neglect duration and transition models, or the Expectation Maximization (EM) training procedure, which is prone to convergence to local optima. Other drawbacks are specific to the application of HMMs to the keyword spotting task. In particular, the scarce occurrence of some keywords in the training corpora often requires ad-hoc modifications of the HMM topology, the transition probabilities or the decoding algorithm. The most acute limitation of HMM-based approaches lies in their training objective. Typically, HMM training aims at maximizing the likelihood of transcribed utterances, and does not provide any guarantees in terms of keyword spotting performance.

The performance of a keyword spotting system is often measured by the ROC curve, that is, a plot of the true positive (spotting a keyword correctly) rate as a function of the false positive (mis-spotting a keyword) rate; see for example Benayed *et al.* (2004), Ketabdar *et al.* (2006), Silaghi and Bourlard (1999). Each point on the ROC curve represents the system performance for a specific trade-off between achieving a high true positive rate and a low false positive rate. Since the preferred trade-off is not always defined in advance, systems are commonly evaluated according to the averaged performance over all operating points. This corresponds to preferring the systems that attain the highest Area Under the ROC Curve (AUC).

In this study, we devise a discriminative large margin approach for learning to spot any given keyword in any given speech utterance. The keyword spotting function gets as input a phoneme sequence representing the keyword and a spoken utterance and outputs a prediction of the time span of the keyword in the spoken utterance and a confidence. If the confidence is above some predefined threshold, the keyword is declared to be spoken in the predicted time span, otherwise the keyword is declared as not spoken. The goal of the training algorithm is to maximize the AUC on the training data and on unseen test data. We call an utterance in the training set in which the keyword is spoken a *positive utterance*, and respectively, an utterance in which the keyword is not spoken a *negative utterance*. Using the Wilcoxon–Mann–Whitney statistics (Cortes and Mohri 2004), we formulate the training as a problem of estimating the model parameters such that the confidence of the correct time span in a positive utterance would be higher than the confidence of any time span in any negative utterance. Formally this problem is stated as a convex optimization problem with constraints. The solution to this optimization problem is a function which is shown analytically to attain high AUC on the training set and is likely to have good generalization properties on unseen test data as well. Moreover, comparing with HMMs, our approach is based on a convex optimization procedure, which converges to the global optima, and it is based on a non-probabilistic framework, which offers greater flexibility in selecting the relative importance of duration modeling with respect to acoustic modeling.

The remainder of this chapter is organized as follows: Section 11.2 describes previous work on keyword spotting, Section 11.3 introduces our discriminative large margin approach, Section 11.4 presents different experiments comparing the proposed model with an HMM-based solution. Finally, Section 11.5 draws some conclusions and delineates possible directions for future research.

# 11.2 Previous Work

The research on keyword spotting has paralleled the development of the automatic speech recognition (ASR) domain in the last thirty years. Like ASR, keyword spotting has first been addressed with models based on Dynamic Time Warping (DTW) (Bridle 1973; Higgins and Wohlford 1985). Then, approaches based on discrete HMMs were introduced (Kawabata *et al.* 1988). Finally, discrete HMMs have been replaced by continuous HMMs (Rabiner and Juang 1993).

The core objective of a keyword spotting system is to discriminate between utterances in which a given keyword is uttered and utterances in which the keyword is not uttered. For this purpose, the first approaches based on DTW proposed to compute the alignment distance between a template utterance representing the target keyword and all possible segments of the test signal (Bridle 1973). In this context, the keyword is considered as detected in a segment of the test utterance whenever the alignment distance is below some predefined threshold. Such approaches are, however, greatly affected by speaker mismatch and varying recording conditions between the template sequence and the test signal. To gain some robustness, it has then been proposed to compute alignment distances not only with respect to the target keyword template, but also with respect to other keyword templates (Higgins and Wohlford 1985). Precisely, given a test utterance, the system identifies the concatenation of templates with the lowest distance to the signal and the keyword is considered as detected if this concatenation contains the target keyword template. Therefore, the keyword alignment distance is not considered as an absolute number, but relatively to the distances to other templates, which increase robustness with respect to changes in the recording conditions.

Along with the development of the speech research, increasingly large amounts of labeled speech data were collected, and DTW-based techniques started showing their shortcomings to leverage from large training sets. In particular, large corpora contain thousands of words and dozens of instances for each word. Considering each instance as a template makes the search for the best template concatenation a prohibitively expensive task. Consequently, discrete HMMs were introduced for ASR (Bahl *et al.* 1986), and then for keyword spotting (Kawabata *et al.* 1988; Wilpon *et al.* 1990). A discrete HMM assumes that the quantized acoustic feature vectors representing the input utterance are independent conditioned on the hidden state variables. This type of model introduces several advantages compared to DTW-based approaches, including an improved robustness to speaker and channel changes, when several training utterances of the targeted keyword are available. However, the most important evolution introduced with the HMM certainly lies in the development of phone or triphone-based modeling (Kawabata *et al.* 1988; Lee and Hon 1988; Rose and Paul 1990), in which a word model is composed of several sub-unit models shared across words. This means that the model of a given word not only benefits from the training utterances containing this word, but also from all the utterances containing its sub-units. Additionally, great scalability is achieved with such an approach since the number of acoustic parameters is proportional to the number of sub-units and does not grow linearly with the corpus size, as opposed to template-based approaches. A further advantage of phone-based modeling is the ability to spot words unavailable at training time, as this paradigm allows one to build a new word model by composing already trained sub-unit models. This aspect is very important, since in most applications the set of test keywords is not known in advance.

Soon after the application of discrete HMMs to speech problems, continuous density HMMs have been introduced in the ASR community (Rabiner and Juang 1993). Continuous HMMs eliminate the need of acoustic vector quantization, as the distributions associated with the HMM states are continuous densities, often modeled by Gaussian Mixture Models (GMMs). The learning of both GMM parameters and the state transition probabilities is performed in a single integrated framework, maximizing the likelihood of the training data given its transcription through the EM algorithm (Bilmes 1998). This approach has been shown to be more effective and allows greater flexibility for speaker or channel adaptation (Rabiner and Juang 1993). It is now the most widely used approach for both ASR and keyword spotting.

In the context of keyword spotting, different strategies based on continuous HMMs have been proposed. In most cases, a sub-unit based HMM is trained over a large corpus of transcribed data and a new model is then built from the sub-unit models. Such a model is composed of two parts, a keyword HMM and a *filler* or *garbage* HMM, which respectively model the keyword and non-keyword parts of the signal. This topology is depicted in Figure 11.1. Given such a model, keyword detection is performed by searching for the sequence of states that yields the highest likelihood for the provided test sequence through Viterbi decoding. Keyword detection is determined by checking whether the Viterbi best-path passes through the keyword model or not. In such a model, the selection of the transition probabilities in the keyword sets the trade-off between low false alarm rate (detecting a keyword when it is not presented), and low false rejection rate (not detecting a keyword when it is indeed presented). Another important aspect of this approach lies in the modeling of non-keyword parts of the signal, and several choices are possible for the garbage HMM. The simplest choice models garbage with an HMM that fully connects all sub-unit models (Rose and Paul 1990), while the most complex choice models garbage with a full-large vocabulary HMM, where the lexicon excludes the keyword (Weintraub 1993). The latter approach obviously yields a better garbage model, using additional linguistic knowledge. This advantage, however, induces a higher decoding cost and requires a larger amount of training data, in particular for language model training. Besides practical concerns, one can conceptually wonder whether an automatic spotting approach should require such a large linguistic knowledge. Of course, several variations of garbage models exist between the two extreme examples pointed out above (see for instance Boite *et al.* (1993)).

Viterbi decoding relies on a sequence of local decisions to determine the best path, which can be fragile with respect to local model mismatch. In the context of HMM-based keyword spotting, a keyword can be missed if only its first phoneme suffers such a mismatch, for instance. To gain some robustness, likelihood ratio approaches have been proposed (Rose and Paul 1990; Weintraub 1995). In this case, the confidence score output from the keyword spotter corresponds to the ratio between the likelihood estimated by an HMM including the occurrence of the target keyword, and the likelihood estimated by an HMM excluding it. These HMM topologies are depicted in Figure 11.2. Detection is then performed by comparing the output scores to a predefined threshold. Different variations on this likelihood ratio approach have then been devised, such as computing the ratio only on the part of the signal where the keyword is assumed to be detected (Junkawitsch *et al.* 1997). Overall, all the above methods are variations over the same HMM paradigm, which consists in training a generative model through likelihood maximization, before introducing different modifications prior to decoding in order to address the keyword spotting task. In other

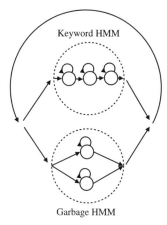

Figure 11.1 HMM topology for keyword spotting with a Viterbi best-path strategy. This approach verifies whether the Viterbi best-path passes through the keyword sub-model.

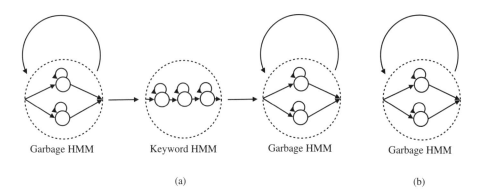

Figure 11.2 HMM topology for keyword spotting with a likelihood ratio strategy. This approach compares the likelihood of the sequence given that the keyword is uttered (a) with the likelihood of the sequence given that the keyword is not uttered (b).

words, these approaches do not propose to train the model so as to maximize the spotting performance, and the keyword spotting task is only introduced in the inference step after training.

Only few studies have proposed discriminative parameter training approaches to circumvent this weakness (Benayed *et al.* 2003; Sandness and Hetherington 2000; Sukkar *et al.* 1996; Weintraub *et al.* 1997). Sukkar *et al.* (1996) proposed to maximize the likelihood ratio between the keyword and garbage models for keyword utterances and to minimize it over a set of false alarms generated by a first keyword spotter. Sandness and Hetherington (2000) proposed to apply Minimum Classification Error (MCE) to the keyword spotting problem.

The training procedure updates the acoustic models to lower the score of non-keyword models in the part of the signal where the keyword is uttered. However, this procedure does not focus on false alarms, and does not aim at lowering the score of the keyword models in parts of the signal where the keyword is not uttered. Other discriminative approaches have been focused on combining different HMM-based keyword detectors. For instance, Weintraub *et al.* (1997) trained a neural network to combine likelihood ratios from different models. Benayed *et al.* (2003) relied on Support Vector Machines (SVMs) to combine different averages of phone-level likelihoods. Both of these approaches propose to minimize the error rate, which equally weights the two possible spotting errors, false positive (or false alarm) and false negative (missed keyword occurrence, often called keyword deletion). This measure is, however, barely used to evaluate keyword spotters, due to the *unbalanced* nature of the problem. Precisely, the targeted keywords generally occur rarely and hence the number of potential false alarms highly exceeds the number of potential missed detections. In this case, the useless model which never predicts the keyword avoids all false alarms and yields a very low error rate, with which it is difficult to compete. For that reason the AUC is more informative and is commonly used to evaluate models. Attaining high AUC would hence be an appropriate learning objective for the discriminative training of a keyword spotter. To the best of our knowledge, only Chang (1995) proposed an approach targeting this goal. This work introduces a methodology to maximize the figure-of-merit (FOM), which corresponds to the AUC over a specific range of false alarm rates. However, the proposed approach relies on various heuristics, such as gradient smoothing and sorting approximations, which does not ensure any theoretical guarantee on obtaining high FOM. Also, these heuristics involve the selection of several hyperparameters, which challenges a practical use.

In the following, we introduce a model that aims at achieving high AUC over a set of training examples, and constitutes a truly discriminative approach to the keyword spotting problem. The proposed model relies on large margin learning techniques for sequence prediction and provides theoretical guarantees regarding the generalization performance. Furthermore, its efficient learning procedure ensures scalability toward large problems and simple practical use.

# 11.3   Discriminative Keyword Spotting

This section formalizes the keyword spotting problem, and introduces the proposed approach. First, we describe the problem of keyword spotting formally. This allows us to introduce a loss derived from the definition of the AUC. Then, we present our model parameterization and the training procedure to minimize efficiently a regularized version of this loss. Finally, we give an analysis of the iterative algorithm, and show that it achieves a high cumulative AUC in the training process and high expected AUC on unseen test data.

## 11.3.1   Problem Setting

In the keyword spotting task, we are provided with a speech signal composed of a sequence of acoustic feature vectors $\bar{\mathbf{x}} = (\mathbf{x}_1, \ldots, \mathbf{x}_T)$, where $\mathbf{x}_t \in \mathcal{X} \subset \mathbb{R}^d$, for all $1 \leq t \leq T$, is a feature vector of length $d$ extracted from the $t$th frame. Naturally, the length of the acoustic signal varies from one signal to another and thus $T$ is not fixed. We denote a keyword by $k \in \mathcal{K}$, where $\mathcal{K}$ is a lexicon of words. Each keyword $k$ is composed of a sequence of phonemes

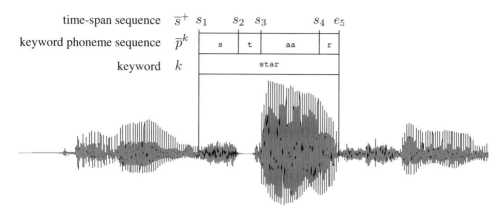

Figure 11.3 Example of our notation. The waveform of the spoken utterance 'a lone star shone...' taken from the TIMIT corpus. The keyword $k$ is the word *star*. The phonetic transcription $\bar{p}^k$ along with the time span sequence $\bar{s}^+$ is depicted in the figure.

$\bar{p}^k = (p_1, \ldots, p_L)$, where $p_l \in \mathcal{P}$ for all $1 \leq l \leq L$ and $\mathcal{P}$ is the domain of the phoneme symbols. The number of phonemes in each keyword may vary from one keyword to another and hence $L$ is not fixed. We denote by $\mathcal{P}^*$ (and similarly $\mathcal{X}^*$) the set of all finite length sequences over $\mathcal{P}$. Let us further define the time span of the phoneme sequence $\bar{p}^k$ in the speech signal $\bar{\mathbf{x}}$. We denote by $s_l \in \{1, \ldots, T\}$ the start time (in frame units) of phoneme $p_l$ in $\bar{\mathbf{x}}$, and by $e_l \in \{1, \ldots, T\}$ the end time of phoneme $p_l$ in $\bar{\mathbf{x}}$. We assume that the start time of any phoneme $p_{l+1}$ is equal to the end time of the previous phoneme $p_l$, that is, $e_l = s_{l+1}$ for all $1 \leq l \leq L - 1$. We define the time span (or segmentation) sequence as $\bar{s}^k = (s_1, \ldots, s_L, e_L)$. An example of our notation is given in Figure 11.3. Our goal is to learn a *keyword spotter*, denoted $f : \mathcal{X}^* \times \mathcal{P}^* \to \mathbb{R}$, which takes as input the pair $(\bar{\mathbf{x}}, \bar{p}^k)$ and returns a real valued score expressing the confidence that the targeted keyword $k$ is uttered in $\bar{\mathbf{x}}$. By comparing this score with a threshold $b \in \mathbb{R}$, we can determine whether $\bar{p}^k$ is uttered in $\bar{\mathbf{x}}$.

In discriminative supervised learning we are provided with a training set of examples and a test set (or evaluation set). Specifically, in the task of discriminative keyword spotting we are provided with a two sets of keywords. The first set $\mathcal{K}_{\text{train}}$ is used for training and the second set $\mathcal{K}_{\text{test}}$ is used for evaluation. Note that the lexicon of keywords is a union of both the training set and the test set, $\mathcal{K} = \mathcal{K}_{\text{train}} \cup \mathcal{K}_{\text{test}}$. Algorithmically, we do not restrict a keyword to be only in one set and a keyword that appears in the training set can appear also in the test set. Nevertheless, in our experiments we picked different keywords for training and test and hence $\mathcal{K}_{\text{train}} \cap \mathcal{K}_{\text{test}} = \emptyset$.

A keyword spotter $f$ is often evaluated using the ROC curve. This curve plots the true positive rate (TPR) as a function of the false positive rate (FPR). The TPR measures the fraction of keyword occurrences correctly spotted, while the FPR measures the fraction of negative utterances yielding a false alarm. The points on the curve are obtained by sweeping the threshold $b$ from the largest value output by the system to the smallest one. These values hence correspond to different trade-offs between the two types of errors a keyword spotter can make, i.e. missing a keyword utterance or raising a false alarm. In order to evaluate a keyword spotter over various trade-offs, it is common to report the AUC as a single value. The AUC hence corresponds to an averaged performance, assuming a flat prior over the different operational settings. Given a keyword $k$, a set of positive utterances $X_k^+$ in which $k$ is uttered, and a set of negative utterances $X_k^-$ in which $k$ is not uttered, the AUC can be written as

$$A_k = \frac{1}{|X_k^+||X_k^-|} \sum_{\bar{\mathbf{x}}^+ \in X_k^+} \sum_{\bar{\mathbf{x}}^- \in X_k^-} \mathbb{1}_{\{f(\bar{p}^k, \bar{\mathbf{x}}^+) > f(\bar{p}^k, \bar{\mathbf{x}}^-)\}},$$

where $|\cdot|$ refers to set cardinality and $\mathbb{1}_{\{\pi\}}$ refers to the indicator function and its value is 1 if the predicate $\pi$ holds and 0 otherwise. The AUC of the keyword $k$, $A_k$, hence estimates the probability that the score assigned to a positive utterance is greater than the score assigned to a negative utterance. This quantity is also referred to as the *Wilcoxon–Mann–Whitney statistics* (Cortes and Mohri 2004; Mann and Whitney 1947; Wilcoxon 1945).

As one is often interested in the expected performance over any keyword, it is common to plot the ROC averaged over a set of evaluation keywords $\mathcal{K}_{\text{test}}$, and to compute the corresponding averaged AUC:

$$A_{\text{test}} = \frac{1}{|\mathcal{K}_{\text{test}}|} \sum_{k \in \mathcal{K}_{\text{test}}} A_k.$$

In this study, we introduce a large-margin approach to learn a keyword spotter $f$ from a training set, which achieves a high averaged AUC.

## 11.3.2   Loss Function and Model Parameterization

In order to build our keyword spotter $f$, we are given training data consisting of a set of training keywords $\mathcal{K}_{\text{train}}$ and a set of training utterances. For each keyword $k \in \mathcal{K}_{\text{train}}$, we denote with $X_k^+$ the set of utterances in which the keyword is spoken and with $X_k^-$ the set of all other utterances, in which the keyword is not spoken. Furthermore, for each positive utterance $\bar{\mathbf{x}}^+ \in X_k^+$, we are also given the timing sequence $\bar{s}^+$ of the keyword phoneme sequence $\bar{p}^k$ in $\bar{\mathbf{x}}^+$. Such a timing sequence provides the start and end points of each of the keyword phonemes, and can either be provided by manual annotators or localized with a forced alignment algorithm, as discussed in Chapter 4. Let us define the training set as $\mathcal{T}_{\text{train}} \equiv \{(p^{k_i}, \bar{\mathbf{x}}_i^+, \bar{s}_i^+, \bar{\mathbf{x}}_i^-)\}_{i=1}^m$. For each keyword in the training set there is only one positive utterance and one negative utterance, hence $|X_k^+| = 1$, $|X_k^-| = 1$ and $|\mathcal{K}_{\text{train}}| = m$, and the AUC over the training set becomes

$$A_{\text{train}} = \frac{1}{m} \sum_{i=1}^m \mathbb{1}_{\{f(\bar{p}^{k_i}, \bar{\mathbf{x}}_i^+) > f(\bar{p}^{k_i}, \bar{\mathbf{x}}_i^-)\}}.$$

The selection of a model maximizing this AUC is equivalent to minimizing the loss:

$$\mathcal{L}^{0/1}(f) = 1 - A_{\text{train}} = \frac{1}{m} \sum_{i=1}^m \mathbb{1}_{\{f(\bar{p}^{k_i}, \bar{\mathbf{x}}_i^+) > f(\bar{p}^{k_i}, \bar{\mathbf{x}}_i^-)\}}.$$

The loss $\mathcal{L}^{0/1}$ is unfortunately not suitable for model training since it is a combinatorial quantity that is difficult to minimize directly. We instead adopt a strategy commonly used in large margin classifiers and employ the convex hinge loss function:

$$\mathcal{L}(f) = \frac{1}{m} \sum_{i=1}^{m} [1 - f(\bar{p}^{k_i}, \bar{\mathbf{x}}_i^{+}) + f(\bar{p}^{k_i}, \bar{\mathbf{x}}_i^{-})]_{+}, \tag{11.1}$$

where $[a]_{+}$ denotes $\max\{0, a\}$. The hinge loss $\mathcal{L}(f)$ upper bounds $\mathcal{L}^{0/1}(f)$: since for any real numbers $a$ and $b$, $[1 - a + b]_{+} \geq \mathbb{1}_{\{a \leq b\}}$, and moreover, when $\mathcal{L}(f) = 0$, then $A_{\text{train}} = 1$, and for any $a$ and $b$, $[1 - a + b]_{+} = 0 \Rightarrow a > b + 1 \Rightarrow a > b$. The hinge loss is related to the ranking loss used in both SVMs for ordinal regression (Herbrich *et al.* 2000) and Ranking SVM (Joachims 2002). These approaches have shown to be successful over highly unbalanced problems, such as information retrieval (Grangier and Bengio 2008; Joachims 2002); using the hinge loss is hence appealing to the keyword spotting problem. The analysis of the relationships between the hinge loss presented in Equation (11.1) and the generalization performance of our keyword spotter is differed to Section 11.3.4, where we show that minimizing the hinge loss yields a keyword spotter likely to attain a high AUC over unseen data.

Our keyword spotter $f$ is parameterized as

$$f_{\mathbf{w}}(\bar{\mathbf{x}}, \bar{p}^k) = \max_{\bar{s}} \mathbf{w} \cdot \boldsymbol{\phi}(\bar{\mathbf{x}}, \bar{p}^k, \bar{s}), \tag{11.2}$$

where $\mathbf{w} \in \mathbb{R}^n$ is a vector of importance weights, $\boldsymbol{\phi}(\bar{\mathbf{x}}, \bar{p}^k, \bar{s})$ is a feature function vector, measuring different characteristics related to the confidence that the phoneme sequence $\bar{p}^k$ representing the keyword $k$ is uttered in $\bar{\mathbf{x}}$ with the time span $\bar{s}$. Formally, $\boldsymbol{\phi}$ is a function defined as $\boldsymbol{\phi} : \mathcal{X}^* \times (\mathcal{P} \times \mathbb{N})^* \to \mathbb{R}^n$. In this study we used seven feature functions ($n = 7$), which are similar to those employed in Chapter 4. These functions are described only briefly for the sake of completeness.

There are four *phoneme transition functions*, which aim at detecting transition between phonemes. For this purpose, they compute the frame distance between the frames before and after a hypothesized transition point. Formally,

$$\forall i = 1, 2, 3, 4, \quad \phi_i(\bar{\mathbf{x}}, \bar{p}^k, \bar{s}) = \frac{1}{L} \sum_{j=2}^{L-1} d(\mathbf{x}_{s_j - i}, \mathbf{x}_{s_j + i}), \tag{11.3}$$

where $d$ refers to the Euclidean distance and $L$ refers to the number of phonemes in keyword $k$.

The *frame based phoneme classifier function* relies on a frame based phoneme classifier to measure the match between each frame and the hypothesized phoneme class,

$$\phi_5(\bar{\mathbf{x}}, \bar{p}^k, \bar{s}) = \frac{1}{L} \sum_{i=1}^{L} \sum_{t=s_i}^{s_{i+1}-1} \frac{1}{s_{i+1} - s_i} g(\mathbf{x}_t, p_i) \tag{11.4}$$

where $g : \mathcal{X} \times \mathcal{P} \to \mathbb{R}$ refers to the phoneme classifier, which returns a confidence that the acoustic feature vector at the $t$th frame, $\mathbf{x}_t$, represents a specific phoneme $p_i$. Different phoneme classifiers might be applied for this feature. In our case, we conduct experiments relying on two alternative solutions. The first assessed classifier is the hierarchical large-margin classifier presented in Dekel *et al.* (2004), while the second classifier is a Bayes classifier with one Gaussian Mixture per phoneme class. In the first case, $g$ is defined as the phoneme confidence output by the classifier, while, in the second case, $g$ is defined as the log posterior of the class $g(\mathbf{x}, p) = \log(P(p|\mathbf{x}))$. The presentation of the training setup, as well as the empirical comparison of both solutions, is deferred to Section 11.4.

The *phoneme duration function* measures the adequacy of the hypothesized segmentation $\bar{s}$, with respect to a duration model,

$$\phi_6(\bar{\mathbf{x}}, \bar{p}^k, \bar{s}) = \frac{1}{L} \sum_{i=1}^{L} \log \mathcal{N}(s_{i+1} - s_i; \mu_{p_i}, \sigma_{p_i}^2), \tag{11.5}$$

where $\mathcal{N}$ denotes the likelihood of a Gaussian duration model, whose mean $\mu_p$ and variance $\sigma_p^2$ parameters for each phoneme $p$ are estimated over the training data.

The *speaking rate function* measures the stability of the speaking rate,

$$\phi_7(\bar{\mathbf{x}}, \bar{p}^k, \bar{s}) = \frac{1}{L} \sum_{i=2}^{L} (r_i - r_{i-1})^2, \tag{11.6}$$

where $r_i$ denotes the estimate of the speaking rate for the $i$th phoneme,

$$r_i = \frac{s_{i+1} - s_i}{\mu_{p_i}}.$$

This set of seven functions has been used in our experiments. Of course, this set can easily be extended to incorporate further features, such as confidences from a triphone frame based classifier or the output of a more refined duration model.

In other words, our keyword spotter outputs a confidence score by maximizing a weighted sum of feature functions over all possible time spans. This maximization corresponds to a search over an exponentially large number of time spans. Nevertheless, it can be performed efficiently by selecting decomposable feature functions, which allows the application of dynamic programming techniques, like those used in HMMs (Rabiner and Juang 1993). Chapter 4 gives a detailed discussion about the efficient computation of Equation (11.2).

### 11.3.3  An Iterative Training Algorithm

In this section we describe an iterative algorithm for finding the weight vector $\mathbf{w}$. We show in the sequel that the weight vector $\mathbf{w}$ found in this process minimizes the loss $\mathcal{L}(f_{\mathbf{w}})$, hence minimizes the loss $\mathcal{L}^{0/1}$ and in turn resulted in a keyword spotting which attains a high AUC over the training set. We also show that the learned weight vector has good generalization properties on the test set.

The procedure starts by initializing the weight vector to be the zero vector, $\mathbf{w}_0 = 0$. Then, at iteration $i \geq 1$, the algorithm examines the $i$th training example $(\bar{p}^{k_i}, \bar{\mathbf{x}}_i^+, \bar{s}_i^+, \bar{\mathbf{x}}_i^-)$. The algorithm first predicts the best time span of the keyword phoneme sequence $\bar{p}^{k_i}$ in the negative utterance $\bar{\mathbf{x}}_i^-$:

$$\bar{s}_i^- = \arg\max_{\bar{s}} \mathbf{w}_{i-1} \cdot \phi(\bar{\mathbf{x}}_i^-, \bar{p}_i^k, \bar{s}). \tag{11.7}$$

Then, the algorithm considers the loss on the $i$th training example and checks that the difference between the score assigned to the positive utterance and the score assigned to the negative example is greater than 1. Formally, define

$$\Delta\phi_i = \phi(\bar{\mathbf{x}}_i^+, \bar{p}_i^k, \bar{s}_i^+) - \phi(\bar{\mathbf{x}}_i^-, \bar{p}_i^k, \bar{s}_i^-).$$

If $\mathbf{w}_{i-1} \cdot \Delta\phi_i \geq 1$ the algorithm keeps the weight vector for the next iteration, namely, $\mathbf{w}_i = \mathbf{w}_{i-1}$. Otherwise, the algorithm updates the weight vector to minimize the following optimization problem:

$$\mathbf{w}_i = \arg\min_{\mathbf{w}} \frac{1}{2}\|\mathbf{w} - \mathbf{w}_{i-1}\|^2 + c[1 - \mathbf{w} \cdot \Delta\phi_i]_+, \tag{11.8}$$

where the hyperparameter $c \geq 1$ controls the trade-off between keeping the new weight vector close to the previous one and satisfying the constraint for the current example. Equation (11.8) can analytically be solved in closed form (Crammer *et al.* 2006), yielding

$$\mathbf{w}_i = \mathbf{w}_{i-1} + \alpha_i \Delta\phi_i,$$

where

$$\alpha_i = \min\left\{c, \frac{[1 - \mathbf{w}_{i-1} \cdot \Delta\phi_i]_+}{\|\Delta\phi_i\|^2}\right\}. \tag{11.9}$$

This update is referred to as *passive–aggressive*, since the algorithm *passively* keeps the previous weight ($\mathbf{w}_i = \mathbf{w}_{i-1}$) if the loss of the current training example is already zero ($[1 - \mathbf{w}_{i-1} \cdot \Delta\phi_i]_+ = 0$), while it *aggressively* updates the weight vector to compensate this loss otherwise. At the end of the training procedure, when all training examples have been visited, the best weight $\mathbf{w}$ among $\{\mathbf{w}_0, \ldots, \mathbf{w}_m\}$ is selected over a set of validation examples $\mathcal{T}_{\text{valid}}$. The hyperparameter $c$ is also selected to optimize performance on the validation data. The pseudo-code of the algorithm is given in Figure 11.4.

## 11.3.4 Analysis

In this section, we derive theoretical bounds on the performance of our keyword spotter. Let us first define the *cumulative AUC* on the training set $\mathcal{T}_{\text{train}}$ as follows:

$$\hat{A}_{\text{train}} = \frac{1}{m} \sum_{i=1}^{m} \mathbb{1}_{\{\mathbf{w}_{i-1}\cdot\phi(\bar{\mathbf{x}}_i^+, \bar{p}^{k_i}, \bar{s}_i^+) > \mathbf{w}_{i-1}\cdot\phi(\bar{\mathbf{x}}_i^-, \bar{p}^{k_i}, \bar{s}_i^-)\}}, \tag{11.10}$$

where $\bar{s}_i^-$ is generated every iteration step according to Equation (11.7). The examination of the cumulative AUC is of great interest as it provides an estimator for the generalization performance. Note that at each iteration step the algorithm receives new example $(\bar{p}^{k_i}, \bar{\mathbf{x}}_i^+, \bar{s}_i^+, \bar{\mathbf{x}}_i^-)$ and predicts the time span of the keyword in the negative instance $\bar{\mathbf{x}}_i^-$ using the previous weight vector $\mathbf{w}_{i-1}$. Only after the prediction is made the algorithm suffers loss by comparing its prediction to the true time span $\bar{s}_i^+$ of the keyword on the positive utterance $\bar{\mathbf{x}}_i^+$. The cumulative AUC is a weighted sum of the performance of the algorithm on the next unseen training example and hence it is a good estimation of the performance of the algorithm on unseen data during training.

Our first theorem states a competitive bound. It compares the cumulative AUC of the weight vectors series, $\{\mathbf{w}_1, \ldots, \mathbf{w}_m\}$, resulted from the iterative algorithm to the best fixed weight vector, $\mathbf{w}^\star$, chosen in hindsight, and essentially proves that, for any sequence of

---

**Input**: Training set $\mathcal{T}_{\text{train}}$, validation set $\mathcal{T}_{\text{valid}}$; parameter $c$.

**Initialize**: $\mathbf{w}_0 = 0$.

**Loop**: for each $(p^{k_i}, \bar{\mathbf{x}}_i^+, \bar{s}_i^+, \bar{\mathbf{x}}_i^-) \in \mathcal{T}_{\text{train}}$

    1. let $\bar{s}_i^- = \arg\max_{\bar{s}} \mathbf{w}_{i-1} \cdot \boldsymbol{\phi}(\bar{\mathbf{x}}_i^-, \bar{p}^{k_i}, \bar{s})$

    2. let $\Delta\boldsymbol{\phi}_i = \boldsymbol{\phi}(\bar{\mathbf{x}}_i^+, \bar{p}^{k_i}, \bar{s}_i^+) - \boldsymbol{\phi}(\bar{\mathbf{x}}_i^-, \bar{p}^{k_i}, \bar{s}_i^-)$

    3. if $\mathbf{w}_{i-1} \cdot \Delta\boldsymbol{\phi}_i < 1$ then

        let $\alpha_i = \min\left\{c, \dfrac{1 - \mathbf{w}_{i-1} \cdot \Delta\boldsymbol{\phi}_i}{\|\Delta\boldsymbol{\phi}_i\|^2}\right\}$

        update $\mathbf{w}_i = \mathbf{w}_{i-1} + \alpha_i \cdot \Delta\boldsymbol{\phi}_i$

**Output**: $\mathbf{w}$ achieving the highest AUC over $\mathcal{T}_{\text{valid}}$:

$$\mathbf{w} = \arg \min_{\mathbf{w} \in \{\mathbf{w}_1, \ldots, \mathbf{w}_m\}} \frac{1}{m_{\text{valid}}} \sum_{j=1}^{m_{\text{valid}}} \mathbb{1}_{\{\max_{\bar{s}^+} \mathbf{w} \cdot \boldsymbol{\phi}(\bar{\mathbf{x}}_j^+, \bar{p}^{k_j}, \bar{s}^+) > \max_{\bar{s}^-} \mathbf{w} \cdot \boldsymbol{\phi}(\bar{\mathbf{x}}_j^-, \bar{p}^{k_j}, \bar{s}^-)\}}$$

---

Figure 11.4   Passive–aggressive training.

examples, our algorithms cannot do much worse than the best fixed weight vector. Formally, it shows that the cumulative area *above* the curve, $1 - \hat{A}_{\text{train}}$, is smaller than the weighted average loss $\mathcal{L}(f_{\mathbf{w}^\star})$ of the best fixed weight vector $\mathbf{w}^\star$ and its weighted complexity, $\|\mathbf{w}^\star\|$. That is, the cumulative AUC of the iterative training algorithm is going to be high, given that the loss of the best solution is small, the complexity of the best solution is small and that the number of training examples, $m$, is sufficiently large.

**Theorem 11.3.1** *Let* $\mathcal{T}_{\text{train}} = \{(\bar{p}^{k_i}, \bar{\mathbf{x}}_i^+, \bar{s}_i^+, \bar{\mathbf{x}}_i^-)\}_{i=1}^m$ *be a set of training examples and assume that for all* $k$, $\bar{\mathbf{x}}$ *and* $\bar{s}$ *we have that* $\|\boldsymbol{\phi}(\bar{\mathbf{x}}, \bar{p}^k, \bar{s})\| \leq 1/\sqrt{2}$. *Let* $\mathbf{w}^\star$ *be the best weight vector selected under some optimization criterion by observing all instances in hindsight. Let* $\mathbf{w}_1, \ldots, \mathbf{w}_m$ *be the sequence of weight vectors obtained by the algorithm in Figure 11.4 given the training set* $\mathcal{T}_{\text{train}}$. *Then,*

$$1 - \hat{A}_{\text{train}} \leq \frac{1}{m} \|\mathbf{w}^\star\|^2 + \frac{2c}{m} \mathcal{L}(f_{\mathbf{w}^\star}) \tag{11.11}$$

*where* $c \geq 1$ *and* $\hat{A}_{\text{train}}$ *is the cumulative AUC defined in Equation 11.10.*

*Proof.* Denote by $\ell_i(\mathbf{w})$ the instantaneous loss the weight vector $\mathbf{w}$ suffers on the $i$th example, that is,

$$\ell_i(\mathbf{w}) = [1 - \mathbf{w} \cdot \boldsymbol{\phi}(\bar{\mathbf{x}}_i^+, \bar{p}^{k_i}, \bar{s}_i^+) + \max_{\bar{s}} \mathbf{w} \cdot \boldsymbol{\phi}(\bar{\mathbf{x}}_i^-, \bar{p}^{k_i}, \bar{s})]_+.$$

The proof of the theorem relies on Lemma 1 and Theorem 4 in Crammer *et al.* (2006). Lemma 1 in Crammer *et al.* (2006) implies that

$$\sum_{i=1}^m \alpha_i (2\ell_i(\mathbf{w}_{i-1}) - \alpha_i \|\Delta\boldsymbol{\phi}_i\|^2 - 2\ell_i(\mathbf{w}^\star)) \leq \|\mathbf{w}^\star\|^2. \tag{11.12}$$

Now if the algorithm makes a prediction mistake and the predicted confidence of the best time span of the keyword in a negative utterance is higher than the confidence of the true time span of the keyword in the positive example, then $\ell_i(\mathbf{w}_{i-1}) \geq 1$. Using the assumption that $\|\boldsymbol{\phi}(\bar{\mathbf{x}}, \bar{p}^k, \bar{s})\| \leq 1/\sqrt{2}$, which means that $\|\Delta\boldsymbol{\phi}(\bar{\mathbf{x}}, \bar{p}^k, \bar{s})\|^2 \leq 1$, and the definition of $\alpha_i$ given in Equation 11.9, when substituting $[1 - \mathbf{w}_{i-1} \cdot \Delta\boldsymbol{\phi}_i]_+$ for $\ell_i(\mathbf{w}_{i-1})$ in its numerator, we conclude that if a prediction mistake occurs then it holds that

$$\alpha_i \ell_i(\mathbf{w}_{i-1}) \geq \min\left\{\frac{\ell_i(\mathbf{w}_{i-1})}{\|\Delta\boldsymbol{\phi}_i\|^2}, c\right\} \geq \min\{1, c\} = 1. \tag{11.13}$$

Summing over all the prediction mistakes made on the entire training set $\mathcal{T}_{\text{train}}$ and taking into account that $\alpha_i \ell_i(\mathbf{w}_{i-1})$ is always non-negative, we have

$$\sum_{i=1}^{m} \alpha_i \ell_i(\mathbf{w}_{i-1}) \geq \sum_{i=1}^{m} \mathbb{1}_{\{\mathbf{w}_{i-1} \cdot \boldsymbol{\phi}(\bar{\mathbf{x}}_i^+, \bar{p}^{k_i}, \bar{s}_i^+) \leq \mathbf{w}_{i-1} \cdot \boldsymbol{\phi}(\bar{\mathbf{x}}_i^-, \bar{p}^{k_i}, \bar{s}_i^-)\}}. \tag{11.14}$$

Again using the definition of $\alpha_i$, we know that $\alpha_i \ell_i(\mathbf{w}^\star) \leq c\ell_i(\mathbf{w}^\star)$ and that $\alpha_i \|\Delta\boldsymbol{\phi}_i\|^2 \leq \ell_i(\mathbf{w}_{i-1})$. Plugging these two inequalities and Equation (11.14) into Equation (11.12) we get

$$\sum_{i=1}^{m} \mathbb{1}_{\{\mathbf{w}_{i-1} \cdot \boldsymbol{\phi}(\bar{\mathbf{x}}_i^+, \bar{p}^{k_i}, \bar{s}_i^+) \leq \mathbf{w}_{i-1} \cdot \boldsymbol{\phi}(\bar{\mathbf{x}}_i^-, \bar{p}^{k_i}, \bar{s}_i^-)\}} \leq \|\mathbf{w}^\star\|^2 + 2c \sum_{i=1}^{m} \ell_i(\mathbf{w}^\star). \tag{11.15}$$

The theorem follows by replacing the sum over prediction mistakes to a sum over prediction hits and plugging the definition of the cumulative AUC given in Equation (11.10). □

The next theorem states that the output of our algorithm is likely to have good generalization, namely, the expected value of the AUC resulted from decoding on unseen test set is likely to be large.

**Theorem 11.3.2** *Under the same conditions of Theorem 11.3.1 assume that the training set $\mathcal{T}_{\text{train}}$ and the validation set $\mathcal{T}_{\text{valid}}$ are both sampled i.i.d. from a distribution $\mathcal{D}$. Denote by $m_{\text{valid}}$ the size of the validation set. With probability of at least $1 - \delta$ we have*

$$1 - A = \mathbb{E}_{\mathcal{D}}[\mathbb{1}_{\{f(\bar{\mathbf{x}}^+, \bar{p}^k) \leq f(\bar{\mathbf{x}}^-, \bar{p}^k)\}}] = \Pr_{\mathcal{D}}[f(\bar{\mathbf{x}}^+, \bar{p}^k) \leq f(\bar{\mathbf{x}}^-, \bar{p}^k)]$$

$$\leq \frac{1}{m}\sum_{i=1}^{m} \ell_i(\mathbf{w}^\star) + \frac{\|\mathbf{w}^\star\|^2}{m} + \frac{\sqrt{2\ln(2/\delta)}}{\sqrt{m}} + \frac{\sqrt{2\ln(2m/\delta)}}{\sqrt{m_{\text{valid}}}}, \tag{11.16}$$

*where A is the mean AUC defined as $A = \mathbb{E}_{\mathcal{D}}[\mathbb{1}_{\{f(\bar{\mathbf{x}}^+, \bar{p}^k) > f(\bar{\mathbf{x}}^-, \bar{p}^k)\}}]$ and*

$$\ell_i(\mathbf{w}) = \left[1 - \mathbf{w} \cdot \boldsymbol{\phi}(\bar{\mathbf{x}}_i^+, \bar{p}^{k_i}, \bar{s}_i^+) + \max_{\bar{s}} \mathbf{w} \cdot \boldsymbol{\phi}(\bar{\mathbf{x}}_i^-, \bar{p}^{k_i}, \bar{s})\right]_+.$$

The proof of the theorem goes along the lines of the proof of Theorem 4.5.2 in Chapter 4. The theorem states that the resulted $\mathbf{w}$ of the iterative algorithm generalizes, with high probability, and is going to have high expected AUC on unseen test data.

## 11.4   Experiments and Results

We started by training the iterative algorithm on the TIMIT training set. We then conducted two types of experiments to evaluate the effectiveness of the proposed discriminative method. First, we compared the performance of the discriminative method with a standard monophone HMM keyword spotter on the TIMIT test set. Second, we compared the robustness of both the discriminative method and the monophone HMM with respect to changing recording conditions by using the models trained on the TIMIT, evaluated on the Wall Street Journal (WSJ) corpus.

### 11.4.1   The TIMIT Experiments

The TIMIT corpus (Garofolo 1993) consists of read speech from 630 American speakers, with 10 utterances per speaker. The corpus provides manually aligned phoneme and word transcriptions for each utterance. It also provides a standard split into training and test data. From the training part of the corpus, we extracted three disjoint sets consisting of 1500, 300 and 200 utterances. The first set was used as the training set of the phoneme classifier and was used by our fifth feature function $\phi_5$. The second set was used as the training set for our discriminative keyword spotter, while the third set was used as the validation set to select the hyperparameter $c$ and the best weight vector $\mathbf{w}$ seen during training. The test set was solely used for evaluation purposes. From each of the last two splits of the training set, 200 words of length greater than or equal to four phonemes were chosen in random. From the test set 80 words were chosen in random as described below.

Mel-frequency cepstral coefficients (MFCC), along with their first ($\Delta$) and second derivatives ($\Delta\Delta$), were extracted every 10 ms. These features were used by the first five feature functions $\phi_1, \ldots, \phi_5$. Two types of phoneme classifier were used for the fifth feature function $\phi_5$, namely, a large margin phoneme classifier (Dekel *et al.* 2004) and a GMM model. Both classifiers were trained to predict 39 phoneme classes (Lee and Hon 1989) over the first part of the training set. The large margin classifier corresponds to a hierarchical classifier with Gaussian kernel, as presented in Dekel *et al.* (2004), where the score assigned to each frame for a given phoneme was used as the function $g$ in Equation (11.4). The GMM model corresponded to a Bayes classifier combining one GMM per class and the phoneme prior probabilities, both learned from the training data. In that case, the log posterior of a phoneme given the frame vector was used as the function $g$ in Equation (11.4). The hyperparameters of both phoneme classifiers were selected to maximize the frame accuracy over part of the training data held out during parameter fitting. In the following, the discriminative keyword spotter relying on the features from the hierarchical phoneme classifier is referred to as *Discriminative/Hier*, while the model relying on the GMM log posteriors is referred to as *Discriminative/GMM*.

We compared the results of both Discriminative/Hier and Discriminative/GMM with a monophone HMM baseline, in which each phoneme was modeled with a left–right HMM of five emitting states. The density of each state was modeled with a 40-Gaussian GMM. Training was performed over the whole TIMIT training set. *Embedded training* was applied, i.e. after an initial training phase relying on the provided phoneme alignment, a second training phase which dynamically determines the most likely alignment was applied. The hyperparameters of this model (the number of states per phoneme, the number of Gaussians

Table 11.1 Area under the ROC curve (AUC) of different models trained on the TIMIT training set and evaluated on the TIMIT test set (the higher the better)

| Model | AUC |
|---|---|
| HMM/Viterbi | 0.942 |
| HMM/Ratio | 0.952 |
| Discriminative/GMM | 0.971 |
| Discriminative/Hier | 0.996 |

per state, as well as the number of expectation-maximization iterations) were selected to maximize the likelihood of a held-out validation set.

The phoneme models of the trained HMM were then used to build a keyword spotting HMM, composed of two sub-models: the keyword model and the garbage model, as illustrated in Figure 11.1. The keyword model was an HMM which estimated the likelihood of an acoustic sequence given that the sequence represented the keyword phoneme sequence. The garbage model was an HMM composed of all phoneme HMMs fully connected to each other, which estimated the likelihood of any phoneme sequence. The overall HMM fully connected the keyword model and the garbage model. The detection of a keyword in a given utterance was performed by checking whether the Viterbi best path passes through the keyword model, as explained in Section 11.2. In this model, the keyword transition probability set the trade-off between the true positive rate and the ROC curve was plotted by varying this probability. This model is referred to as *HMM/Viterbi*.

We also experimented with an alternative decoding strategy, in which the system output the ratio of the likelihood of the acoustic sequence knowing the keyword was uttered versus the likelihood of the sequence knowing the keyword was *not* uttered, as discussed in Section 11.2. In this case, the first likelihood was determined by an HMM forcing an occurence of the keyword, and the second likelihood was determined by the garbage model, as illustrated in Figure 11.2. This likelihood-ratio strategy is referred to as *HMM/Ratio* in the following.

The evaluation of discriminative and HMM-based models was performed over 80 keywords, randomly selected among the words occurring in the TIMIT test set. This random sampling of the keyword set aimed at evaluating the expected performance over any keyword. For each keyword $k$, we considered a spotting problem, which consisted of a set of positive utterances $X_k^+$ and a set of negative utterance $X_k^-$. Each positive set $X_k^+$ contained between 1 and 20 sequences, depending on the number of occurrences of $k$ in the TIMIT test set. Each negative set contained 20 sequences, randomly sampled among the utterances of TIMIT which do not contain $k$. This setup represented an unbalanced problem, with only 10% of the sequences being labeled as positive.

Table 11.1 reports the average AUC results of the 80 test keywords, for different models trained on the TIMIT training set and evaluated on the TIMIT test set. These results show the advantage of our discriminative approach. The two discriminative models outperform the two HMM-based models. The improvement introduced by our discriminative model algorithm can be observed when comparing the performance of Discriminative/GMM with the performance of the HMM spotters. In that case, both spotters rely on GMMs to estimate the frame likelihood given a phoneme class. In our case we use that probability to compute the feature $\phi_5$, while the HMM uses it as the state emission probability.

Table 11.2 The distribution of the 80 keywords among the models which better spotted them. Each row in the table represents the keywords for which the model written at the beginning of the row received the highest area under the ROC curve. The models were trained on the TIMIT training set and evaluated on the TIMIT test set

| Best model | Keywords |
| --- | --- |
| Discriminative/Hier | absolute admitted apartments apparently argued controlled depicts dominant drunk efficient followed freedom introduced millionaires needed obvious radiation rejected spilled street superb sympathetically weekday (**23 keywords**) |
| HMM/Ratio | materials (**1 keyword**) |
| No differences | aligning anxiety bedrooms brand camera characters cleaning climates creeping crossings crushed decaying demands dressy episode everything excellent experience family firing forgiveness fulfillment functional grazing henceforth ignored illnesses imitate increasing inevitable January mutineer package paramagnetic patiently pleasant possessed pressure recriminations redecorating secularist shampooed solid spreader story strained streamlined stripped stupid surface swimming unenthusiastic unlined urethane usual walking (**56 keywords**) |

Moreover, our keyword spotter can benefit from non-probabilistic frame based classifiers, as illustrated with *Discriminative/Hier*. This model relies on the output of a large margin classifier, which outperforms all other models, and reaches a mean AUC of 0.996. In order to verify whether the differences observed on averaged AUC could be due only to a few keywords, we applied the Wilcoxon test (Rice 1995) to compare the results of both HMM approaches (HMM/Viterbi and HMM/Ratio) with the results of both discriminative approaches (Discriminative/GMM and Discriminative/Hier). At the 90% confidence level, the test rejected this hypothesis, showing that the performance gain of the discriminative approach is consistent over the keyword set.

Table 11.2 further presents the performance per keyword and compares the results of the best HMM configuration, HMM/Ratio with the performance of the best discriminative configuration, Discriminative/Hier. Out of a total of 80 keywords, 23 keywords were better spotted with the discriminative model, one keyword was better spotted with the HMM, and both models yielded the same spotting accuracy for 56 keywords. The discriminative model seems to be better for shorter keywords, as it outperforms the HMM for most of the keywords of five phonemes or less (e.g. *drunk, spilled, street*).

## 11.4.2    The WSJ Experiments

WSJ (Paul and Baker 1992) is a large corpus of American English. It consists of read and spontaneous speech corresponding to the reading and the dictation of articles from the *Wall Street Journal*. In the following, all models were trained on the TIMIT training set and evaluated on the `si_tr_s` subset of WSJ. This subset corresponds to the recordings of 200 speakers. Compared with TIMIT, this subset introduces several variations, both regarding the type of sentences recorded and the recording conditions (Paul and Baker 1992).

Table 11.3 Area under the ROC curve (AUC) of different models trained on the TIMIT training set and evaluated on the `si_tr_s` subset of WSJ (the higher the better)

| Model | AUC |
|---|---|
| HMM/Viterbi | 0.868 |
| HMM/Ratio | 0.884 |
| Discriminative/GMM | 0.922 |
| Discriminative/Hier | 0.914 |

These experiments hence evaluate the robustness of the different approaches when they encounter differing conditions for training and testing. As for TIMIT, the evaluation is performed over 80 keywords randomly selected from the corpus transcription. For each keyword $k$, the evaluation was performed over a set $X_k^+$, containing between 1 and 20 positive sequences, and a $X_k^-$, containing 20 randomly selected negative sequences. This setup also represents an unbalanced problem, with 27% of the sequences being labeled as positive.

Table 11.3 reports the average AUC results of the 80 test keywords, for different models trained on the TIMIT training set and evaluated on the `si_tr_s` subset of WSJ. Overall, the results show that the differences between the TIMIT training conditions and the WSJ test conditions affect the performance of all models. However, the measured performance still yields acceptable performance in all cases (AUC of 0.868 in the worst case). Comparing the individual model performance, the WSJ results confirm the conclusions of TIMIT experiments and the discriminative spotters outperform the HMM-based alternatives. For the HMM models, HMM/Ratio outperforms HMM/Viterbi as in the TIMIT experiments. For the discriminative spotters, Discriminative/GMM outperforms Discriminative/Hier, which was not the case over TIMIT. Since these two models only differ in the frame based classifier used as the feature function $\phi_5$, this result certainly indicates that the hierarchical frame based classifier on which Discriminative/Hier relies is less robust to the acoustic condition changes than the GMM alternative. As with TIMIT, we checked whether the differences observed on the whole set could be due to a few keywords. The Wilcoxon test rejected this hypothesis at the 90% confidence level, for the four tests comparing Discriminative/GMM and Discriminative/Hier with HMM/Viterbi and HMM/Hier.

We further compared the best discriminative spotter, Discriminative/GMM, and the best HMM spotter HMM/Ratio over each keyword. These results are summarized in Table 11.4. Out of the 80 keywords, the discriminative model outperforms the HMM for 50 keywords, the HMM outperforms the discriminative model for 20 keywords and both models yield the same results for 10 keywords. As with the TIMIT experiments, the discriminative model is shown to be especially advantageous for short keywords, with five phonemes or less (e.g. *Adams, kings, serving*).

Overall, the experiments over both WSJ and TIMIT highlight the advantage of our discriminative learning method.

# 11.5 Conclusions

This chapter introduced a discriminative method for the keyword spotting problem. In this task, the model receives a keyword and a spoken utterance as input and should decide whether

Table 11.4 The distribution of the 80 keywords among the models which better spotted them. Each row in the table represents the keywords for which the model written at the beginning of the row received the highest area under the ROC curve. The models were trained on the TIMIT training set but evaluated on the `si_tr_s` subset of WSJ

| Best model | Keywords |
| --- | --- |
| Discriminative/Hier | Adams additions Allen Amerongen apiece buses Bushby Colombians consistently cracked dictate drop fantasy fills gross Higa historic implied interact kings list lobby lucrative measures Melbourne millions Munich nightly observance owning plus proudly queasy regency retooling Rubin scramble Seidler serving significance sluggish strengthening Sutton's tariffs Timberland today truths understands withhold Witter's **(50 keywords)** |
| HMM/Ratio | artificially Colorado elements Fulton itinerary longer lunchroom merchant mission multilateral narrowed outlets Owens piper replaced reward sabotaged shards spurt therefore **(20 keywords)** |
| No differences | aftershocks Americas farms Flamson hammer homosexual philosophically purchasers sinking steel-makers **(10 keywords)** |

the keyword is uttered in the utterance. Keyword spotting corresponds to an unbalanced detection problem, since, in standard setups, most tested utterances do not contain the targeted keyword. In that unbalanced context, the AUC is generally used for evaluation. This work proposed a learning algorithm, which aims at maximizing the AUC over a set of training spotting problems. Our strategy is based on a large margin formulation of the task, and relies on an efficient iterative training procedure. The resulting model contrasts with standard approaches based on HMMs, for which the training procedure does not rely on a loss directly related to the spotting task. Compared with such alternatives, our model is shown to yield significant improvements over various spotting problems on the TIMIT and the WSJ corpus. For instance, the best HMM configuration over TIMIT reaches an AUC of 0.953, compared with an AUC of 0.996 for the best discriminative spotter.

Several potential directions of research can be identified from this work. In its current configuration, our keyword spotter relies on the output of a pre-trained frame based phoneme classifier. It would be of great interest to learn the frame based classifier and the keyword spotter jointly, so that all model parameters are selected to maximize the performance on the final spotting task.

Also, our work currently represents keywords as sequence of phonemes, without considering the neighboring context. Possible improvement might result from the use of phonemes in context, such as triphones. We hence plan to investigate the use of triphones in a discriminative framework, and to compare the resulting model with triphone-based HMMs. More generally, our model parameterization offers greater flexibility to incorporate new features compared with probabilistic approaches such as HMMs. Therefore, in addition to triphones, features extracted from the speaker identity, the channel characteristics or the linguistic context could possibly be included to improve performance.

# Acknowledgments

This research was partly performed while David Grangier was visiting Google Inc. (Mountain View, USA), and while Samy Bengio was with the IDIAP Research Institute (Martigny, Switzerland). This research was supported by the European PASCAL Network of Excellence and the DIRAC project.

# References

Bahl LR, Brown PF, de Souza P and Mercer RL 1986 Maximum mutual information estimation of hidden Markov model parameters for speech recognition. *Proceedings of the International Conference on Acoustics, Speech, and Signal Processing (ICASSP)*. IEEE Computer Society.

Benayed Y, Fohr D, Haton JP and Chollet G 2003 Confidence measures for keyword spotting using support vector machines. *Proceedings of the International Conference on Acoustics, Speech, and Signal Processing (ICASSP)*. IEEE Computer Society.

Benayed Y, Fohr D, Haton JP and Chollet G 2004 Confidence measure for keyword spotting using support vector machines. *Proceedings of the International Conference on Audio, Speech and Signal Processing*, pp. 588–591.

Bilmes JA 1998 A gentle tutorial of the EM algorithm and its application to parameter estimation for gaussian mixture and hidden Markov models. International Computer Science Institute, Berkeley, CA, USA. *Technical Report TR-97-021*.

Boite JM, Bourlard H, D'hoore B and Haesen M 1993 Keyword recognition using template concatenation. *Proceedings of the European Conference on Speech and Communication Technologies (EUROSPEECH)*. International Speech Communication Association.

Bridle JS 1973 An efficient elastic-template method for detecting given words in running speech. *Proceedings of the British Acoustic Society Meeting*. British Acoustic Society.

Chang E 1995 Improving word spotting performance with limited training. Data Massachusetts Institute of Technology (MIT), PhD thesis.

Cortes C and Mohri M 2004 Confidence intervals for the area under the ROC curve. *Advances in Neural Information Processing Systems (NIPS)*. MIT Press.

Crammer K, Dekel O, Keshet J, Shalev-Shwartz S and Singer Y 2006 Online passive aggressive algorithms. *Journal of Machine Learning Research* **7**.

Dekel O, Keshet J and Singer Y 2004 Online algorithm for hierarchical phoneme classification. *Workshop on Multimodal Interaction and Related Machine Learning Algorithms; Lecture Notes in Computer Science*. Springer-Verlag, pp. 146–159.

Garofolo JS 1993 TIMIT acoustic-phonetic continuous speech corpus. Linguistic Data Consortium, Philadelphia, PA, USA. *Technical Report LDC93S1*.

Grangier D and Bengio S 2008 A discriminative kernel-based model to rank images from text queries. *IEEE Transactions on Pattern Analysis and Machine Intelligence (TPAMI)*.

Herbrich R, Graepel T and Obermayer K 2000 Large margin rank boundaries for ordinal regression, in *Advances in Large Margin Classifiers* (eds. A Smola, B Schölkopf and D Schuurmans). MIT Press.

Higgins AL and Wohlford RE 1985 Keyword recognition using template concatenation. *Proceedings of the International Conference on Acoustics, Speech, and Signal Processing (ICASSP)*. IEEE Computer Society.

Joachims T 2002 Optimizing search engines using clickthrough data. *Proceedings of the ACM Conference on Knowledge Discovery and Data Mining (KDD)*. Association for Computing Machinery.

Junkawitsch J, Ruske G and Hoege H 1997 Efficient methods in detecting keywords in continuous speech. *Proceedings of the European Conference on Speech and Communication Technologies (EUROSPEECH)*. International Speech Communication Association.

Kawabata T, Hanazawa T and Shikano K 1988 Word spotting method based on HMM phoneme recognition. *Journal of the Acoustical Society of America (JASA)* **1**(84).

Ketabdar H, Vepa J, Bengio S and Bourlard H 2006 Posterior based keyword spotting with a priori thresholds *Proceedings of the InterSpeech Conference*. International Speech Communication Association.

Lee KF and Hon HF 1988 Large-vocabulary speaker-independent continuous speech recognition using HMM. *Proceedings of the International Conference on Acoustics, Speech, and Signal Processing (ICASSP)*. IEEE Computer Society.

Lee KF and Hon HW 1989 Speaker independent phone recognition using hidden Markov models. *IEEE Transactions on Acoustics, Speech and Signal Processing (TASSP)* **11**(37).

Mann H and Whitney D 1947 On a test of whether one of two random variables is stochastically larger than the other. *Annals of Mathematical Statistics* **1**(18).

Paul D and Baker J 1992 The design for the Wall Street Journal-based CSR corpus. *Proceedings of the Human Language Technology Conference (HLT)*. Morgan Kaufmann.

Rabiner L and Juang B 1993 *Fundamentals of Speech Recognition*. Prentice-Hall, Upper Saddle River, NJ, USA.

Rice J 1995 *Rice, Mathematical Statistics and Data Analysis*. Duxbury Press.

Rose RC and Paul DB 1990 A hidden Markov model based keyword recognition system. *Proceedings of the International Conference on Acoustics, Speech, and Signal Processing (ICASSP)*. IEEE Computer Society.

Sandness ED and Hetherington IL 2000 Keyword-based discriminative training of acoustic models. *Proceedings of the International Conference on Spoken Language Processing (ICSLP)*. IEEE Computer Society.

Silaghi MC and Bourlard H 1999 Iterative posterior-based keyword spotting without filler models. *Proceedings of the IEEE Automatic Speech Recognition and Understanding Workshop*, Keystone, USA. IEEE Computer Society, pp. 213–216.

Sukkar RA, Seltur AR, Rahim MG and Lee CH 1996 Utterance verification of keyword strings using word-based minimum verification error training. *Proceedings of the International Conference on Acoustics, Speech, and Signal Processing (ICASSP)*. IEEE Computer Society.

Weintraub M 1993 Keyword spotting using SRI's DECIPHER large vocabulary speech recognition system. *Proceedings of the International Conference on Acoustics, Speech, and Signal Processing (ICASSP)*. IEEE Computer Society.

Weintraub M 1995 LVCSR log-likelihood ratio scoring for keyword spotting. *Proceedings of the International Conference on Acoustics, Speech, and Signal Processing (ICASSP)*. IEEE Computer Society.

Weintraub M, Beaufays F, Rivlin Z, Konig Y and Stolcke A 1997 Neural-network based measures of confidence for word recognition. *Proceedings of the International Conference on Acoustics, Speech, and Signal Processing (ICASSP)*. IEEE Computer Society.

Wilcoxon F 1945 Individual comparisons by ranking methods. *Biometrics Bulletin* **1**.

Wilpon JG, Rabiner LR, Lee CH and Goldman ER 1990 Automatic recognition of keywords in unconstrained speech using hidden Markov models. *IEEE Transactions on Acoustics, Speech and Signal Processing (TASSP)* **38**(11).

# 12

# Kernel-based Text-independent Speaker Verification

## Johnny Mariéthoz, Samy Bengio and Yves Grandvalet

The goal of a person authentication system is to certify/attest the claimed identity of a user. When this authentication is based on the voice of the user, without respect to what the user exactly said, the system is called a text-independent speaker verification system.

Speaker verification systems are increasingly often used to secure personal information, particularly for mobile phone based applications. Furthermore, text-independent versions of speaker verification systems are most used for their simplicity, as they do not require complex speech recognition modules. The most common approach to this task is based on Gaussian Mixture Models (GMMs) (Reynolds *et al.* 2000), which do not take into account any temporal information. GMMs have been intensively used thanks to their good performance, especially with the use of the Maximum a posteriori (MAP) (Gauvain and Lee 1994) adaptation algorithm. This approach is based on the density estimation of an impostor data distribution, followed by its adaptation to a specific client data set. Note that the estimation of these densities is not the final goal of speaker verification systems, which is rather to discriminate the client and impostor classes; hence discriminative approaches might appear good candidates for this task as well.

As a matter of fact, Support Vector Machine (SVM) based systems have been the subject of several recent publications in the speaker verification community, in which they obtain performance similar to or even better than GMMs on several text-independent speaker verification tasks. In order to use SVMs or any other discriminant approaches for speaker verification, several modifications of the classical techniques need to be performed. The purpose of this chapter is to present an overview of discriminant approaches that have been used successfully for the task of text-independent speaker verification, to analyze their differences from and their similarities to each other and to classical generative approaches

*Automatic Speech and Speaker Recognition: Large Margin and Kernel Methods*   Joseph Keshet and Samy Bengio
© 2009 John Wiley & Sons, Ltd

based on GMMs. An open-source version of the C++ source code used to performed all experiments described in this chapter can be found at http://speaker.abracadoudou.com.

## 12.1   Introduction

Person authentication systems are in general designed in order to let genuine clients access a given service while preventing impostors' access. This can be seen as a two-class classification problem suitable for machine learning approaches.

A number of specificities make speaker verification different from a standard binary classification problem. First, the input data consists of sentences whose lengths depend on its phonetic content and the speaking rate of the underlying speaker.

Second, only few client training examples are available: in most real applications, it is not possible to ask a client to speak during several hours or days in order to capture the entire variability of his/her voice. There are typically between one and three utterances for each client.

Third, the impostor distribution is not known and not even well defined: we have no idea of what an impostor is in a 'real' application. In order to simulate impostor accesses, one usually considers other speakers in the database. This ignorance is somewhat remedied by evaluating the models with impostor identities that are not available when creating the models. This incidentally means that plenty of impostor accesses are usually available, often more than 1000 times the number of client accesses, which makes the problem highly unbalanced.

The distribution of impostors being only loosely defined, the prior probability of each class is unknown, and the cost of each type of error is usually not known beforehand. Thus, one usually selects a model that gives reasonable performance for several possible cost trade-offs.

Finally, the recording conditions change over time. The speaker can be located in several kinds of place: office, street, train station, etc. The device used to perform the authentication can also change between authentication attempts: land line phone, mobile phone, laptop microphone, etc.

That being said, the problem of accepting or rejecting someone's identity claim can be formally stated as a binary classification task. Let $S$ be a set of clients and $s_i \in S$ be the $i$th client of that set. We look for a discriminant function $f(\cdot; \vartheta_i)$ and a decision threshold $\Delta$ such that

$$f(\bar{\mathbf{x}}; \vartheta_i) > \Delta \qquad\qquad (12.1)$$

if and only if sentence $\bar{\mathbf{x}}$ was pronounced by speaker $s_i$.

The parameters $\vartheta_i$ are typically determined by optimizing an empirical criterion computed on a set of $L_i$ sentences, called either the training or the learning set $\mathcal{L}_i = \{(\bar{\mathbf{x}}_l, y_l)\}_{l=1}^{L_i}$, where $\bar{\mathbf{x}}_l \in \mathbb{R}^{d \times T_l}$ is an input waveform sequence encoded as $T_l$ $d$-dimensional frames, and $y_l \in \{-1, 1\}$ is the corresponding target, where 1 stands for a true client sequence and $-1$ for an impostor access. The search space is defined as the set of functions $f : \mathbb{R}^{d \times T_l} \mapsto \mathbb{R}$ parameterized by $\vartheta_i$, and $\vartheta_i$ is identified by minimizing the mean loss on the training set, where the loss $\ell(\cdot)$ returns low values when $f(\bar{\mathbf{x}}; \vartheta_i)$ is near $y$ and high values otherwise:

$$\vartheta_i = \arg\min_{\theta} \sum_{(\bar{\mathbf{x}}, y) \in \mathcal{L}_i} \ell(f(\bar{\mathbf{x}}; \theta), y).$$

Note that the overall goal is not to obtain zero error on $\mathcal{L}_i$ but rather on unseen examples drawn from the same probability distribution. This objective is monitored by measuring the classification performance on an independent test set $\mathcal{T}_i$, in order to provide an unbiased estimate of performance on the population.

A standard taxonomy of machine learning algorithms sets apart discriminant models, which directly estimate the function $f(\cdot; \boldsymbol{\vartheta}_i)$, from generative models, where $f(\cdot; \boldsymbol{\vartheta}_i)$ is defined through the estimation of the conditional distribution of sequences knowing the speaker. We briefly present hereafter the classical generative approach that encompasses the very popular GMM, which will provide a baseline in the experimental section. All the other methods presented in this chapter are kernel-based systems that belong to the discriminative approach.

# 12.2   Generative Approaches

The state-of-the-art generative approaches for speaker verification use atypical models in the sense that they do not model the joint distribution of inputs and outputs. This is due to the fact that we have no clue of what the prior probability of having client $s_i$ speaking should be, since the distribution of impostors is only loosely defined and the proportion of client accesses in the training set may not be representative of the proportion in future accesses. Although the model is not complete, a decision function is computed using the rationale described below.

## 12.2.1   Rationale

The system has to decide whether a sentence $\bar{\mathbf{x}}$ was pronounced by speaker $s_i$ or by any other person $s_0$. It should accept a claimed speaker as a *client* if and only if:

$$P(s_i|\bar{\mathbf{x}}) > \alpha_i P(s_0|\bar{\mathbf{x}}), \tag{12.2}$$

where $\alpha_i$ is a trade-off parameter that accounts for the loss of false acceptance of an impostor access versus false rejection of a genuine client access.

Using Bayes' theorem, we rewrite Equation (12.2) as follows:

$$\frac{p(\bar{\mathbf{x}}|s_i)}{p(\bar{\mathbf{x}}|s_0)} > \alpha_i \frac{P(s_0)}{P(s_i)} = \Delta_i = \Delta, \tag{12.3}$$

where $\Delta_i$ is proportional to the ratio of the prior probabilities of being or not being the client. This ratio being unknown, $\Delta_i$ is replaced by a client independent decision threshold $\Delta$. This corresponds to having different (unknown) settings for the trade-off parameters $\alpha_i$.

The left ratio in Equation (12.3) plays the role of $f(\bar{\mathbf{x}}; \boldsymbol{\vartheta}_i)$ in Equation (12.1), where the set of parameters $\boldsymbol{\vartheta}_i$ is decomposed as follows:

$$f(\bar{\mathbf{x}}; \boldsymbol{\vartheta}_i) = \frac{p(\bar{\mathbf{x}}|s_i, \boldsymbol{\theta}_i)}{p(\bar{\mathbf{x}}|s_0, \boldsymbol{\theta}_0)},$$

with $\boldsymbol{\vartheta}_i = \{\boldsymbol{\theta}_i, \boldsymbol{\theta}_0\}$. The loss function used to estimate $\boldsymbol{\theta}_0$ is the negative log-likelihood

$$\boldsymbol{\theta}_0 = \arg\min_{\boldsymbol{\theta}} \sum_{(\bar{\mathbf{x}}, y) \in \mathcal{L}_i^-} -\log p(\bar{\mathbf{x}}|s_0, \boldsymbol{\theta}),$$

where $\mathcal{L}_i^-$ is the subset of pairs $(\bar{\mathbf{x}}, y)$ in the learning set $\mathcal{L}_i$ for which $y = -1$. As generally few positive examples are available, the loss function used to estimate $\theta_i$ is based on a MAP adaptation scheme (Gauvain and Lee 1994) and can be written as follows:

$$\theta_i = \arg\min_\theta \sum_{(\bar{\mathbf{x}}, y) \in \mathcal{L}_i^+} -\log(p(\bar{\mathbf{x}}|s_i, \theta)p(\theta))$$

where $\mathcal{L}_i^+$ is the subset of pairs $(\bar{\mathbf{x}}, y)$ in $\mathcal{L}_i$ for which $y = 1$. This MAP approach puts some prior on $\theta$ to constrain these parameters to some reasonable values. In practice, they are constrained to be near $\theta_0$, which represents reasonable parameters for any unknown person. See for instance Reynolds *et al.* (2000) for a practical implementation.

## 12.2.2    Gaussian Mixture Models

State-of-the-art systems compute the density of a sentence $\bar{\mathbf{x}}$ by a rough estimate that assumes independence of the $T$ frames that encode $\bar{\mathbf{x}}$. The density of the frames themselves is assumed to be independent of the sequence length, and is estimated by a GMM with diagonal covariance matrices, as follows:

$$p(\bar{\mathbf{x}}|s, \theta) = P(T)p(\bar{\mathbf{x}}|T, s, \theta)$$

$$= P(T) \prod_{t=1}^{T} p(\mathbf{x}^t|T, s, \theta)$$

$$= P(T) \prod_{t=1}^{T} p(\mathbf{x}^t|s, \theta)$$

$$= P(T) \prod_{t=1}^{T} \sum_{m=1}^{M} \pi_m \mathcal{N}(\mathbf{x}^t|\mu_m, \sigma_m), \qquad (12.4)$$

where $P(T)$ is the probability distribution[1] of the length of sequence $\bar{\mathbf{x}}$, $\mathbf{x}^t$ is the $t$th frame of $\bar{\mathbf{x}}$, and $M$ is the number of mixture components. The parameters $\theta$ comprise the means $\{\mu_m\}_{m=1}^{M}$, standard deviations $\{\sigma_m\}_{m=1}^{M}$ and mixing weights $\{\pi_m\}_{m=1}^{M}$ for all Gaussian components. The Gaussian density is defined as follows:

$$\mathcal{N}(\mathbf{x}|\mu, \sigma) = \frac{1}{(2\pi)^{d/2}|\Sigma|} \exp\left(-\frac{1}{2}(\mathbf{x} - \mu)^T \Sigma^{-2}(\mathbf{x} - \mu)\right),$$

where $d$ is the dimension of $\mathbf{x}$, $\Sigma$ is the diagonal matrix with diagonal elements $\Sigma_{ii} = \sigma_i$, and $|\Sigma|$ denotes the determinant of $\Sigma$.

As stated in the previous section, we first train an impostor model $p(\bar{\mathbf{x}}|s_0, \theta_0)$, called *world model* or *universal background model* when it is common to all speakers $s_i$. For this purpose, we use the expectation maximization (EM) algorithm to maximize the likelihood of the negative examples in the training set. Note that in order to obtain state-of-the-art performance, the variances of all Gaussian components are constrained to be higher than

---

[1]Under the reasonable assumption that the distributions of sentence length are identical for each speaker, this distribution does not play any discriminating role and can be left unspecified.

some threshold, normally selected on a separate development set. This process, often called *variance flooring* (Melin *et al.* 1998), can be seen as a way to control the capacity of the overall model.

For each client $s_i$, we use a variant of MAP adaptation (Reynolds *et al.* 2000) to estimate a client model $p(\bar{\mathbf{x}}|s_i, \boldsymbol{\theta}_i)$ that only departs partly from the world model $p(\bar{\mathbf{x}}|s_0, \boldsymbol{\theta}_0)$. In this setting, only the mean parameters of the world model are adapted to each client, using the following update rule:

$$\boldsymbol{\mu}_m^i = \tau_{i,m} \, \widehat{\boldsymbol{\mu}}_m^i + (1 - \tau_{i,m})\boldsymbol{\mu}_m^0,$$

where $\boldsymbol{\mu}_m^0$ is the vector of means of Gaussian $m$ of the world model, $\widehat{\boldsymbol{\mu}}_m^i$ is the corresponding vector estimated by maximum likelihood on the sequences available for client $s_i$ and $\tau_i$ is the adaptation factor that represents the faith we have in the client data. The latter is defined as follows (Reynolds *et al.* 2000):

$$\tau_{i,m} = \frac{n_{i,m}}{n_{i,m} + r} \tag{12.5}$$

where $n_{i,m}$ is the effective number of frames used to compute $\widehat{\boldsymbol{\mu}}_m^i$, that is, the sum of memberships to component $m$ for all the frames of the training sequence(s) uttered by client $s_i$ (see Section 12.5.2 for details). The MAP relevant factor $r$ is chosen by cross-validation.

Finally, when all GMMs have been estimated, one can instantiate Equation (12.3) to take a decision for a given access as follows:

$$\frac{1}{T} \sum_{t=1}^{T} \log \frac{\sum_{m=1}^{M} \pi_m \, \mathcal{N}(\mathbf{x}^t; \boldsymbol{\mu}_m^i, \boldsymbol{\sigma}_m)}{\sum_{m=1}^{M} \pi_m \, \mathcal{N}(\mathbf{x}^t; \boldsymbol{\mu}_m^0, \boldsymbol{\sigma}_m)} > \log \Delta,$$

where $\boldsymbol{\theta}_0 = \{\boldsymbol{\mu}_m^0, \boldsymbol{\sigma}_m, \pi_m\}_{m=1}^{M}$ are the GMM parameters for the world model, and $\boldsymbol{\theta}_i = \{\boldsymbol{\mu}_m^i, \boldsymbol{\sigma}_m, \pi_m\}_{m=1}^{M}$ are the GMM parameters for the client model. Note that $1/T$ does not follow from Equation (12.3) and is an empirical normalization factor added to yield a threshold $\Delta$ that is independent of the length of the sentence.

## 12.3 Discriminative Approaches

SVMs (Vapnik 2000) are now a standard tool in numerous applications of machine learning, such as in text or vision (Joachims 2002; Pontil and Verri 1998). While GMM is the mainstream generative model in speaker verification, SVMs are prevailing in the discriminative approach. This section provides a basic description of SVMs that introduces the kernel trick that relates feature expansions to kernels, on which we will focus in Section 12.5.

### 12.3.1 Support Vector Machines

In the context of binary classification problems, the SVM decision function is defined by the sign of

$$f(\mathbf{x}; \boldsymbol{\vartheta}) = \mathbf{w} \cdot \Phi(\mathbf{x}) + b, \tag{12.6}$$

where $\mathbf{x}$ is the current example, $\boldsymbol{\vartheta} = \{\mathbf{w}, b\}$ are the model parameters and $\Phi(\cdot)$ is a mapping, chosen 'a priori', that associates a possibly high dimensional feature to each input data.

The SVM training problem consists in solving the following problem:

$$
\begin{cases}
(\mathbf{w}^*, b^*) = \arg\min_{(\mathbf{w}, b)} \dfrac{1}{2}\|\mathbf{w}\|^2 + C \sum_{l=1}^{L} \xi_l \\
\text{s.t.} \qquad y_l(\mathbf{w} \cdot \mathbf{x}_l + b) \geq 1 - \xi_l \quad \forall l \\
\qquad\qquad \xi_l \geq 0 \quad \forall l,
\end{cases}
\tag{12.7}
$$

where $L$ is the number of training examples, the target class label $y_l \in \{-1, 1\}$ corresponds to $\mathbf{x}_l$, and $C$ is a hyper-parameter that trades off the minimization of classification error (upper-bounded by $\xi_l$) and the maximization of the margin, which provides generalization guarantees (Vapnik 2000).

Solving Equation (12.7) leads to a discriminant function expressed as a linear combination of training examples in the feature space $\Phi(\cdot)$. We can thus rewrite Equation (12.6) as follows:

$$
f(\mathbf{x}; \vartheta) = \sum_{l=1}^{L} \alpha_l y_l \Phi(\mathbf{x}_l) \cdot \Phi(\mathbf{x}) + b,
$$

where most training examples do not enter this combination ($\alpha_l = 0$); the training examples for which $\alpha_l \neq 0$ are called *support vectors*.

As the feature mapping $\Phi(\cdot)$ only appears in dot products, the SVM solution can be expressed as follows:

$$
f(\mathbf{x}; \vartheta) = \sum_{l=1}^{L} \alpha_l y_l k(\mathbf{x}_l, \mathbf{x}) + b,
$$

where $k(\cdot, \cdot)$ is the dot product $\Phi(\cdot) \cdot \Phi(\cdot)$. More generally, $k(\cdot, \cdot)$ can be any kernel function that fulfills the Mercer conditions (Burges 1998), which ensure that, for any possible training set, the optimization problem is convex.

## 12.3.2 Kernels

A usual problem in machine learning is to extract features that are relevant for the classification task. For SVMs, choosing the features and choosing the kernel are equivalent problems, thanks to the so-called 'kernel trick' mentioned above. The latter also permits to map $\mathbf{x}_l$ into potentially infinite dimensional feature spaces by avoiding the explicit computation of $\Phi(\mathbf{x}_l)$; it also reduces the computational load for mappings in finite but high dimension.

The two most well known kernels are the radial basis function (RBF) kernel:

$$
k(\mathbf{x}_l, \mathbf{x}_{l'}) = \exp\left(\frac{-\|\mathbf{x}_l - \mathbf{x}_{l'}\|^2}{2\sigma^2}\right)
\tag{12.8}
$$

and the polynomial kernel:

$$
k(\mathbf{x}_l, \mathbf{x}_{l'}) = (a\mathbf{x}_l \cdot \mathbf{x}_{l'} + b)^p,
\tag{12.9}
$$

where $\sigma, p, b, a$ are hyper-parameters that define the feature space.

Several SVM-based approaches have been proposed recently to tackle the speaker verification problem (Campbell *et al.* 2006; Wan and Renals 2003). These approaches rely on constructing an ad-hoc kernel for the problem at hand. These kernels will be presented and evaluated after the following section, which describes the details of the experimental methodology and the data that will be used to compare the various methods.

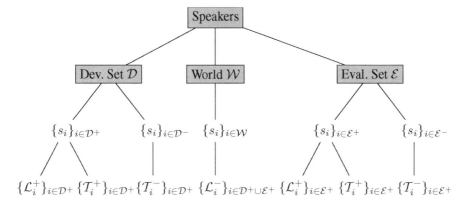

Figure 12.1 Split of the speaker population into three subsets, with the final decomposition in learning and test sets.

## 12.4 Benchmarking Methodology

In this section, we describe the methodology and the data used in all the experiments reported in this chapter. We first present the data splitting strategy that is used to imitate a realistic use of speaker verification systems. Then, we discuss the measures evaluating the performances of learning algorithms. Finally, we detail the database used to benchmark these algorithms, and the pre-processing that builds sequences of frames from waveform signals.

### 12.4.1 Data Splitting for Speaker Verification

A speaker verification problem is not a standard classification problem, since the objective is not to certify accesses from a pre-defined set of clients. Instead, we want to be able to authenticate new clients when they subscribe to the service, that is, we want to learn how to build new classifiers on the fly. Hence, a speaker verification system is evaluated by its ability to produce new classifiers with small test error. This is emulated by the data splitting process depicted in Figure 12.1.

The root level gathers the population of speakers, which is split into three sub-populations, defined by their role in building classifiers: the development set $\mathcal{D}$, the world set $\mathcal{W}$ and the evaluation set $\mathcal{E}$. All accesses from the speakers of $\mathcal{W}$ will be used as the set of negative examples $\mathcal{L}_i^-$ for training the models responsible for authenticating client $s_i$, where $s_i$ may belong either to the development set $\mathcal{D}$ or to the evaluation set $\mathcal{E}$. The sets $\mathcal{D}$ and $\mathcal{E}$ are further split into clients (respectively $\mathcal{D}^+$ and $\mathcal{E}^+$) and impostors (respectively $\mathcal{D}^-$ and $\mathcal{E}^-$) at the second level of the tree. The clients and the test impostors hence differ between the development and the evaluation sets.

The impostor accesses in $\mathcal{D}^-$ and $\mathcal{E}^-$ form the set of negative test examples $\mathcal{T}_i^-$, that is, 'attempt data' from out-of-training impostors claiming identity $s_i$, where $s_i$ belongs respectively to $\mathcal{D}^+$ and $\mathcal{E}^+$. Finally, at the third level of the tree, the accesses of client $s_i$ are split to form the positive examples of the training set $\mathcal{L}_i^+$ (also known as the 'enrollment data', usually a single access), and the set of positive 'attempt data' $\mathcal{T}_i^+$ that play the role of out-of-training client accesses requiring authentication.

To summarize, the development set $\mathcal{D}$ is used jointly with $\mathcal{W}$ to train models and select their various hyper-parameters (such as the number of Gaussians, the MAP adaptation factor, kernel parameters, etc.). For each hyper-parameter, we define a range of possible values, and for each value, each client model is trained using the enrollment data $\mathcal{L}_i^+$ and the world data $\mathcal{L}_i^-$, before being evaluated with the positive and negative attempt data $\mathcal{T}_i^+$ and $\mathcal{T}_i^-$. We then select the value of the hyper-parameters that optimizes a given performance measure (the equal error rate described below) on $\{\mathcal{T}_i^+ \cup \mathcal{T}_i^-\}$. Finally, the evaluation set $\mathcal{E}$ is used to train new client models using these hyper-parameters, and to measure the performance of the system on these new clients.

## 12.4.2 Performance Measures

The classification error rate is the most common performance measure in the machine learning literature, but it is not well suited to the type of problems encountered in speaker verification, where class priors are unknown and misclassification losses are unbalanced. Hence, a weighted version of the misclassification rate is used, where one distinguishes two kinds of error: *false rejection* (FR), which consists in rejecting a genuine client, and *false acceptance* (FA), which consists in accepting an impostor. All the measures used in this chapter are based on the corresponding error rates: the *false acceptance rate* (FAR) is the number of FAs divided by the number of client accesses, and the *false rejection rate* (FRR) is the number of FRs divided by the number of impostor accesses.

As stated in the previous section, in practice, we aim at building a single system that is able to take decisions for all future users. The performance is measured globally, on the set of speakers of the evaluation set, by averaging the performance over all trials independently of the claimed identity.

In the speaker verification literature, a point often overlooked is that most of the results are reported with 'a posteriori' measures, in the sense that the decision threshold $\Delta$ in Equation (12.1) is selected such that it optimizes some criterion on the evaluation set. We believe that this is unfortunate, and, in order to obtain unbiased results, we will use 'a priori' measures, where the decision threshold $\Delta$ is selected on a development set, before seeing the evaluation set, and then applied to the evaluation data.

Common a posteriori measures include the equal error rate (EER), where the threshold $\Delta$ is chosen such that (FAR = FRR), and the detection error trade-off (DET) curve (Martin *et al.* 1997), which depicts FRR as a function of FAR when $\Delta$ varies. Note that the DET curve is a nonlinear transformation of the receiver operating characteristic (ROC) curve (Van Trees 1968). The nonlinearity is in fact a normal deviate, coming from the hypothesis that the scores of client accesses and impostor accesses follow a Gaussian distribution. These measures are perfectly legitimate for exploratory analysis or for tuning hyper-parameters on the development set and they are used for this purpose here. To avoid confusion with proper test results, we will only report DET curves computed on the development set. For test performance, we will use a priori measures: the half total error rate (HTER $= \frac{1}{2}$(FAR($\Delta$) + FRR($\Delta$))) and the expected performance curve (EPC) (Bengio *et al.* 2005), which depicts the evaluation set HTER as a function of a trade-off parameter $\alpha$. The latter defines a decision threshold, computed on the development set, by minimizing the following convex combination of development FAR and FRR:

$$\Delta^* = \arg \min_{\Delta}(\alpha \cdot \text{FAR}(\Delta) + (1 - \alpha) \cdot \text{FRR}(\Delta)). \qquad (12.10)$$

We will provide confidence intervals around HTER and EPC. In this chapter, we report confidence intervals computed at the 5% significance level, using an adapted version of the standard proportion test (Bengio and Mariéthoz 2004).

### 12.4.3   NIST Data

The NIST database is a subset of the database that was used for the *NIST 2005 and 2006 Speaker Recognition Evaluation*, which comes from the second release of the cellular switchboard corpus (Switchboard Cellular – Part 2) of the Linguistic Data Consortium. This data was used as development and evaluation sets while the training (negative) examples come from previous NIST campaigns. For both development and evaluation clients, there are about 2 minutes of telephone speech available to train the models and each test access was less than 1 minute long. Only male speakers were used. The development population consisted of 264 speakers, while the evaluation set contained 349 speakers. Two hundred and nineteen different records were used as negative examples for the discriminant models. The total number of accesses in the development population is 13 596 and 22 131 for the evaluation set population with a proportion of 10% of true target accesses.

### 12.4.4   Pre-processing

To extract input features, the original waveforms are sampled every 10 ms with a window size of 20 ms. Each sentence is parameterized using 24 triangular band-pass filters with a DCT transformation of order 16, complemented by their first derivative (delta) and the 10th second derivative (delta-delta), the log-energy, the delta-log-energy and delta-delta-log-energy, for a total of 51 coefficients. The NIST database being telephone-based, the signal is band-pass filtered between 300 and 3400 Hz.

A simple silence detector, based on a two-component GMM, is used to remove all silence frames. The model is first learned on a random recording with land line microphone and adapted for each new sequence using the MAP adaptation algorithm. The sequences are then normalized in order to have zero mean and unit variance on each feature.

While the log-energy is important in order to remove the silence frames, it is known to be inappropriate to discriminate between clients and impostors. This feature is thus eliminated after silence removal, while its first derivative is kept. Hence, the speaker verification models are trained with 50 (51 − 1) features.

## 12.5   Kernels for Speaker Verification

One particularity of speaker verification is that patterns are sequences. An SVM based classification thus requires a kernel handling variable size sequence. Most solutions proposed in the literature use a procedure that converts the sequences into fixed size vectors that are processed by a linear SVM. Other sequence kernels allow embeddings in infinite-dimensional feature spaces (Mariéthoz and Bengio 2007). However, compared with the mainstream approach, this type of kernel is computationally too demanding for long sequences. It will not be applied here, since the NIST database contains long sequences.

In the following we describe several approaches using sequence kernels. The most promising are then compared in Section 12.8.

## 12.5.1   Mean Operator Sequence Kernels

For kernel methods, a simple approach to tackling variable length sequences considers the following kernel between two sequences:

$$K(\bar{\mathbf{x}}_i, \bar{\mathbf{x}}_j) = \frac{1}{T_i T_j} \sum_{t=1}^{T_i} \sum_{u=1}^{T_j} k(\mathbf{x}_i^t, \mathbf{x}_j^u), \tag{12.11}$$

where we denote by $K(\cdot, \cdot)$ a sequence kernel, $\bar{\mathbf{x}}_i$ is a sequence of size $T_i$ and $\mathbf{x}_i^t$ is a frame of $\bar{\mathbf{x}}_i$. We thus apply a frame based kernel $k(\cdot, \cdot)$ to all possible pairs of frames coming from the two input sequences $\bar{\mathbf{x}}_i$ and $\bar{\mathbf{x}}_j$.

As the kernel $K$ represents the average similarity between all possible pairs of frames, it will be referred to as the mean operator sequence kernel. This kind of kernel has been applied successfully in other domains such as object recognition (Boughorbel *et al.* 2004). Provided that $k(\cdot, \cdot)$ is positive-definite, the resulting kernel $K(\cdot, \cdot)$ is also positive-definite.

The sequences in the NIST database typically consist of several thousands of frames, hence the double summation in Equation (12.11) is very costly. As the number of operations for each sequence kernel evaluation is proportional to the product of sequence lengths, such a computation typically requires the order of a million operations. We thus will consider factorizable kernels $k(\cdot, \cdot)$, such that the mean operator sequence kernel Equation (12.11) can be expressed as follows:

$$K(\bar{\mathbf{x}}_i, \bar{\mathbf{x}}_j) = \frac{1}{T_i T_j} \sum_{t=1}^{T_i} \sum_{u=1}^{T_j} \phi(\mathbf{x}_i^t) \cdot \phi(\mathbf{x}_j^u)$$

$$= \left[ \frac{1}{T_i} \sum_{t=1}^{T_i} \phi(\mathbf{x}_i^t) \right] \cdot \left[ \frac{1}{T_j} \sum_{u=1}^{T_j} \phi(\mathbf{x}_j^u) \right]. \tag{12.12}$$

When the dimension of the feature space is not too large, computing the dot product explicitly is not too demanding, and replacing the double summation by two single ones may result in a significant reduction of computing time.

Explicit polynomial expansions have been used in Campbell (2002), Campbell *et al.* (2006) and Wan and Renals (2003). In practice, the average feature vectors within brackets in Equation (12.12) are used as input to a linear SVM. The generalized linear discriminant sequence (GLDS) kernel of Campbell departs slightly from a raw polynomial expansion, by using a normalization in the feature space:

$$K(\bar{\mathbf{x}}_i, \bar{\mathbf{x}}_j) = \frac{1}{T_i T_j} \Phi(\bar{\mathbf{x}}_i) \mathbf{\Gamma}^{-1} \Phi(\bar{\mathbf{x}}_j), \tag{12.13}$$

where $\boldsymbol{\Gamma}$ defines a metric in the feature space. Typically, this is a diagonal approximation of the Mahalanobis metric, that is, $\boldsymbol{\Gamma}$ is a diagonal matrix whose diagonal elements $\gamma_k$ are the empirical variances[2] for each feature, computed over the training data.

The polynomial expansion sends $d$-dimensional frames to a feature space of dimension $(d + p)!/d!p! - 1$, where $p$ is the degree of the polynomial. With our 50 input features, and for a polynomial of degree $p = 3$, the dimension of the feature space is 23 425. For higher polynomial degrees and for other feature spaces of higher dimension, the computational advantage of the decomposition of Equation (12.12) disappears, and it is better to use explicit kernel in the form of Equation (12.11). We empirically show below that, for the usual representation of frames described in Section 12.4.4, the GLDS normalization in Equation (12.13) is embedded in the standard polynomial kernel.

Let us define $k(\mathbf{x}_i, \mathbf{x}_j)$ as a polynomial kernel of the form $(\mathbf{x}_i \cdot \mathbf{x}_j + 1)^p$, where $p$ is the degree of the polynomial. After removing the constant term, the explicit expansion of this standard polynomial kernel involves $(d + p)!/d!p! - 1$ terms that can be indexed by $\mathbf{r} = (r_1, r_2, \ldots, r_d)$, such that

$$\phi_{\mathbf{r}}(\mathbf{x}) = \sqrt{c_{\mathbf{r}}} x_1^{r_1} x_2^{r_2} \ldots x_d^{r_d},$$

$$\text{where} \quad \sum_{i=1}^{d} r_i = p, \quad r_i \geq 0, \quad \text{and} \quad c_{\mathbf{r}} = \frac{p!}{r_1! r_2! \ldots r_{d+1}!}.$$

In the above equations, $\sqrt{c_{\mathbf{r}}}$ has exactly the same role as the $1/\sqrt{\gamma_k}$ coefficients on the diagonal of $\boldsymbol{\Gamma}^{-1/2}$ in Equation (12.13). In Figure 12.2, we compare these coefficient values, where the normalization factors $1/\sqrt{\gamma_k}$ are estimated on two real datasets, after a polynomial expansion of degree 3. The values are very similar, with highs and lows on the same monomial. In fact, the performance of the two approaches obtained on the development set of NIST are about the same, as shown by the DET curves given in Figure 12.3.

Even if this approach is simple and easy to use, the accuracy can be improved by introducing priors. In fact, to train a client model very few positive examples are available. Thus, if we can put pieces of information collected on a large set of speakers into the SVM model, as done for the GMM system, we can expect an improvement. One can for example try to include the world model in the kernel function as proposed in the next section.

## 12.5.2 Fisher Kernels

Jaakkola and Haussler (1998) proposed a principled means for building kernel functions from generative models: the Fisher kernel. In this framework, which has been applied to speaker verification by Wan and Renals (2005a), the generative model is used to specify the similarity between pairs of examples, instead of the usual practice where it is used to provide a likelihood score, which measures how well the example fits the model. Put another way, a Fisher kernel utilizes a generative model to measure the differences in the generative process between pairs of examples instead of the differences in posterior probabilities.

The key ingredient of the Fisher kernel is the vector of Fisher scores:

$$\mathbf{u}_{\bar{\mathbf{x}}} = \nabla_{\boldsymbol{\theta}} \log p(\bar{\mathbf{x}}|\boldsymbol{\theta}),$$

---

[2]The constant feature is removed from the feature space prior to normalization.

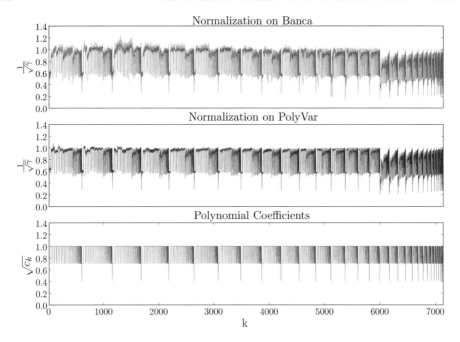

Figure 12.2 Coefficient values of polynomial terms, as computed on two different datasets (Banca and PolyVar), compared with the $c_k$ polynomial coefficients.

where $\theta$ denotes here the parameters of the generative model, and $\nabla_\theta$ is the gradient with respect to $\theta$. The Fisher scores quantify how much each parameter contributes to the generation of example $\bar{\mathbf{x}}$.

The Fisher kernel itself is given by

$$K(\bar{\mathbf{x}}_i, \bar{\mathbf{x}}_j) = \mathbf{u}_{\bar{\mathbf{x}}_i}{}^T \mathbf{I}(\theta)^{-1} \mathbf{u}_{\bar{\mathbf{x}}_j}, \qquad (12.14)$$

where $\mathbf{I}(\theta)$ is the Fisher information matrix at $\theta$, that is, the covariance matrix of Fisher scores:

$$\mathbf{I}(\theta) = \mathbb{E}_{\bar{\mathbf{x}}}(\mathbf{u}_{\bar{\mathbf{x}}}\mathbf{u}_{\bar{\mathbf{x}}}{}^T), \qquad (12.15)$$

where we have used that $\mathbb{E}_{\bar{\mathbf{x}}}(\mathbf{u}_{\bar{\mathbf{x}}}) = 0$. The Fisher kernel of Equation (12.14) can thus be interpreted as a Mahalanobis distance between two Fisher scores.

Another interpretation of the Fisher kernel is based on the representation of a parametric class of generative models as a Riemannian manifold (Jaakkola and Haussler 1998). Here, the vector of Fisher scores defines a tangent direction at a given location, that is, at a given model parameterized by $\theta$. The Fisher information matrix is the local metric at this given point, which defines the distance between the current model $p(\bar{\mathbf{x}}|\theta)$ and its neighbors $p(\bar{\mathbf{x}}|\theta + \delta)$. The (squared) distance $d(\theta, \theta + \delta) = \frac{1}{2}\delta^T \mathbf{I}\delta$ approximates the Kullback–Leibler divergence between the two models. Note that, unlike the Kullback–Leibler divergence, the Fisher kernel of Equation (12.14) is symmetric. It is also positive-definite since the Fisher information matrix $\mathbf{I}(\theta)$ is obviously positive-definite at $\theta$.

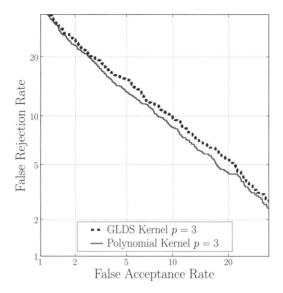

Figure 12.3 Detection error trade-off curves on the development set of the NIST database comparing the explicit polynomial expansion (noted as 'GLDS kernel $p = 3$, in the legend), and the principled polynomial kernel (noted 'Polynomial kernel $p = 3$').

**Fisher Kernels for GMMs**

In the MAP framework, the family of generative models we consider is the set of Gaussian mixtures of Equation (12.4) that differ in their mean vectors $\boldsymbol{\mu}_m$. Hence, a relevant dissimilarity between examples will be measured by the Fisher scores computed on these vectors: $\mathbf{u}_{\bar{\mathbf{x}}} = (\nabla_{\boldsymbol{\mu}_1}^T \log p(\bar{\mathbf{x}}|\boldsymbol{\theta}), \ldots, \nabla_{\boldsymbol{\mu}_M}^T \log p(\bar{\mathbf{x}}|\boldsymbol{\theta}))^T$, where

$$
\begin{aligned}
\nabla_{\boldsymbol{\mu}_m} \log p(\bar{\mathbf{x}}|\boldsymbol{\theta}) &= \sum_{t=1}^{T} \nabla_{\boldsymbol{\mu}_m} \log \sum_{m'=1}^{M} \pi_{m'} \mathcal{N}(\mathbf{x}^t|\boldsymbol{\mu}_{m'}, \boldsymbol{\sigma}_{m'}) \\
&= \sum_{t=1}^{T} P(m|\mathbf{x}^t) \nabla_{\boldsymbol{\mu}_m} \left( -\frac{1}{2}(\mathbf{x}^t - \boldsymbol{\mu}_m)^T \boldsymbol{\Sigma}_m^{-2} (\mathbf{x}^t - \boldsymbol{\mu}_m) \right) \\
&= \sum_{t=1}^{T} P(m|\mathbf{x}^t) \boldsymbol{\Sigma}_m^{-2} (\mathbf{x}^t - \boldsymbol{\mu}_m).
\end{aligned}
\tag{12.16}
$$

Using the definition of Equation (12.15), the Fisher information matrix can be expressed block-wise, with $M \times M$ blocks of size $d \times d$:

$$
\mathbf{I} = (\mathbf{I}_{m,m'})_{1 \leq m \leq M, 1 \leq m' \leq M},
$$

with

$$\mathbf{I}_{m,m'} = \mathbb{E}_{\bar{\mathbf{x}}}\left[\sum_{t=1}^{T}\sum_{u=1}^{T} P(m|\mathbf{x}^t)P(m'|\mathbf{x}^u)\boldsymbol{\Sigma}_m^{-2}(\mathbf{x}^t - \boldsymbol{\mu}_m)(\mathbf{x}^u - \boldsymbol{\mu}_{m'})^T \boldsymbol{\Sigma}_{m'}^{-2}\right]. \qquad (12.17)$$

There is no simple analytical expression of this expectation, due, among other things, to the product $P(m|\mathbf{x}^t)P(m'|\mathbf{x}^u)$. Hence, several options are possible:

1. Ignore the information matrix in the computation of the Fisher kernel of Equation (12.14). This option, mentioned by Jaakkola and Haussler as a simpler suitable substitute, is often used in the application of Fisher kernels.

2. Approximate the expectation in the definition of Fisher information by Monte Carlo sampling.

3. Approximate the product $P(m|\mathbf{x}^t)P(m'|\mathbf{x}^u)$ by a simpler expression in Equation (12.17). For example, if we assume that the considered GMM performs hard assignments of frames to mixture components, then $P(m|\mathbf{x}^t)P(m'|\mathbf{x}^u)$ is null if $m \neq m'$. Furthermore, this product is also null for $m = m'$ when $\mathbf{x}^t$ or $\mathbf{x}^u$ is generated from another component of the mixture distribution, otherwise, we have $P(m|\mathbf{x}^t)P(m'|\mathbf{x}^u) = 1$. Let $g_m$ denote the function such that $g_m(\mathbf{x},\mathbf{y}) = \boldsymbol{\Sigma}_m^{-2}(\mathbf{x} - \boldsymbol{\mu}_m)(\mathbf{y} - \boldsymbol{\mu}_{m'})^T \boldsymbol{\Sigma}_{m'}^{-2}$ if $P(m|\mathbf{x}) = P(m|\mathbf{y}) = 1$ and $g_m(\mathbf{x},\mathbf{y}) = 0$ otherwise. With this notation and the above approximations, Equation (12.17) reads

$$\mathbf{I}_{m,m'} \simeq 0 \text{ if } m \neq m'$$

$$\mathbf{I}_{m,m} \simeq \mathbb{E}_{\bar{\mathbf{x}}}\left[\sum_{t=1}^{T}\sum_{u=1}^{T} g_m(\mathbf{x}^t, \mathbf{x}^u)\right]$$

$$\simeq \mathbb{E}_{\mathbf{x}}[g_m(\mathbf{x}, \mathbf{x})]\mathbb{E}_T[T]$$

$$\simeq \boldsymbol{\Sigma}_m^{-2}\mathbb{E}_T[T].$$

The unknown constant $\mathbb{E}_T[T]$ is not relevant and can been dropped from the implementation of this approximation to the Fisher kernel.

We now introduce some definitions with the following scenario. Suppose that we trained the GMM world model on a large set of speakers, resulting in parameters $\boldsymbol{\theta}_0 = \{\boldsymbol{\mu}_m^0, \boldsymbol{\sigma}_m, \pi_m\}_{m=1}^{M}$. We then use this GMM as an initial guess for the model for client $s_i$. If, as in the MAP framework, the client model differs from the world model in the mean vector only, then, after one EM update, the training sequence $\bar{\mathbf{x}}_i$ will result in the following estimates:

$$\boldsymbol{\mu}_m^i = \frac{1}{n_{i,m}}\sum_{t=1}^{T_i}\mathbf{x}_i^t P(m|\mathbf{x}_i^t),$$

$$\text{where} \quad n_{i,m} = \sum_{t=1}^{T_i} P(m|\mathbf{x}_i^t).$$

Hence, $n_{i,m}$ is the effective number of frames used to compute $\boldsymbol{\mu}_m^i$, that is, the sum of the membership of all frames of $\bar{\mathbf{x}}_i$ to component $m$. These definitions of $\boldsymbol{\mu}_m^i$ and $n_{i,m}$ are convenient for expressing Fisher scores, when the reference generative model is the world model parameterized by $\theta_0$:

$$\nabla_{\mu_m} \log p(\bar{\mathbf{x}}_i|\theta)|_{\theta=\theta_0} = \Sigma_m^{-2} \sum_{t=1}^{T_i} P(m|\mathbf{x}_i^t)(\mathbf{x}_i^t - \mu_m)$$

$$= n_{i,m} \Sigma_m^{-2}(\mu_m^i - \mu_m^0). \tag{12.18}$$

With the approximations of the Fisher information discussed above, the kernel is expressed as:

1. For the option where the Fisher information matrix is ignored:

$$K(\bar{\mathbf{x}}_i, \bar{\mathbf{x}}_j) = \mathbf{u}_{\bar{\mathbf{x}}_i}^T \mathbf{u}_{\bar{\mathbf{x}}_j}$$

$$= \sum_{m=1}^{M} (n_{i,m} \Sigma_m^{-2}(\mu_m^i - \mu_m^0))^T (n_{j,m} \Sigma_m^{-2}(\mu_m^j - \mu_m^0)).$$

2. For the option where the Fisher information matrix is approximated by Monte Carlo integration: here, for computational reasons, we only consider a block-diagonal approximation $\widehat{\mathbf{I}}$, where

$$\widehat{\mathbf{I}} = (\widehat{\mathbf{I}}_{m,m'})_{1 \le m \le M, 1 \le m' \le M},$$

with

$$\widehat{\mathbf{I}}_{m,m'} = 0 \text{ if } m \neq m'$$

$$\widehat{\mathbf{I}}_{m,m} = \frac{1}{n} \sum_t P(m|\mathbf{x}^t)^2 \Sigma_m^{-2}(\mathbf{x}^t - \mu_m^0)(\mathbf{x}^t - \mu_m^0)^T \Sigma_m^{-2},$$

where $n$ is the number of random draws of $\mathbf{x}^t$ generated from the world model.

We then have:

$$K(\bar{\mathbf{x}}_i, \bar{\mathbf{x}}_j) = \sum_{m=1}^{M} (n_{i,m} \Sigma_m^{-2}(\mu_m^i - \mu_m^0))^T \widehat{\mathbf{I}}_{m,m}^{-1} (n_{j,m} \Sigma_m^{-2}(\mu_m^j - \mu_m^0)).$$

3. For the option where the Fisher information matrix is approximated analytically:

$$K(\bar{\mathbf{x}}_i, \bar{\mathbf{x}}_j) = \sum_{m=1}^{M} (n_{i,m} \Sigma_m^{-1}(\mu_m^i - \mu_m^0))^T (n_{j,m} \Sigma_m^{-1}(\mu_m^j - \mu_m^0)).$$

These three variants of the Fisher kernel are compared in Figure 12.4, which compares the DET curves obtained on the development set of the NIST database. The three curves almost overlap, confirming that ignoring the information matrix in the Fisher kernel is not harmful in our setup.

Table 12.1 Equal error rates (EERs) (the lower the better) on the development set of the NIST database, comparing Fisher kernel (approximation 3) with the normalized Fisher kernel

|                  | Fisher | Normalized Fisher |
|------------------|--------|-------------------|
| EER (%)          | 9.3    | 8.2               |
| 95% confidence   | ±0.9   | ±0.8              |
| # support vectors| 37     | 32                |

### 12.5.3 Beyond Fisher Kernels

The previous experimental results confirm that the main component of the Fisher kernel is the Fisher score. The latter is based on a probabilistic model viewed through the log-likelihood function. We can depart from the original setup described above, by using other models and/or score. Some alternative approaches have already been investigated, for example Wan and Renals (2005b) use scores based on a log-likelihood ratio between the world model and the adapted client model. We describe below a very simple modification of the scoring function that brings noticeable improvements in performance.

#### (a) Normalized Fisher Scores

We saw in Section 12.2.2 that the scores used for classifying examples are normalized, in order to counterbalance the exponential decrease of likelihoods with sequence lengths. Using the normalized likelihood leads to the following Fisher-like kernel:

$$K(\bar{\mathbf{x}}_i, \bar{\mathbf{x}}_j) = \frac{1}{T_i T_j} \mathbf{u}_{\bar{\mathbf{x}}_i}^T \mathbf{u}_{\bar{\mathbf{x}}_j}$$

$$= \sum_{m=1}^{M} \left( \frac{n_{i,m}}{T_i} \boldsymbol{\Sigma}_m^{-2} (\boldsymbol{\mu}_m^i - \boldsymbol{\mu}_m^0) \right)^T \left( \frac{n_{j,m}}{T_j} \boldsymbol{\Sigma}_m^{-2} (\boldsymbol{\mu}_m^j - \boldsymbol{\mu}_m^0) \right).$$

Here also one may consider several options for approximating the Fisher information matrix, but the results displayed in Figure 12.4 suggest it is not worth pursuing this further. Table 12.1 and Figure 12.5 compare empirically the Fisher kernel (approximation 3) with the normalized Fisher kernel. Including a normalization seems to have a positive impact on the accuracy. Thus other kinds of scores should be explored.

#### (b) GMM Supervector Linear Kernel

The Fisher kernel is a similarity based on the differences in the generation of examples. In this matter, it is related to the GMM supervector linear kernel (GSLK) proposed by Campbell *et al.* (2006).

The GSLK approximates the Kullback–Leibler divergence, which measures the dissimilarity between two GMMs, each one being obtained by adapting the world model to one example of the pair $(\bar{\mathbf{x}}_i, \bar{\mathbf{x}}_j)$. Hence, instead of looking at how a single generative process differs for each example of the $(\bar{\mathbf{x}}_i, \bar{\mathbf{x}}_j)$ pair, GSLK looks at the difference between pairs of

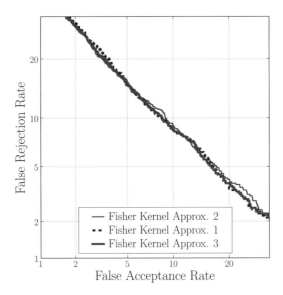

Figure 12.4 Detection error trade-off curves on the development set of the NIST database comparing the three different approximations of the Fisher information matrix.

generative models. The GSLK is given by:

$$K(\bar{\mathbf{x}}_i, \bar{\mathbf{x}}_j) = \sum_{m=1}^{M} (\sqrt{\pi_m} \Sigma_m^{-1} (\tau_{i,m} \boldsymbol{\mu}_m^i + (1 - \tau_{i,m}) \boldsymbol{\mu}_m^0))^T$$
$$\cdot (\sqrt{\pi_m} \Sigma_m^{-1} (\tau_{j,m} \boldsymbol{\mu}_m^j + (1 - \tau_{j,m}) \boldsymbol{\mu}_m^0)),$$

where $\tau_{i,m}$ is the adaptation factor for the mixture component $m$ adapted with sequence $\bar{\mathbf{x}}_i$, as defined in Equation (12.5). The MAP relevant factor $r$ is chosen by cross-validation, as in GMM based text-independent speaker verification systems (Reynolds *et al.* 2000).

The Fisher kernel and GSLK are somewhat similar scalar products, with the most noticeable difference being that the Fisher similarity is based on difference from the reference $\boldsymbol{\mu}^0$ whereas the GSLK kernel above is based on a convex combination of the observations and the reference $\boldsymbol{\mu}^0$ that has no obvious interpretation. Both are an approximation of the Kullback–Leibler divergence as mentioned in Section 12.5.2. The difference is that GSLK compares two adapted distributions whereas the Fisher kernel compares the world model and the updated model using the access data.

Table 12.2 and Figure 12.6 compare empirically GSLK with the normalized Fisher kernel. There is no significant difference between GSLK and the normalized Fisher kernel.

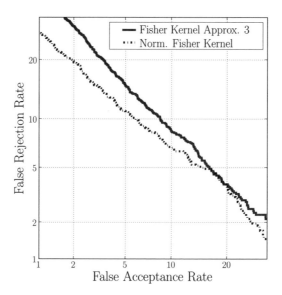

Figure 12.5 Detection error trade-off curves on the development set of the NIST database for Fisher kernel (approximation 3) and normalized Fisher kernel.

Table 12.2 Equal error reports (EERs) (the lower the better) on the development set of the NIST database, comparing GMM supervector linear kernel (GSLK) and the normalized Fisher kernel

|                  | GSLK | Normalized Fisher |
|------------------|------|-------------------|
| EER (%)          | 7.9  | 8.2               |
| 95% confidence   | ±0.8 | ±0.8              |
| # support vectors| 34   | 32                |

## 12.6 Parameter Sharing

The text-independent speaker verification problem is actually a set of several binary classification problems, one for each client of the system. Although few positive examples are available for each client, the overall number of available positive examples may be large. Hence, techniques that share information between classification problems should be beneficial. We already mentioned such a technique: the MAP adaptation scheme that trains a single world model on a common data set, and uses it as a prior distribution over the parameters to train a GMM for each client. Here, the role of the world model is to bias each client model towards a reference speaker model. This bias amounts to a soft sharing of parameters.

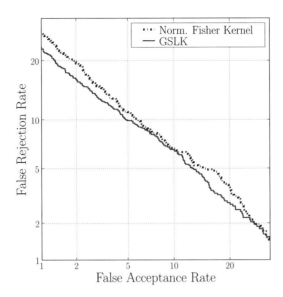

Figure 12.6  Detection error trade-off (DET) curves on the development set of the NIST database for GMM supervector linear kernel (GSLK) and normalized Fisher kernel.

Additional parameter sharing techniques are now used in discriminant approaches. In the following, we discuss one of them, the *Nuisance Attribute Projection* (NAP).

## 12.6.1  Nuisance Attribute Projection

The NAP approach (Solomonoff *et al.* 2004) looks for a linear subspace such that similar accesses (that is, accesses coming from the same client or from the same channel, etc) are near each other. In order to avoid finding an obvious bad solution, the dimension of the target subspace is controlled by cross-validation. This transformation is learned on a large set of clients (similarly to learning a generic GMM in the generative approach). After this step is performed, a standard linear SVM is usually trained for each new client over the transformed access data. This approach provided very good performance in recent NIST evaluations.

More specifically, assume each access sequence $\bar{\mathbf{x}}$ is mapped into a fixed-size feature space through some transformation $\Phi(\bar{\mathbf{x}})$ such as the one used in the GLDS kernel. Let $\mathbf{W}^c$ be a proximity matrix encoding, for each pair of accesses $(\bar{\mathbf{x}}_i, \bar{\mathbf{x}}_j)$, that these sequences were recorded over the same channel ($W_{i,j}^c = 0$) or not ($W_{i,j}^c = 1$). The NAP approach then consists in finding a projection matrix $\mathbf{P}^\star$ such that

$$\mathbf{P}^\star = \arg \min_{\mathbf{P}} \sum_{i,j} W_{i,j}^c \|\mathbf{P}(\Phi(\bar{\mathbf{x}}_i) - \Phi(\bar{\mathbf{x}}_j))\|^2 \qquad (12.19)$$

among orthonormal projection matrices of a given rank. Hence $\mathbf{P}^\star$ minimizes the average difference between accesses from differing channels, in the feature space. Similarly, a second matrix $\mathbf{W}^s$ could encode the fact that two accesses come from the same speaker.

A combination of this prior knowledge could be encoded as follows:

$$\mathbf{W} = \alpha \mathbf{W}^c - \gamma \mathbf{W}^s, \tag{12.20}$$

with $\alpha$ and $\gamma$ hyper-parameters to tune, and $\mathbf{P}^\star$ found to minimize Equation (12.19) with $\mathbf{W}$ instead of $\mathbf{W}^c$.

As stated earlier, $\mathbf{P}^\star$ is then used to project each access $\Phi(\bar{\mathbf{x}})$ into a feature subspace where, for each client, a linear SVM is used to discriminate client and impostor accesses. As shown in Table 12.3 and Figure 12.7, NAP brings significant improvement when combined with the GSLK kernel. On the other hand, the number of support vectors also grows significantly. This can be interpreted that now all accesses are in the same space and are independent of the channel and thus more training impostors are good candidates.

Table 12.3  Equal error rates (EERs) (the lower the better) on the development set of the NIST database, comparing a Support Vector Machine classifier with GMM supervector linear kernel (GSLK) with and without Nuisance Attribute Projection (NAP) (polynomial kernel of degree 3)

|                   | GSKL  | GSLK with NAP |
| ----------------- | ----- | ------------- |
| EER (%)           | 7.9   | 5.8           |
| 95% confidence    | ±0.8  | ±0.6          |
| # support vectors | 34    | 59            |

Although the approach has been shown to yield very good performance results, we believe that there is still room for improvement, since $\mathbf{P}^\star$ is not selected using the criterion that is directly related to the task. Minimizing the average squared distance between accesses of the same client (or accesses of a different channel) is likely to help classification, but it would also be relevant to do something about accesses from different clients, such as moving them away.

## 12.6.2   Other Approaches

Another recent approach that goes in the same direction and that obtains state-of-the-art performance similar to the NAP approach is the Bayesian Factor Analysis approach (Kenny *et al.* 2005). In this case, one assumes that the mean vector of a client model is a linear combination of a generic mean vector, the mean vector of the available training data for that client and the mean vector of the particular channel used in this training data. Once again, the linear combination parameters are trained on a large amount of access data, involving a large number of clients. While this approach is nicely presented theoretically (and obtains very good empirical performance), it does not try to find the optimal parameters of client models and linear combination by taking into account the global cost function.

Another very promising line of research that has emerged in machine learning relates to the general problem of learning a similarity metric (Chopra *et al.* 2005; Lebanon 2006; Weinberger *et al.* 2005). In this setting, where the learning algorithm relies on the comparison of two examples, one can set aside some training examples to actually learn what would be

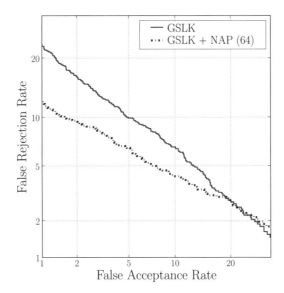

Figure 12.7 Detection error trade-off curves on the development set of the NIST database with GMM supervector linear kernel (GSLK) with and without Nuisance Attribute Projection (NAP).

a good metric to compare pairs of examples. Obviously, in the SVM world, this relates to learning the kernel itself (Crammer *et al.* 2002; Lanckriet *et al.* 2004).

In the context of discriminant approaches to speaker verification, none of these techniques has been tried, to the best of our knowledge. Using a large base of accesses for which one knows the correct identity, one could for instance train a parametric similarity measure that would assess whether two accesses are coming from the same person or not. That could be done efficiently by stochastic gradient descent using a scheme similar to the so-called *Siamese neural network* (Chopra *et al.* 2005) and a margin criterion with proximity constraints.

## 12.7 Is the Margin Useful for This Problem?

The scarcity of positive training examples in speaker verification explains the great improvements that pertain to parameter sharing techniques. In this section, we question whether this specificity also hinders large margin methods to improve upon more simple approaches.

The K-Nearest Neighbors (KNN) algorithm (Duda and Hart 1973) is probably the simplest and the best known non-parametric classifier. Instead of learning a decision boundary, decisions are computed on-the-fly for each test access by using the $k$ nearest labelled sequences in the database as 'experts,' whose votes are aggregated to make up the decision on the current access.

Table 12.4 Equal error rates (ERRs) (the lower the better) on the development set of the NIST database, comparing the Fisher normalized kernel with Nuisance Attribute Projection (NAP) (250) for K-Nearest Neighbors (KNN) and Support Vector Machine (SVM)

|                  | SVM  | KNN  |
| ---------------- | ---- | ---- |
| EER (%)          | 6.7  | 5.3  |
| 95% confidence   | ±0.7 | ±0.7 |
| # support vectors | 47   | –    |

In the weighted KNN (Dudani 1986) variant, the votes of the nearest neighbors are weighted according to their distance to the query:

$$f(\bar{\mathbf{x}}_j) = \sum_{i=1}^{k} y_i w_i, \text{ with } w_i = \begin{cases} 1 & \text{if } d(j, k) = d(j, 1), \\ \dfrac{d(j, k) - d(j, i)}{d(j, k) - d(j, 1)} & \text{otherwise}, \end{cases} \tag{12.21}$$

where the sum runs over the $k$ neighbors of the query $\bar{\mathbf{x}}_j$, $y_i \in \{-1, 1\}$ determines whether the neighbor's access is from a client ($y_i = 1$) or an impostor ($y_1 = -1$), and $d(j, i)$ is the distance from $\bar{\mathbf{x}}_j$ to its $i$th neighbor.

One can then use kernels to define distances, as follows:

$$d(i, j) = \sqrt{K(\bar{\mathbf{x}}_i, \bar{\mathbf{x}}_i) - 2K(\bar{\mathbf{x}}_i, \bar{\mathbf{x}}_j) + K(\bar{\mathbf{x}}_j, \bar{\mathbf{x}}_j)}, \tag{12.22}$$

but it is often better to normalize the data also in the feature space so that they have unit norm, as follows:

$$K_{\text{norm}}(\bar{\mathbf{x}}_i, \bar{\mathbf{x}}_j) = \frac{K(\bar{\mathbf{x}}_i, \bar{\mathbf{x}}_j)}{\sqrt{K(\bar{\mathbf{x}}_i, \bar{\mathbf{x}}_i) K(\bar{\mathbf{x}}_j, \bar{\mathbf{x}}_j)}}, \tag{12.23}$$

which leads to the final distance measure used in the experiments:

$$d_{\text{norm}}(i, j) = \sqrt{2 - 2\frac{K(\bar{\mathbf{x}}_i, \bar{\mathbf{x}}_j)}{\sqrt{K(\bar{\mathbf{x}}_i, \bar{\mathbf{x}}_i) K(\bar{\mathbf{x}}_j, \bar{\mathbf{x}}_j)}}}. \tag{12.24}$$

Table 12.4 and Figure 12.8 compare the normalized Fisher score with the NAP approach followed by either an SVM or a KNN, and as can be seen the KNN approach yields similar if not better performance than the SVM approach. Furthermore, the KNN has several advantages with respect to SVMs: there is no training session and KNN can easily approximate posterior probabilities and does not rely on potentially constraining Mercer conditions to work. On the other hand, the test session might be longer as finding nearest neighbors needs to be efficiently implemented.

## 12.8   Comparing all Methods

As a final experiment, we have compared all the proposed approaches and now report the results on the evaluation set. Figure 12.9 compares a state-of-the-art diagonal GMM with an

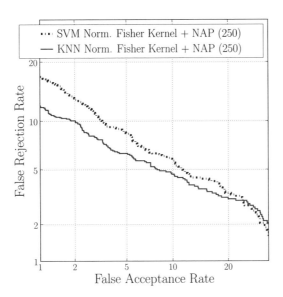

Figure 12.8  Detection error trade-off curves on the development set of the NIST database comparing Fisher normalized kernel with Nuisance Attribute Projection (NAP) for K-Nearest Neighbors (KNN) and Support Vector Machine (SVM).

SVM using a GSLK kernel with NAP, and also with a KNN based on the normalized Fisher kernel with NAP.

In this experiment, the following set of hyper-parameters was tuned according to the EER obtained on the development set:

- The number of neighbors $K$ in the KNN approach was varied between 20 and 200, with optimal value: 100.

- The size of **P**, the transformed space in NAP for the GSLK kernel, was varied between 40 and 250, with optimal value: 64.

- The size of **P**, the transformed space in NAP for the Fisher kernel, was varied between 40 and 400, with optimal value: 250.

- The number of Gaussians in the GMM used for the GSLK and Fisher kernel approaches was varied between 100 and 500, with optimal value: 200.

- All other parameters of the state-of-the-art diagonal GMM baseline were taken from previously published experiments.

The GMM yields the worst performance, probably partly because no channel compensation method is used (while the others use NAP). KNN and SVM performances do not differ significantly, hence the margin does not appear to be at all necessary for speaker verification.

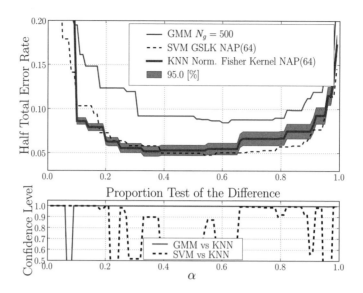

Figure 12.9 Expected performance curve (the lower, the better) on the evaluation set of the NIST database comparing Gaussian Mixture Model (GMM) with T-norm, Support Vector Machine (SVM) with a GMM supervector linear kernel (GSLK) kernel and Nuisance Attribute Projection (NAP), SVM Fisher normalized kernel with NAP and K-nearest Neighbors (KNN) with a Fisher normalized kernel with NAP.

Table 12.5  Final results on the evaluation set of the NIST database

|  | GMM | SVM GSKL NAP | KNN Normalized Fisher NAP |
|---|---|---|---|
| Half total error rate (%) | 10.2 | 5.4 | 5.5 |
| 95% confidence | ±0.7 | ±0.5 | ±0.5 |

GMM, Gaussian Mixture Model; SVM, Support Vector Machine; GSLK, GMM supervector linear kernel; NAP, Nuisance Attribute Projection; KNN, K-Nearest Neighbors.

## 12.9    Conclusion

In this chapter, we have presented the task of text independent speaker verification. We have shown that the traditional method to approach this task is through a generative approach based on GMM.

We have then presented a discriminative framework for this task, and presented several recent approaches in this framework, mainly based on SVM. We have presented various kernels adapted to the task, including the GLDS, GSLK and Fisher kernels. While many of the proposed kernels in the literature were proposed in some heuristic way, including the GLDS and GSLK kernels, we have shown the relation between the principled polynomial kernel and the GLDS kernel, as well as the relation between the principled Fisher kernel and the GSLK kernel. We have then shown that in order for SVMs to perform at a state-of-the-art level, parameter sharing in one way or another was necessary. Approaches such as NAP or Bayesian Factor Analysis were designed for that purpose and indeed helped SVMs to reach better performance.

Finally, we have questioned the main purpose of using SVMs, which maximize the margin in the feature space. We have tried instead a plain KNN approach, which yielded similar performance. This simple experiment shows that future research should concentrate more on better modelling of the distance measure than on maximizing the margin.

A drawback of the current approaches is that they consist in various blocks (feature extraction, feature normalization, distance measure, etc) which were all trained using a separate *ad hoc* criterion. Ultimately, a system that would train all these steps in a single framework to optimize the final objective should perform better, but more research is necessary to reach that goal.

In order to foster more research in this domain, an open-source version of the C++ source code used to performed all experiments described in this chapter has been made available at http://speaker.abracadoudou.com.

# References

Bengio S and Mariéthoz J 2004 A statistical significance test for person authentication. *Proceedings of Odyssey 2004: The Speaker and Language Recognition Workshop* (eds. J Ortega-García *et al.*). ISCA Archive, pp. 237–240.

Bengio S, Mariéthoz J and Keller M 2005 The expected performance curve. *Proceedings of the International Conference on Machine Learning, ICML, Workshop on ROC Analysis in Machine Learning*.

Boughorbel S, Tarel JP and Fleuret F 2004 Non-Mercer kernel for SVM object recognition. *Proceedings of the British Machine Vision Conference*.

Burges C 1998 A tutorial on support vector machines for pattern recognition. *Knowledge Discovery and Data Mining* **2**(2), 121–167.

Campbell W 2002 Generalized linear discriminant sequence kernels for speaker recognition *Proc IEEE International Conference on Audio Speech and Signal Processing*, pp. 161–164.

Campbell W, Campbell J, Reynolds D, Singer E and Torres-Carrasquillo P 2006 Support vector machines for speaker and language recognition. *Computer Speech and Language* **20**(2–3), 125–127.

Campbell W, Sturim D and Reynolds D 2006 Support vector machines using gmm supervectors for speaker verification. *IEEE Signal Processing Letters* **13**(5), 308–311.

Chopra S, Hadsell R and LeCun Y 2005 Learning a similarity metric discriminatively, with application to face verification. *Proceedings of the IEEE Computer Society Conference on Computer Vision and Pattern Recognition (CVPR)*.

Crammer K, Keshet J and Singer Y 2002 Kernel design using boosting *Advances in Neural Information Processing Systems, NIPS*, vol. 15 (eds. S Becker, S Thrum and K Obermayer). MIT Press, Cambridge, MA, pp. 537–544.

Duda RO and Hart PE 1973 *Pattern Classification and Scene Analysis*. John Wiley & Sons, New York.

Dudani SA 1986 The distance-weighted k-nearest neighbor rule. *IEEE Transactions on Systems, Man and Cybernetics* **6**(4), 325–327.

Gauvain JL and Lee CH 1994 Maximum a posteriori estimation for multivariate gaussian mixture observation of Markov chains *IEEE Transactions on Speech and Audio Processing*, **2**, 291–298.

Jaakkola T and Haussler D 1998 Exploiting generative models in discriminative classifiers. *Advances in Neural Information Processing* **11**, 487–493.

Joachims T 2002 *Learning to Classify Text using Support Vector Machines*. Kluwer Academic Publishers, Dordrecht, Netherlands.

Kenny P, Boulianne G and Dumouchel P 2005 Eigenvoice modeling with sparse training data. *IEEE Transactions on Speech and Audio Processing* **13**(3), 345–354.

Lanckriet GRG, Cristianini N, Bartlett P, Ghaoui LE and Jordan MI 2004 Learning the kernel matrix with semidefinite programming. *Journal of Machine Learning Research* **5**, 27–72.

Lebanon G 2006 Metric learning for text documents. *IEEE Transactions on Pattern Analysis and Machine Intelligence* **28**, 497–508.

Mariéthoz J and Bengio S 2007 A kernel trick for sequences applied to text-independent speaker verification systems. *Pattern Recognition*. IDIAP-RR 05-77.

Martin A, Doddington G, Kamm T, Ordowski M and Przybocki M 1997 The DET curve in assessment of detection task performance. *Proceedings of EuroSpeech'97*, Rhodes, Greece, pp. 1895–1898.

Melin H, Koolwaaij J, Lindberg J and Bimbot F 1998 A comparative evaluation of variance flooring techniques in hmm-based speaker verification. *Proceedings of the International Conference on Spoken Language Processing (ICSLP) 1998*, pp. 1903–1906.

Pontil M and Verri A 1998 Support vector machines for 3-d object recognition. *IEEE Transactions on Pattern Analysis and Machine Intelligence* **20**, 637–646.

Reynolds DA, Quatieri TF and Dunn RB 2000 Speaker verification using adapted gaussian mixture models. *Digital Signal Processing* **10**, 19–41.

Solomonoff A, Quillen C and Campbell W 2004 Channel compensation for SVM speaker recognition. *Proceedings of Odyssey 2004: The Speaker and Language Recognition Workshop*, pp. 57–62.

Van Trees HL 1968 *Detection, Estimation and Modulation Theory*, vol. 1. John Wiley & Sons, New York.

Vapnik VN 2000 *The Nature of Statistical Learning Theory*, 2nd edn. Springer.

Wan V and Renals S 2003 Support vector machine speaker verification methodology. *Proceedings of the IEEE International Conference on Acoustic, Speech, and Signal Processing, ICASSP*, pp. 221–224.

Wan V and Renals S 2005a Speaker verification using sequence discriminant support vector machines. *IEEE Transactions on Speech and Audio Processing* **13**(2), 203–210.

Wan V and Renals S 2005b Speaker verification using sequence discriminant support vector machines. *IEEE Transactions on Speech and Audio Processing* **12**(2), 203–210.

Weinberger KQ, Blitzer J and Saul LK 2005 Distance metric learning for large margin nearest neighbor classification. *Advances in Neural Information Processing Systems, NIPS*, vol. 18 (eds. Y Weiss, B Schölkopf and J Platt). MIT Press, Cambridge, MA, pp. 1473–1480.

# 13

# Spectral Clustering for Speech Separation

## Francis R. Bach and Michael I. Jordan

Spectral clustering refers to a class of recent techniques which rely on the eigenstructure of a similarity matrix to partition points into disjoint clusters, with points in the same cluster having high similarity and points in different clusters having low similarity. In this chapter, we introduce the main concepts and algorithms together with recent advances in learning the similarity matrix from data. The techniques are illustrated on the blind one-microphone speech separation problem, by casting the problem as one of segmentation of the spectrogram.

## 13.1 Introduction

Clustering has many applications in machine learning, exploratory data analysis, computer vision and speech processing. Many algorithms have been proposed for this task of grouping data into several subsets that share some common structure (see, e.g., Hastie *et al.* (2001); Mitchell (1997)). Two distinct approaches are very popular: (1) Linkage algorithms are based on thresholding pairwise distances and are best suited for complex elongated structures, but are very sensitive to noise in the data; (2) $K$-means algorithms, on the other hand, are very robust to noise but are best suited for rounded linearly separable clusters. Spectral clustering is aimed at bridging the gap between these approaches, providing a methodology for finding elongated clusters while being more robust to noise than linkage algorithms.

Spectral clustering relies on the eigenstructure of a similarity matrix to partition points into disjoint clusters, with points in the same cluster having high similarity and points in different clusters having low similarity. As presented in Section 13.2, spectral clustering can be cast as a relaxation of a hard combinatorial problem based on *normalized cuts*.

Most clustering techniques explicitly or implicitly assume a metric or a similarity structure over the space of configurations, which is then used by clustering algorithms. The success of such algorithms depends heavily on the choice of the metric, but this choice is generally not treated as part of the learning problem. Thus, time-consuming manual feature selection and weighting is often a necessary precursor to the use of spectral methods.

Several recent papers have considered ways to alleviate this burden by incorporating prior knowledge into the metric, either in the setting of $K$-means clustering (Bar-Hillel *et al.* 2003; Wagstaff *et al.* 2001; Xing *et al.* 2003) or spectral clustering (Kamvar *et al.* 2003; Yu and Shi 2002). In this chapter, we consider a complementary approach, providing a general framework for learning the similarity matrix for spectral clustering from examples. We assume that we are given sample data with known partitions and are asked to build similarity matrices that will lead to these partitions when spectral clustering is performed. This problem is motivated by the availability of such datasets for at least two domains of application: in vision and image segmentation, databases of hand-labeled segmented images are now available (Martin *et al.* 2001), while for the blind separation of speech signals via partitioning of the time-frequency plane (Brown and Cooke 1994), training examples can be created by mixing previously captured signals.

Another important motivation for our work is the need to develop spectral clustering methods that are robust to irrelevant features. Indeed, as we show in Section 13.4.5, the performance of current spectral methods can degrade dramatically in the presence of such irrelevant features. By using our learning algorithm to learn a diagonally-scaled Gaussian kernel for generating the similarity matrix, we obtain an algorithm that is significantly more robust.

Our work is based on a cost function that characterizes how close the eigenstructure of a similarity matrix $\mathbf{W}$ is to a partition $\mathbf{E}$. We derive this cost function in Section 13.2. As we show in Section 13.2.5, minimizing this cost function with respect to the partition $\mathbf{E}$ leads to a new clustering algorithm that takes the form of weighted $K$-means algorithms. Minimizing them with respect to $\mathbf{W}$ yields a theoretical framework for learning the similarity matrix, as we show in Section 13.3. Section 13.3.3 provides foundational material on the approximation of the eigensubspace of a symmetric matrix that is needed for Section 13.4, which presents learning algorithms for spectral clustering.

We highlight one other aspect of the problem here – the major computational challenge involved in applying spectral methods to domains such as vision or speech separation. Indeed, in image segmentation, the number of pixels in an image is usually greater than hundreds of thousands, leading to similarity matrices of potential huge sizes, while, for speech separation, four seconds of speech sampled at 5.5 kHz yields 22 000 samples and thus a naive implementation would need to manipulate similarity matrices of dimension at least 22 000 × 22 000. Thus a major part of our effort to apply spectral clustering techniques to speech separation has involved the design of numerical approximation schemes that exploit the different time scales present in speech signals. In Section 13.4.4, we present numerical techniques that are appropriate for generic clustering problems, while in Section 13.6.3, we show how these techniques specialize to speech. The results presented in this chapter are taken from Bach and Jordan (2006).

## 13.2 Spectral Clustering and Normalized Cuts

In this section, we present our spectral clustering framework. Following Shi and Malik (2000) and Gu *et al.* (2001), we derive the spectral relaxation through normalized cuts. Alternative frameworks, based on Markov random walks (Meila and Shi 2002), on different definitions of the normalized cut (Meila and Xu 2003), or on constrained optimization (Higham and Kibble 2004) lead to similar spectral relaxations.

### 13.2.1 Similarity Matrices

Spectral clustering refers to a class of techniques for clustering that are based on pairwise similarity relations among data points. Given a dataset $\mathcal{I}$ of $P$ points in a space $\mathbf{X}$, we assume that we are given a $P \times P$ 'similarity matrix' $\mathbf{W}$ that measures the similarity between each pair of points: $\mathbf{W}_{pp'}$ is large when points indexed by $p$ and $p'$ are preferably in the same cluster, and is small otherwise. The goal of clustering is to organize the dataset into disjoint subsets with high intra-cluster similarity and low inter-cluster similarity.

Throughout this chapter we always assume that the elements of $\mathbf{W}$ are non-negative ($\mathbf{W} \geqslant 0$) and that $\mathbf{W}$ is symmetric ($\mathbf{W} = \mathbf{W}^\top$). Moreover, we make the assumption that the diagonal elements of $\mathbf{W}$ are strictly positive. In particular, contrary to most work on kernel based algorithms, our theoretical framework makes no assumptions regarding the positive semidefiniteness of the matrix (a symmetric matrix $\mathbf{W}$ is positive semidefinite if and only if for all vectors $\mathbf{u} \in \mathbb{R}^P$, $\mathbf{u}^\top \mathbf{W} \mathbf{u} \geqslant 0$). If in fact the matrix is positive semidefinite this can be exploited in the design of efficient approximation algorithms (see Section 13.4.4). But the spectral clustering algorithms presented in this chapter are not limited to positive semidefinite matrices. In particular, it differs from the usual assumption made in kernel methods. For similar clustering approaches based on kernel methods, see Filippone *et al.* (2008), Bie and Cristianini (2006) and Bach and Harchaoui (2008).

A classical similarity matrix for clustering in $\mathbb{R}^d$ is the diagonally-scaled Gaussian similarity, defined between pairs of points $(\mathbf{x}, \mathbf{y}) \in \mathbb{R}^\mathbf{d} \times \mathbb{R}^\mathbf{d}$ as:

$$\mathbf{W}(\mathbf{x}, \mathbf{y}) = \exp(-(\mathbf{x} - \mathbf{y})^\top \operatorname{Diag}(\mathbf{a})(\mathbf{x} - \mathbf{y})),$$

where $\mathbf{a} \in \mathbb{R}^d$ is a vector of positive parameters, and $\operatorname{Diag}(\mathbf{a})$ denotes the $d \times d$ diagonal matrix with diagonal $\mathbf{a}$. It is also very common to use such similarity matrices after transformation to a set of 'features,' where each feature can depend on the entire dataset $(\mathbf{x}_i)_{i=1,\ldots,P}$ or a subset thereof (see, e.g., Shi and Malik 2000, for an example from computational vision and see Section 13.5 of the current chapter for examples from speech separation). In Figure 13.1, we present a toy example in two dimensions with the Gaussian similarity.

In the context of *graph partitioning* where data points are vertices of an undirected graph and $\mathbf{W}_{ij}$ is defined to be 1 if there is an edge between $i$ and $j$, and zero otherwise, $\mathbf{W}$ is often referred to as an 'affinity matrix' (Chung 1997).

### 13.2.2 Normalized Cuts

We let $V = \{1, \ldots, P\}$ denote the index set of all data points. We wish to find $R$ disjoint clusters, $A = (A_r)_{r \in \{1,\ldots,R\}}$, where $\bigcup_r A_r = V$, that optimize a certain cost function. In this

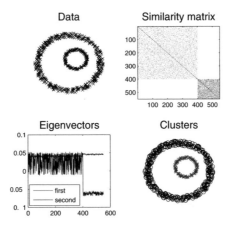

Figure 13.1  Toy examples in two dimensions.

chapter, we consider the $R$-way normalized cut, $C(A, \mathbf{W})$, defined as follows (Gu *et al.* 2001; Shi and Malik 2000). For two subsets $A$, $B$ of $V$, define the total weight between $A$ and $B$ as $W(A, B) = \sum_{i \in A} \sum_{j \in B} \mathbf{W}_{ij}$. Then the normalized cut is equal to

$$C(A, \mathbf{W}) = \sum_{r=1}^{R} \frac{W(A_r, V \backslash A_r)}{W(A_r, V)}. \tag{13.1}$$

Noting that $W(A_r, V) = W(A_r, A_r) + W(A_r, V \backslash A_r)$, we see that the normalized cut is small if for all $r$, the weight between the $r$th cluster and the remaining data points is small compared with the weight within that cluster. The normalized cut criterion thus penalizes unbalanced partitions, while non-normalized criteria do not and often lead to trivial solutions (e.g. a cluster with only one point) when applied to clustering. In addition to being more immune to outliers, the normalized cut criterion and the ensuing spectral relaxations have a simpler theoretical asymptotic behavior when the number of data points tends to infinity (von Luxburg *et al.* 2005).

Let $\mathbf{e}_r$ be the indicator vector in $\mathbb{R}^P$ for the $r$th cluster, i.e. $\mathbf{e}_r \in \{0, 1\}^P$ is such that $\mathbf{e}_r$ has a non-zero component only for points in the $r$th cluster. Knowledge of $\mathbf{E} = (\mathbf{e}_1, \ldots, \mathbf{e}_R) \in \mathbb{R}^{P \times R}$ is equivalent to knowledge of $A = (A_1, \ldots, A_R)$ and, when referring to partitions, we will use the two formulations interchangeably. A short calculation reveals that the normalized cut is then equal to

$$C(E, \mathbf{W}) = \sum_{r=1}^{R} \frac{\mathbf{e}_r^\top (\mathbf{D} - \mathbf{W}) \mathbf{e}_r}{\mathbf{e}_r^\top \mathbf{D} \mathbf{e}_r}, \tag{13.2}$$

where $\mathbf{D}$ denotes the diagonal matrix whose $i$th diagonal element is the sum of the elements in the $i$th row of $\mathbf{W}$, i.e. $\mathbf{D} = \mathrm{Diag}(\mathbf{W1})$, where $\mathbf{1}$ is defined as the vector in $\mathbb{R}^P$ composed of 1s. Since we have assumed that all similarities are non-negative, the matrix $\mathbf{L} = \mathbf{D} - \mathbf{W}$, usually referred to as the 'Laplacian matrix,' is a positive-semidefinite matrix (Chung 1997). In addition, its smallest eigenvalue is always zero, with eigenvector $\mathbf{1}$. Also, we have assumed that the diagonal of $\mathbf{W}$ is strictly positive, which implies that $\mathbf{D}$ is positive-definite. Finally, in the next section, we also consider the normalized Laplacian matrix defined as $\tilde{\mathbf{L}} = \mathbf{I} - \mathbf{D}^{-1/2}\mathbf{W}\mathbf{D}^{-1/2}$. This matrix is also positive-definite with zero as its smallest eigenvalue, associated with eigenvector $\mathbf{D}^{1/2}\mathbf{1}$.

Minimizing the normalized cut is an NP-hard problem (Meila and Xu 2003; Shi and Malik 2000). Fortunately, tractable relaxations based on eigenvalue decomposition can be found.

## 13.2.3 Spectral Relaxation

The following proposition, which extends a result of Shi and Malik (2000) for two clusters to an arbitrary number of clusters, gives an alternative description of the clustering task, and leads to a spectral relaxation.

**Proposition 13.2.1** *For all partitions* $\mathbf{E}$ *into* $R$ *clusters, the* $R$-*way normalized cut* $C(\mathbf{W}, \mathbf{E})$ *is equal to* $R - \mathrm{tr}\, \mathbf{Y}^\top \mathbf{D}^{-1/2}\mathbf{W}\mathbf{D}^{-1/2}\mathbf{Y}$ *for any matrix* $\mathbf{Y} \in \mathbb{R}^{P \times R}$ *such that:*
*(a) the columns of* $\mathbf{D}^{-1/2}\mathbf{Y}$ *are piecewise constant with respect to the clusters* $\mathbf{E}$,
*(b)* $\mathbf{Y}$ *has orthonormal columns* ($\mathbf{Y}^\top \mathbf{Y} = \mathbf{I}$).

*Proof.* The constraint (*a*) is equivalent to the existence of a matrix $\mathbf{\Lambda} \in \mathbb{R}^{R \times R}$ such that $\mathbf{D}^{-1/2}\mathbf{Y} = \mathbf{E}\mathbf{\Lambda}$. The constraint (*b*) is thus written as $\mathbf{I} = \mathbf{Y}^\top \mathbf{Y} = \mathbf{\Lambda}^\top \mathbf{E}^\top \mathbf{D}\mathbf{E}\mathbf{\Lambda}$. The matrix $\mathbf{E}^\top \mathbf{D}\mathbf{E}$ is diagonal, with elements $\mathbf{e}_r^\top \mathbf{D}\mathbf{e}_r$ and is thus positive and invertible. The $R \times R$ matrix $\mathbf{M} = (\mathbf{E}^\top \mathbf{D}\mathbf{E})^{1/2}\mathbf{\Lambda}$ satisfies $\mathbf{M}^\top \mathbf{M} = \mathbf{I}$, i.e. $\mathbf{M}$ is orthogonal, which implies $\mathbf{I} = \mathbf{M}\mathbf{M}^\top = (\mathbf{E}^\top \mathbf{D}\mathbf{E})^{1/2}\mathbf{\Lambda}\mathbf{\Lambda}^\top(\mathbf{E}^\top \mathbf{D}\mathbf{E})^{1/2}$.

This immediately implies that $\mathbf{\Lambda}\mathbf{\Lambda}^\top = (\mathbf{E}^\top \mathbf{D}\mathbf{E})^{-1}$. Thus we have

$$R - \mathrm{tr}\, \mathbf{Y}^\top(\mathbf{D}^{-1/2}\mathbf{W}\mathbf{D}^{-1/2})\mathbf{Y} = R - \mathrm{tr}\, \mathbf{\Lambda}^\top \mathbf{E}^\top \mathbf{D}^{1/2}(\mathbf{D}^{-1/2}\mathbf{W}\mathbf{D}^{-1/2})\mathbf{D}^{1/2}\mathbf{E}\mathbf{\Lambda}$$

$$= R - \mathrm{tr}\, \mathbf{\Lambda}^\top \mathbf{E}^\top \mathbf{W}\mathbf{E}\mathbf{\Lambda}$$

$$= R - \mathbf{E}^\top \mathbf{W}\mathbf{E}\mathbf{\Lambda}\mathbf{\Lambda}^\top = \mathrm{tr}\, \mathbf{E}^\top \mathbf{W}\mathbf{E}(\mathbf{E}^\top \mathbf{D}\mathbf{E})^{-1}$$

$$= C(\mathbf{W}, \mathbf{E}),$$

which completes the proof. $\qquad\square$

By removing the constraint (*a*), we obtain a relaxed optimization problem, whose solutions involve the eigenstructure of $\mathbf{D}^{-1/2}\mathbf{W}\mathbf{D}^{-1/2}$ and which leads to the classical lower bound on the optimal normalized cut (Chan *et al.* 1994; Zha *et al.* 2002). The following proposition gives the solution obtained from the spectral relaxation.

**Proposition 13.2.2** *The maximum of* $\mathrm{tr}\, \mathbf{Y}^\top \mathbf{D}^{-1/2}\mathbf{W}\mathbf{D}^{-1/2}\mathbf{Y}$ *over matrices* $\mathbf{Y} \in \mathbb{R}^{P \times R}$ *such that* $\mathbf{Y}^\top \mathbf{Y} = \mathbf{I}$ *is the sum of the* $R$ *largest eigenvalues of* $\mathbf{D}^{-1/2}\mathbf{W}\mathbf{D}^{-1/2}$. *It is attained at all* $\mathbf{Y}$ *of the form* $\mathbf{Y} = \mathbf{U}\mathbf{B}_1$ *where* $\mathbf{U} \in \mathbb{R}^{P \times R}$ *is any orthonormal basis of the* $R$th *principal subspace of* $\mathbf{D}^{-1/2}\mathbf{W}\mathbf{D}^{-1/2}$ *and* $\mathbf{B}_1$ *is an arbitrary orthogonal matrix in* $\mathbb{R}^{R \times R}$.

*Proof.* Let $\tilde{\mathbf{W}} = \mathbf{D}^{-1/2}\mathbf{W}\mathbf{D}^{-1/2}$. The proposition is equivalent to the classical variational characterization of the sum of the $R$ largest eigenvalues $\lambda_1(\tilde{\mathbf{W}}) \geq \cdots \geq \lambda_R(\tilde{\mathbf{W}})$ of $\tilde{\mathbf{W}}$ – a result known as Ky Fan's theorem (Overton and Womersley 1993):

$$\lambda_1(\tilde{\mathbf{W}}) + \cdots + \lambda_R(\tilde{\mathbf{W}}) = \max\{\mathrm{tr}\, \mathbf{Y}^\top \tilde{\mathbf{W}}\mathbf{Y}, \mathbf{Y} \in \mathbb{R}^{P \times R}, \mathbf{Y}^\top \mathbf{Y} = \mathbf{I}\},$$

where the maximum is attained for all matrices $\mathbf{Y}$ of the form $\mathbf{Y} = \mathbf{UB}_1$, where $\mathbf{U} \in \mathbb{R}^{P \times R}$ is any orthonormal basis of the $R$th principal subspace of $\widetilde{\mathbf{W}}$ and $\mathbf{B}_1$ is an arbitrary orthogonal matrix in $\mathbb{R}^{R \times R}$. Note that the $R$th principal subspace is uniquely defined if and only if $\lambda_R \neq \lambda_{R+1}$ (i.e. there is a positive eigengap). $\qquad\square$

The solutions found by this relaxation will not in general be piecewise constant, i.e. they will not in general satisfy constraint ($a$) in Proposition 13.2.1, and thus the relaxed solution has to be projected back to the constraint set defined by ($a$), an operation we refer to as 'rounding,' due to the similarity with the rounding performed after a linear programming relaxation of an integer programming problem (Bertsimas and Tsitsiklis 1997).

(Tighter relaxations that exploit the non-negativity of cluster indicators can be obtained (Xing and Jordan 2003). These lead to convex relaxations, but their solution cannot be simply interpreted in terms of eigenvectors).

## 13.2.4 Rounding

Our rounding procedure is based on the minimization of a metric between the relaxed solution and the entire set of discrete allowed solutions. Different metrics lead to different rounding schemes. In this section, we present two different metrics that take into account the known invariances of the problem.

Solutions of the relaxed problem are defined up to an orthogonal matrix, i.e. $\mathbf{Y}_{\text{eig}} = \mathbf{UB}_1$, where $\mathbf{U} \in \mathbb{R}^{P \times R}$ is any orthonormal basis of the $R$th principal subspace of $\mathbf{M}$ and $\mathbf{B}_1$ is an arbitrary orthogonal matrix. The set of matrices $\mathbf{Y}$ that correspond to a partition $\mathbf{E}$ and that satisfy constraints ($a$) and ($b$) are of the form $\mathbf{Y}_{\text{part}} = \mathbf{D}^{1/2}\mathbf{E}(\mathbf{E}^\top \mathbf{DE})^{-1/2}\mathbf{B}_2$, where $\mathbf{B}_2$ is an arbitrary orthogonal matrix.

Since both matrices are defined up to an orthogonal matrix, it makes sense to compare the *subspaces* spanned by their columns. A common way to compare subspaces is to compare the orthogonal projection operators on those subspaces (Golub and Loan 1996), that is, to compute the Frobenius norm between $\mathbf{Y}_{\text{eig}}\mathbf{Y}_{\text{eig}}^\top = \mathbf{UU}^\top$ and the orthogonal projection operator $\mathbf{\Pi}_0(\mathbf{W}, \mathbf{E})$ on the subspace spanned by the columns of $\mathbf{D}^{1/2}\mathbf{E} = \mathbf{D}^{1/2}(\mathbf{e}_1, \ldots, \mathbf{e}_r)$, equal to:

$$
\begin{aligned}
\mathbf{\Pi}_0(\mathbf{W}, \mathbf{E}) &= \mathbf{Y}_{\text{part}}\mathbf{Y}_{\text{part}}^\top \\
&= \mathbf{D}^{1/2}\mathbf{E}(\mathbf{E}^\top \mathbf{DE})^{-1}\mathbf{E}^\top \mathbf{D}^{1/2} \\
&= \sum_r \frac{\mathbf{D}^{1/2}\mathbf{e}_r \mathbf{e}_r^\top \mathbf{D}^{1/2}}{\mathbf{e}_r^\top \mathbf{De}_r}.
\end{aligned}
$$

We thus define the following cost function:

$$
J_1(\mathbf{W}, \mathbf{E}) = \tfrac{1}{2}\|\mathbf{U}(\mathbf{W})\mathbf{U}(\mathbf{W})^\top - \mathbf{\Pi}_0(\mathbf{W}, \mathbf{E})\|_F^2. \tag{13.3}
$$

Other cost functions could be derived using different metrics between linear subspaces, but as shown in Section 13.2.5, the Frobenius norm between orthogonal projections has the appealing feature that it leads to a weighted $K$-means algorithm. Another possibility has been followed by (Yu and Shi 2003).[1]

Using the fact that both $\mathbf{U}(\mathbf{W})\mathbf{U}(\mathbf{W})^\top$ and $\boldsymbol{\Pi}_0(\mathbf{W}, \mathbf{E})$ are orthogonal projection operators on linear subspaces of dimension $R$, we have:

$$J_1(\mathbf{W}, \mathbf{E}) = \frac{1}{2} \operatorname{tr} \mathbf{U}(\mathbf{W})\mathbf{U}(\mathbf{W})^\top + \frac{1}{2} \operatorname{tr} \boldsymbol{\Pi}_0(\mathbf{W}, \mathbf{E})\boldsymbol{\Pi}_0(\mathbf{W}, \mathbf{E})^\top - \operatorname{tr} \mathbf{U}(\mathbf{W})\mathbf{U}(\mathbf{W})^\top \boldsymbol{\Pi}_0(\mathbf{W}, \mathbf{E})$$

$$= \frac{R}{2} + \frac{R}{2} - \operatorname{tr} \mathbf{U}(\mathbf{W})\mathbf{U}(\mathbf{W})^\top \boldsymbol{\Pi}_0(\mathbf{W}, \mathbf{E})$$

$$= R - \sum_r \frac{\mathbf{e}_r^\top \mathbf{D}^{1/2}\mathbf{U}(\mathbf{W})\mathbf{U}(\mathbf{W})^\top \mathbf{D}^{1/2}\mathbf{e}_r}{\mathbf{e}_r^\top \mathbf{D}\mathbf{e}_r}.$$

Note that if the similarity matrix $\mathbf{W}$ has rank equal to $R$, then our cost function $J_1(\mathbf{W}, \mathbf{E})$ is exactly equal to the normalized cut $C(\mathbf{W}, \mathbf{E})$.

**Alternative normalization of eigenvectors**   By construction of the orthonormal basis $\mathbf{U}$ of the $R$-dimensional principal subspace of $\widetilde{\mathbf{W}} = \mathbf{D}^{-1/2}\mathbf{W}\mathbf{D}^{-1/2}$, the $P$ $R$-dimensional rows $\mathbf{u}_1, \ldots, \mathbf{u}_P \in \mathbb{R}^R$ are already *globally* normalized, i.e. they satisfy $\mathbf{U}^\top \mathbf{U} = \sum_{i=1}^P \mathbf{u}_i \mathbf{u}_i^\top = \mathbf{I}$. Additional renormalization of those eigenvectors has proved worthwhile in clustering applications (Ng *et al.* 2002; Scott and Longuet-Higgins 1990; Weiss 1999), as can be seen in the idealized situation in which the similarity is zero between points that belong to different clusters and strictly positive between points in the same clusters. In this situation, the eigenvalue 1 has multiplicity $R$, and $\mathbf{D}^{1/2}\mathbf{E}$ is an orthonormal basis of the principal subspace. Thus, any basis $\mathbf{U}$ of the principal subspace has rows which are located on orthogonal *rays* in $\mathbb{R}^R$, where the distance from the $i$th row $\mathbf{u}_i$ to the origin is simply $\mathbf{D}_{ii}^{1/2}$. By normalizing each row by the value $\mathbf{D}_{ii}^{1/2}$ or by its norm $\|\mathbf{u}_i\|$, the rows become orthonormal points in $\mathbb{R}^R$ (in the idealized situation) and thus are trivial to cluster. Ng *et al.* (2002) have shown that when the similarity matrix is 'close' to this idealized situation, the properly normalized rows tightly cluster around an orthonormal basis.

Our cost function characterizes the ability of the matrix $\mathbf{W}$ to produce the partition $\mathbf{E}$ when using its eigenvectors. Minimizing with respect to $\mathbf{E}$ leads to new clustering algorithms, which we now present. Minimizing with respect to the matrix $\mathbf{W}$ for a given partition $\mathbf{E}$ leads to algorithms for learning the similarity matrix, as we show in Section 13.3 and Section 13.4.

## 13.2.5   Spectral Clustering Algorithms

In this section, we provide a variational formulation of our cost function. Those variational formulations lead naturally to $K$-means and weighted $K$-means algorithms for minimizing this cost function with respect to the partition. While $K$-means is often used heuristically as a post-processor for spectral clustering (Meila and Shi 2002; Ng *et al.* 2002), our approach provides a mathematical foundation for the use of $K$-means.

The following theorem, inspired by the spectral relaxation of $K$-means presented by Zha *et al.* (2002), shows that the cost function can be interpreted as a weighted distortion measure:

---

[1]Another natural possibility is to compare directly $\mathbf{U}$ (or a normalized version thereof) with the indicator matrix $\mathbf{E}$, up to an orthogonal matrix, which then has to be estimated. This approach leads to an alternating minimization scheme similar to $K$-means.

---

**Input**: Similarity matrix $\mathbf{W} \in \mathbb{R}^{P \times P}$.

**Algorithm**:
1. Compute first $R$ eigenvectors $\mathbf{U}$ of $\mathbf{D}^{-1/2}\mathbf{W}\mathbf{D}^{-1/2}$ where $\mathbf{D} = \mathrm{diag}(\mathbf{W}\mathbf{1})$.
2. Let $\mathbf{U} = (\mathbf{u}_1, \ldots, \mathbf{u}_P)^\top \in \mathbb{R}^{P \times R}$ and $d_p = \mathbf{D}_{pp}$.
3. Initialize partition A.
4. Weighted $K$-means: While partition $A$ is not stationary,
      a. For all $r$, $\boldsymbol{\mu}_r = \sum_{p \in A_r} d_p^{1/2} \mathbf{u}_p / \sum_{p \in A_r} d_p$
      b. For all $p$, assign $p$ to $A_r$ where $r = \arg\min_{r'} \|\mathbf{u}_p d_p^{-1/2} - \boldsymbol{\mu}_{r'}\|$

**Output**: partition $A$, distortion measure $\sum_r \sum_{p \in A_r} d_p \|\mathbf{u}_p d_p^{-1/2} - \boldsymbol{\mu}_r\|^2$

---

Figure 13.2  Spectral clustering algorithm that minimizes $J_1(\mathbf{W}, \mathbf{E})$ with respect to $\mathbf{E}$ with weighted $K$-mean. See Section 13.2.5 for the initialization of the partition $A$.

**Theorem 13.2.3** *Let $\mathbf{W}$ be a similarity matrix and let $\mathbf{U} = (\mathbf{u}_1, \ldots, \mathbf{u}_P)^\top$, where $\mathbf{u}_p \in \mathbb{R}^R$, be an orthonormal basis of the Rth principal subspace of $\mathbf{D}^{-1/2}\mathbf{W}\mathbf{D}^{-1/2}$, and $d_p = \mathbf{D}_{pp}$ for all p. For any partition $\mathbf{E} \equiv A$, we have*

$$J_1(\mathbf{W}, \mathbf{E}) = \min_{(\boldsymbol{\mu}_1, \ldots, \boldsymbol{\mu}_R) \in \mathbb{R}^{R \times R}} \sum_r \sum_{p \in A_r} d_p \|\mathbf{u}_p d_p^{-1/2} - \boldsymbol{\mu}_r\|^2.$$

*Proof.* Let $D(\boldsymbol{\mu}, A) = \sum_r \sum_{p \in A_r} d_p \|\mathbf{u}_p d_p^{-1/2} - \boldsymbol{\mu}_r\|^2$. Minimizing $D(\boldsymbol{\mu}, A)$ with respect to $\mu$ is a decoupled least-squares problem and we get

$$\min_{\boldsymbol{\mu}} D(\boldsymbol{\mu}, A) = \sum_r \sum_{p \in A_r} \mathbf{u}_p^\top \mathbf{u}_p - \sum_r \left\| \sum_{p \in A_r} d_p^{1/2} \mathbf{u}_p \right\|^2 \Big/ \left( \sum_{p \in A_r} d_p \right)$$

$$= \sum_p \mathbf{u}_p^\top \mathbf{u}_p - \sum_r \sum_{p, p' \in A_r} d_p^{1/2} d_{p'}^{1/2} \mathbf{u}_p^\top \mathbf{u}_{p'} / (\mathbf{e}_r^\top \mathbf{D} \mathbf{e}_r)$$

$$= R - \sum_r \mathbf{e}_r^\top \mathbf{D}^{1/2} \mathbf{U} \mathbf{U}^\top \mathbf{D}^{1/2} \mathbf{e}_r / (\mathbf{e}_r^\top \mathbf{D} \mathbf{e}_r) = J_1(\mathbf{W}, \mathbf{E}). \qquad \square$$

This theorem has an immediate algorithmic implication – to minimize the cost function $J_1(\mathbf{W}, \mathbf{E})$ with respect to the partition $\mathbf{E}$, we can use a weighted $K$-means algorithm. The resulting algorithm is presented in Figure 13.2.

The rounding procedures that we propose in this chapter are similar to those in other spectral clustering algorithms (Ng *et al.* 2002; Yu and Shi 2003). Empirically, all such rounding schemes usually lead to similar partitions. The main advantage of our procedure – which differs from the others in being derived from a cost function – its that it naturally leads to an algorithm for learning the similarity matrix from data, presented in Section 13.3.

**Initialization**   The $K$-means algorithm can be interpreted as a coordinate descent algorithm and is thus subject to problems of local minima. Thus good initialization is crucial for the practical success of the algorithm in Figure 13.2.

A similarity matrix $\mathbf{W}$ is said to be *perfect with respect to a partition* $\mathbf{E}$ with $R$ clusters if the cost function $J_1(\mathbf{W}, \mathbf{E})$ is exactly equal to zero. This is true in at least two potentially distinct situations: (1) when the matrix $\mathbf{W}$ is block-constant, where the block structure follows the partition $\mathbf{E}$, and, as seen earlier, (2) when the matrix $\mathbf{W}$ is such that the similarity between points in different clusters is zero, while the similarity between points in the same clusters is strictly positive (Ng *et al.* 2002; Weiss 1999).

In both situations, the $R$ cluster centroids are orthogonal vectors, and Ng *et al.* (2002) have shown that when the similarity matrix is 'close' to the second known type of perfect matrices, those centroids are close to orthogonal. This leads to the following natural initialization of the partition $A$ for the $K$-means algorithm in Figure 13.2 (Ng *et al.* 2002): select a point $\mathbf{u}_p$ at random, and successively select $R - 1$ points whose directions are most orthogonal to the previously chosen points; then assign each data point to the closest of the $R$ chosen points.

### 13.2.6   Variational Formulation for the Normalized Cut

In this section, we show that there is a variational formulation of the normalized cut similar to Theorem 13.2.3 for positive semidefinite similarity matrices, i.e. for matrices that can be factorized as $\mathbf{W} = \mathbf{G}\mathbf{G}^\top$ where $\mathbf{G} \in \mathbb{R}^{P \times M}$, where $M \leqslant P$. Indeed we have the following theorem, whose proof is almost identical to the proof of Theorem 13.2.3.

**Theorem 13.2.4**  *If* $\mathbf{W} = \mathbf{G}\mathbf{G}^\top$, *where* $\mathbf{G} \in \mathbb{R}^{P \times M}$, *then for any partition* $\mathbf{E}$, *we have*

$$C(\mathbf{W}, \mathbf{E}) = \min_{(\boldsymbol{\mu}_1, \ldots, \boldsymbol{\mu}_R) \in \mathbb{R}^{R \times R}} \sum_r \sum_{p \in A_r} \mathbf{d}_p \| \mathbf{g}_p d_p^{-1} - \boldsymbol{\mu}_r \|^2 + R - \operatorname{tr} \mathbf{D}^{-1/2}\mathbf{W}\mathbf{D}^{-1/2}. \quad (13.4)$$

This theorem shows that for positive semidefinite matrices, the normalized cut problem is equivalent to the minimization of a weighted distortion measure. However, the dimensionality of the space involved in the distortion measure is equal to the rank of the similarity matrices, and thus can be very large (as large as the number of data points). Consequently, this theorem does not lead straightforwardly to an efficient algorithm for minimizing normalized cuts, since a weighted $K$-means algorithm in very high dimensions is subject to severe local minima problems (see, e.g., Meila and Heckerman 2001). See Dhillon *et al.* (2004) for further algorithms based on the equivalence between normalized cuts and weighted $K$-means.

## 13.3   Cost Functions for Learning the Similarity Matrix

Given a similarity matrix $\mathbf{W}$, the steps of a spectral clustering algorithm are (1) normalization, (2) computation of eigenvalues and (3) partitioning of the eigenvectors using (weighted) $K$-means to obtain a partition $\mathbf{E}$. In this section, we assume that the partition $\mathbf{E}$ is given, and we develop a theoretical framework and a set of algorithms for learning a similarity matrix $\mathbf{W}$.

It is important to note that if we put no constraints on $\mathbf{W}$, then there is a trivial solution, namely any perfect similarity matrix with respect to the partition $\mathbf{E}$; in particular, any matrix that is block-constant with the appropriate blocks. For our problem to be meaningful, we thus must consider a setting in which there are several datasets to partition and we have a parametric form for the similarity matrix. The objective is to learn parameters that generalize to unseen datasets with a similar structure. We thus assume that the similarity matrix is a function of a vector variable $\boldsymbol{\alpha} \in \mathbb{R}^F$, and develop a method for learning $\boldsymbol{\alpha}$.

Given a distance between partitions, a naive algorithm would simply minimize the distance between the true partition $\mathbf{E}$ and the output of the spectral clustering algorithm. However, the $K$-means algorithm that is used to cluster eigenvectors is a non-continuous map and the naive cost function would be non-continuous and thus hard to optimize. In this section, we first show that the cost function we have presented is an upper bound on the naive cost function; this upper bound has better differentiability properties and is amenable to gradient-based optimization. The function that we obtain is a function of eigensubspaces and we provide numerical algorithms to efficiently minimize such functions in Section 13.3.3.

## 13.3.1 Distance Between Partitions

Let $\mathbf{E} = (\mathbf{e}_r)_{r=1,\dots,R}$ and $\mathbf{F} = (\mathbf{f}_s)_{s=1,\dots,S}$ be two partitions of $P$ data points with $R$ and $S$ clusters, represented by the indicator matrices of sizes $P \times R$ and $P \times S$, respectively. We use the following distance between the two partitions (Hubert and Arabie 1985):

$$d(\mathbf{E}, \mathbf{F}) = \frac{1}{\sqrt{2}} \|\mathbf{E}(\mathbf{E}^\top \mathbf{E})^{-1}\mathbf{E}^\top - \mathbf{F}(\mathbf{F}^\top \mathbf{F})^{-1}\mathbf{F}^\top\|$$

$$= \frac{1}{\sqrt{2}} \left\| \sum_r \frac{\mathbf{e}_r \mathbf{e}_r^\top}{\mathbf{e}_r^\top \mathbf{e}_r} - \sum_s \frac{\mathbf{f}_s \mathbf{f}_s^\top}{\mathbf{f}_s^\top \mathbf{f}_s} \right\|_F$$

$$= \frac{1}{\sqrt{2}} \left( R + S - 2 \sum_{r,s} \frac{(\mathbf{e}_r^\top \mathbf{f}_s)^2}{(\mathbf{e}_r^\top \mathbf{e}_r)(\mathbf{f}_s^\top \mathbf{f}_s)} \right)^{1/2}. \tag{13.5}$$

The term $\mathbf{e}_r^\top \mathbf{f}_s$ simply counts the number of data points which belong to the $r$th cluster of $\mathbf{E}$ and the $s$th cluster of $\mathbf{F}$. The function $d(\mathbf{E}, \mathbf{F})$ is a distance for partitions, i.e. it is non-negative and symmetric, it is equal to zero if and only if the partitions are equal, and it satisfies the triangle inequality. Moreover, if $\mathbf{F}$ has $S$ clusters and $\mathbf{E}$ has $R$ clusters, we have $0 \leq d(\mathbf{E}, \mathbf{F}) \leq ((R + S)/2 - 1)^{1/2}$. In simulations, we compare partitions using the squared distance.

## 13.3.2 Cost Functions as Upper Bounds

We let $\mathbf{E}_1(\mathbf{W})$ denote the clustering obtained by minimizing the cost function $J_1(\mathbf{W}, \mathbf{E})$ with respect to $\mathbf{E}$. The following theorem shows that our cost function is an upper bound on the distance between a partition and the output of the spectral clustering algorithm.

**Theorem 13.3.1** *Let* $\eta(\mathbf{W}) = \max_p \mathbf{D}_{pp}/ \min_p \mathbf{D}_{pp} \geq 1$. *If* $\mathbf{E}_1(\mathbf{W}) = \arg\min_R J_1(\mathbf{W}, \mathbf{E})$, *then for all partitions* $\mathbf{E}$, *we have*

$$d(\mathbf{E}, \mathbf{E}_1(\mathbf{W}))^2 \leq 4\eta(\mathbf{W}) J_1(\mathbf{W}, \mathbf{E}). \tag{13.6}$$

The previous theorem shows that minimizing our cost function is equivalent to minimizing an upper bound on the true cost function. This bound is tight at zero; consequently, if we are able to produce a similarity matrix $\mathbf{W}$ with small $J_1(\mathbf{W}, \mathbf{E})$ cost, then the matrix will provably lead to a partition that is close to $\mathbf{E}$. Note that the bound in Equation (13.6) contains a constant term dependent on $\mathbf{W}$ and the framework can be slightly modified to take this into account (Bach and Jordan 2006). In Section 13.3.4, we compare our cost function with previously proposed cost functions.

### 13.3.3 Functions of Eigensubspaces

Our cost function, as defined in Equation (13.3), depends on the $R$th principal eigensubspace, i.e. the subspace spanned by the first $R$ eigenvectors, $\mathbf{U} \in \mathbb{R}^{P \times R}$, of $\widetilde{\mathbf{W}} = \mathbf{D}^{-1/2}\mathbf{W}\mathbf{D}^{-1/2}$. In this section, we review classical properties of eigensubspaces, and present optimization techniques to minimize functions of eigensubspaces. In this section, we focus mainly on the cost function $J_1(\mathbf{W}, \mathbf{E})$ which is defined in terms of the projections onto the principal subspace of $\widetilde{\mathbf{W}} = \mathbf{D}^{-1/2}\mathbf{W}\mathbf{D}^{-1/2}$. In this section, we first assume that all considered matrices are positive-semidefinite, so that all eigenvalues are non-negative, postponing the treatment of the general case to Section 13.3.3.

#### (a) Properties of Eigensubspaces

Let $\mathbf{M}_{P,R}$ be the set of symmetric matrices such that there is a positive gap between the $R$th largest eigenvalue and the $(R+1)$th largest eigenvalue. The set $\mathbf{M}_{P,R}$ is open (Magnus and Neudecker 1999), and for any matrix in $\mathbf{M}_{P,R}$, the $R$th principal subspace $E_R(\mathbf{M})$ is uniquely defined and the orthogonal projection $\mathbf{\Pi}_R(\mathbf{M})$ on that subspace is an unique identifier of that subspace. If $\mathbf{U}_R(\mathbf{M})$ is an orthonormal basis of eigenvectors associated with the $R$ largest eigenvalues, we have $\mathbf{\Pi}_R(\mathbf{M}) = \mathbf{U}_R(\mathbf{M})\mathbf{U}_R(\mathbf{M})^{\top}$, and the value is independent of the choice of the basis $\mathbf{U}_R(\mathbf{M})$. Note that the $R$th eigensubspace is well defined even if some eigenvalues larger than the $R$th eigenvalue coalesce (in which case, the $R$ eigenvectors are not well defined but the $R$th principal eigensubspace is).

The computation of eigenvectors and eigenvalues is a well-studied problem in numerical linear algebra (see, e.g., Golub and Loan 1996). The two classical iterative techniques to obtain a few eigenvalues of a symmetric matrix are the *orthogonal iterations* (a generalization of the power method for one eigenvalue) and the *Lanczös method*.

The method of orthogonal iterations starts with a random matrix $\mathbf{V}$ in $\mathbb{R}^{P \times R}$, successively multiplies $\mathbf{V}$ by the matrix $\mathbf{M}$ and orthonormalizes the result with the QR decomposition. For almost all $\mathbf{V}$, the orthogonal iterations converge to the principal eigensubspace, and the convergence is linear with rate $\lambda_{R+1}(\mathbf{M})/\lambda_R(\mathbf{M})$, where $\lambda_1(\mathbf{M}) \geqslant \cdots \geqslant \lambda_{R+1}(\mathbf{M})$ are the $R+1$ largest eigenvalues of $\mathbf{M}$. The complexity of performing $q$ steps of the orthogonal iterations is $qR$ times the complexity of the matrix-vector product with the matrix $\mathbf{M}$. If $\mathbf{M}$ has no special structure, the complexity is thus $O(qRP^2)$. As discussed in Section 13.4.4, if special structure is present in $\mathbf{M}$ it is possible to reduce this to linear in $P$. The number of steps to obtain a given precision depends directly on the multiplicative eigengap $\varepsilon_R(\mathbf{M}) = \lambda_{R+1}(\mathbf{M})/\lambda_R(\mathbf{M}) \leqslant 1$; indeed this number of iterations is $O(1/(1 - \varepsilon_R(\mathbf{M})))$.

The Lanczös method is also an iterative method, one which makes better use of the available information to obtain more rapid convergence. Indeed the number of iterations is only $O(1/(1 - \varepsilon_R(\mathbf{M}))^{1/2})$, i.e. the square root of the number of iterations for the orthogonal iterations (Golub and Loan 1996). Note that it is usual to perform subspace iterations on more than the desired number of eigenvalues in order to improve convergence (Bathe and Wilson 1976).

Finally, in our setting of learning the similarity matrix, we can speed up the eigenvalue computation by initializing the power or Lanczös method with the eigensubspace of previous iterations. Other techniques are also available that can provide a similar speed-up by efficiently tracking the principal subspace of slowly varying matrices (Comon and Golub 1990; Edelman *et al.* 1999).

## (b) Approximation of Eigensubspace and its Differential

When learning the similarity matrix, the cost function and its derivatives are computed many times and it is thus worthwhile to use an efficient approximation of the eigensubspace as well as its differential. A very natural solution is to stop the iterative methods for computing eigenvectors at a fixed iteration $q$. The following proposition shows that for the method of power iterations, for almost all starting matrices $\mathbf{V} \in \mathbb{R}^{P \times R}$, the projection obtained by early stopping is an infinitely differentiable function:

**Proposition 13.3.2** *Let* $\mathbf{V} \in \mathbb{R}^{P \times R}$ *be such that* $\eta = \max_{\mathbf{u} \in E_R(\mathbf{M})^\perp, \mathbf{v} \in \mathrm{range}(\mathbf{V})} \cos(\mathbf{u}, \mathbf{v})$ $< 1$. *Then if we let* $\mathbf{V}_q(\mathbf{M})$ *denote the results of* $q$ *orthogonal iterations, the function* $\mathbf{V}_q(\mathbf{M})\mathbf{V}_q(\mathbf{M})^\top$ *is infinitely differentiable in a neighborhood of* $\mathbf{M}$, *and we have:* $\|\mathbf{V}_q(\mathbf{M})\mathbf{V}_q(\mathbf{M})^\top - \mathbf{\Pi}_R(\mathbf{M})\|_2 \le (\eta/(1-\eta^2)^{1/2})(|\lambda_{R+1}(\mathbf{M})|/|\lambda_R(\mathbf{M})|)^q.$

*Proof.* Golub and Loan (1996) show that for all $q$, $\mathbf{M}^q\mathbf{V}$ always has rank $R$. When only the projection on the column space is sought, the result of the orthogonal iterations does not depend on the chosen method of orthonormalization (usually the $QR$ decomposition), and the final result is theoretically equivalent to orthonormalizing at the last iteration. Thus $\mathbf{V}_q(\mathbf{M})\mathbf{V}_q(\mathbf{M})^\top = \mathbf{M}^q\mathbf{V}(\mathbf{V}^\top\mathbf{M}^{2q}\mathbf{V})^{-1}\mathbf{V}^\top\mathbf{M}^q$. $\mathbf{V}_q(\mathbf{M})\mathbf{V}_q(\mathbf{M})^\top$ is $C^\infty$ since matrix inversion and multiplication are $C^\infty$. The bound is proved in Golub and Loan (1996) for the $QR$ orthogonal iterations, and since the subspaces computed by the two methods are the same, the bound also holds here. The derivative can easily be computed using the chain rule. □

Note that numerically taking powers of matrices without care can lead to disastrous results (Golub and Loan 1996). By using successive $QR$ iterations, the computations can be made stable and the same technique can be used for the computation of the derivatives.

## (c) Potentially Hard Eigenvalue Problems

In most of the literature on spectral clustering, it is taken for granted that the eigenvalue problem is easy to solve. It turns out that in many situations, the (multiplicative) eigengap is very close to 1, making the eigenvector computation difficult (examples are given in the following section).

When the eigengap is close to 1, a large power is necessary for the orthogonal iterations to converge. In order to avoid those situations, we regularize the approximation of the cost function based on the orthogonal iterations by a term which is large when the matrix $\mathbf{D}^{-1/2}\mathbf{W}\mathbf{D}^{-1/2}$ is expected to have a small eigengap, and small otherwise. We use the function $n(\mathbf{W}) = \mathrm{tr}\,\mathbf{W}/\mathrm{tr}\,\mathbf{D}$, which is always between 0 and 1, and is equal to 1 when $\mathbf{W}$ is diagonal (and thus has no eigengap).

We thus use the cost function defined as follows. Let $\mathbf{V} \in \mathbb{R}^{P \times R}$ be defined as $\mathbf{D}^{1/2}\mathbf{F}$, where the $r$th column of $\mathbf{F}$ is the indicator matrix of a random subset of the $r$th cluster normalized by the number of points in that cluster. This definition of $\mathbf{W}$ ensures that when

**W** is diagonal, the cost function is equal to $R - 1$, i.e. if the power iterations are likely not to converge, then the value is the maximum possible true value of the cost.

Let $\mathbf{B}(\mathbf{W})$ be an approximate orthonormal basis of the projections on the $R$th principal subspace of $\mathbf{D}^{-1/2}\mathbf{W}\mathbf{D}^{-1/2}$, based on orthogonal iterations starting from $\mathbf{V}$.[2]

The cost function that we use to approximate $J_1(\mathbf{W}, \mathbf{E})$ is

$$F_1(\mathbf{W}, \mathbf{E}) = \tfrac{1}{2}\|\mathbf{B}(\mathbf{W})\mathbf{B}(\mathbf{W})^\top - \mathbf{\Pi}_0(\mathbf{W}, \mathbf{E})\|_F^2 - \kappa \log(1 - n(\mathbf{W})). \qquad (13.7)$$

**(d) Negative Eigenvalues**

The spectral relaxation in Proposition 13.2.2 involves the largest eigenvalues of the matrix $\widetilde{\mathbf{W}} = \mathbf{D}^{-1/2}\mathbf{W}\mathbf{D}^{-1/2}$. The vector $\mathbf{D}^{1/2}\mathbf{1}$ is an eigenvector with eigenvalue 1; since we have assumed that $\mathbf{W}$ is pointwise non-negative, 1 is the largest eigenvalue of $\widetilde{\mathbf{W}}$. Given any symmetric matrices (not necessarily positive-semidefinite) orthogonal iterations will converge to eigensubspaces corresponding to eigenvalues which have largest magnitude, and it may well be the case that some negative eigenvalues of $\widetilde{\mathbf{W}}$ have larger magnitude than the largest (positive) eigenvalues, thus preventing the orthogonal iterations from converging to the desired eigenvectors. When the matrix $\mathbf{W}$ is positive-semidefinite this is not possible. However, in the general case, eigenvalues have to be shifted so that they are all non-negative. This is done by adding a multiple of the identity matrix to the matrix $\widetilde{\mathbf{W}}$, which does not modify the eigenvectors but simply potentially changes the signs of the eigenvalues. In our context adding exactly the identity matrix is sufficient to make the matrix positive; indeed, when $\mathbf{W}$ is pointwise non-negative, then both $\mathbf{D} + \mathbf{W}$ and $\mathbf{D} - \mathbf{W}$ are *diagonally dominant* with non-negative diagonal entries, and are thus positive semidefinite (Golub and Loan 1996), which implies that $-\mathbf{I} \preccurlyeq \widetilde{\mathbf{W}} \preccurlyeq \mathbf{I}$, and thus $\mathbf{I} + \widetilde{\mathbf{W}}$ is positive semidefinite.

## 13.3.4 Empirical Comparisons Between Cost Functions

In this section, we study the ability of the various cost functions we have proposed to track the gold standard error measure in Equation (13.5) as we vary the parameter $\alpha$ in the similarity matrix $\mathbf{W}_{pp'} = \exp(-\alpha\|\mathbf{x}_p - \mathbf{x}_{p'}\|^2)$. We study the cost function $J_1(\mathbf{W}, \mathbf{E})$ as well as their approximations based on the power method presented in Section 13.3.3. We also present results for two existing approaches, one based on a Markov chain interpretation of spectral clustering (Meila and Shi 2002) and one based on the alignment (Cristianini *et al.* 2002) of $\mathbf{D}^{-1/2}\mathbf{W}\mathbf{D}^{-1/2}$ and $\mathbf{\Pi}_0$. Our experiment is based on the simple clustering problem shown in Figure 13.3(a). This apparently simple toy example captures much of the core difficulty of spectral clustering – nonlinear separability and thinness/sparsity of clusters (any point has very few near neighbors belonging to the same cluster, so that the weighted graph is sparse). In particular, in Figure 13.3(b) we plot the eigengap of the similarity matrix as a function of $\alpha$, noting that for all optimum values of $\alpha$, this gap is very close to 1, and thus the eigenvalue problem is hard to solve. Worse, for large values of $\alpha$, the eigengap becomes so small that the eigensolver starts to diverge. It is thus essential to prevent our learning algorithm from yielding parameter settings that lead to a very small eigengap. In Figure 13.3(e), we plot our approximation of the cost function based on the power method,

---

[2]The matrix $\mathbf{D}^{-1/2}\mathbf{W}\mathbf{D}^{-1/2}$ always has the same largest eigenvalue 1 with eigenvector $\mathbf{D}^{1/2}\mathbf{1}$ and we could consider instead the $(R - 1)$th principal subspace of $\mathbf{D}^{-1/2}\mathbf{W}\mathbf{D}^{-1/2} - \mathbf{D}^{1/2}\mathbf{1}\mathbf{1}^\top\mathbf{D}^{1/2}/(\mathbf{1}^\top\mathbf{D}\mathbf{1})$.

and we see that, even without the additional regularization presented in Section 13.3.3, our approximate cost function avoids a very small eigengap. The regularization presented in Section 13.3.3 strengthens this behavior.

In Figure 13.3(c) and (d), we plot the four cost functions against the gold standard. The gold standard curve shows that the optimal $\alpha$ lies above 2.5 on a log scale, and as seen in Figure 13.3(c) and (e), the minima of the new cost function and its approximation lie among these values. As seen in Figure 13.3(d), on the other hand, the alignment and Markov-chain-based cost functions show a poor match to the gold standard, and yield minima far from the optimum.

The problem with the latter cost functions is that these functions essentially measure the distance between the similarity matrix $\mathbf{W}$ (or a normalized version of $\mathbf{W}$) and a matrix $T$ which (after permutation) is block-diagonal with constant blocks. Spectral clustering does work with matrices which are close to block-constant; however, one of the strengths of spectral clustering is its ability to work effectively with similarity matrices which are not block-constant, and which may exhibit strong variations among each block.

Indeed, in examples such as that shown in Figure 13.3, the optimal similarity matrix is very far from being block-diagonal with constant blocks. Rather, given that data points that lie in the same ring are in general far apart, the blocks are very sparse – not constant and full. Methods that try to find constant blocks cannot find the optimal matrices in these cases. In the language of spectral graph partitioning, where we have a weighted graph with weights $\mathbf{W}$, each cluster is a connected but very sparse graph. The power $\mathbf{W}^q$ corresponds to the $q$th power of the graph; i.e. the graph in which two vertices are linked by an edge if and only if they are linked by a path of length no more than $q$ in the original graph. Thus taking powers can be interpreted as 'thickening' the graph to make the clusters more apparent, while not changing the eigenstructure of the matrix (taking powers of symmetric matrices only changes the eigenvalues, not the eigenvectors). Note that other clustering approaches based on taking powers of similarity matrices have been studied by Tishby and Slonim (2001) and Szummer and Jaakkola (2002); these differ from our approach in which we only take powers to approximate the cost function used for learning the similarity matrix.

# 13.4   Algorithms for Learning the Similarity Matrix

We now turn to the problem of learning the similarity matrix from data. We assume that we are given one or more sets of data for which the desired clustering is known. The goal is to design a 'similarity map,' that is, a mapping from datasets of elements in $\mathbf{X}$ to the space of symmetric matrices with non-negative elements. In this chapter, we assume that this space is parameterized. In particular, we consider diagonally-scaled Gaussian kernel matrices (for which the parameters are the scales of each dimension), as well as more complex parameterized matrices for speech separation in Section 13.5. In general we assume that the similarity matrix is a function of a vector variable $\mathbf{a} \in \mathbb{R}^F$. We also assume that the parameters are in one-to-one correspondence with the features; setting one of these parameters to zero is equivalent to ignoring the corresponding feature.

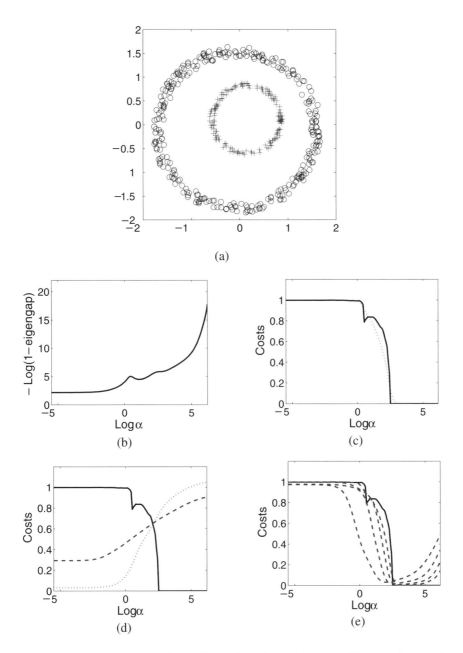

Figure 13.3 Empirical comparison of cost functions. (a) Data with two clusters (crosses and circles). (b) Eigengap of the similarity matrix as a function of $\alpha$. (c) Gold standard clustering error (solid), spectral cost function $J_1$ (dotted). (d) Gold standard clustering error (solid), the alignment (dashed), and a Markov-chain-based cost, divided by 20 (dotted). (e) Approximations based on the power method, with increasing power $q$: 2 4 16 32.

## 13.4.1    Learning Algorithm

We assume that we are given several related datasets with known partitions and our objective is to learn parameters of similarity matrices adapted to the overall problem. This 'supervised' setting is not uncommon in practice. In particular, as we show in Section 13.5, labeled datasets are readily obtained for the speech separation task by artificially combining separately-recorded samples. Note also that in the image segmentation domain, numerous images have been hand-labeled and a dataset of segmented natural images is available (Martin *et al.* 2001).

More precisely, we assume that we are given $N$ datasets $\mathbf{D}_n$, $n \in \{1, \ldots, N\}$, of points in $\mathbf{X}$. Each dataset $\mathbf{D}_n$ is composed of $P_n$ points $\mathbf{x}_{np}$, $p \in \{1, \ldots, P_n\}$. Each dataset is segmented; that is, for each $n$ we know the partition $\mathbf{E}_n$. For each $n$ and each $\alpha$, we have a similarity matrix $\mathbf{W}_n(\mathbf{a})$. The cost function that we use is $H(\mathbf{a}) = (1/N) \sum_n F(\mathbf{W}_n(\mathbf{a}), \mathbf{E}_n) + C \sum_{f=1}^{F} |\mathbf{a}_f|$. The $\ell_1$ penalty serves as a feature selection term, tending to make the solution sparse. The learning algorithm is the minimization of $H(\mathbf{a})$ with respect to $\mathbf{a} \in \mathbb{R}^F$, using the method of steepest descent.

Given that the complexity of the cost function increases with $q$, we start the minimization with small $q$ and gradually increase $q$ up to its maximum value. We have observed that for small $q$, the function to optimize is smoother and thus easier to optimize – in particular, the long plateaus of constant values are less pronounced. In some cases, we may end the optimization with a few steps of steepest descent using the cost function with the true eigenvectors, i.e. for $q = \infty$; this is particularly appropriate when the eigengaps of the optimal similarity matrices happen to be small.

## 13.4.2    Related Work

Several other frameworks aim at learning the similarity matrices for spectral clustering or related procedures. Closest to our own work is the algorithm of Cour *et al.* (2005) which optimizes directly the eigenvectors of the similarity matrix, rather than the eigensubpaces, and is applied to image segmentation tasks. Although differently motivated, the frameworks of Meila and Shi (2002) and Shental *et al.* (2003) lead to similar convex optimization problems. The framework of Meila and Shi (2002) directly applies to spectral clustering, but we have shown in Section 13.3.4 that the cost function, although convex, may lead to similarity matrices that do not perform well. The probabilistic framework of Shental *et al.* (2003) is based on the model granular magnet of Blatt *et al.* (1997) and applies recent graphical model approximate inference techniques to solve the intractable inference required for the clustering task. Their framework leads to a convex maximum likelihood estimation problem for the similarity parameters, which is based on the same approximate inference algorithms. Among all those frameworks, ours has the advantage of providing theoretical bounds linking the cost function and the actual performance of spectral clustering.

## 13.4.3    Testing Algorithm

The output of the learning algorithm is a vector $\mathbf{a} \in \mathbb{R}^F$. In order to cluster previously unseen datasets, we compute the similarity matrix $\mathbf{W}$ and use the algorithm of Figure 13.2. In order to further enhance testing performance, we also adopt an idea due to Ng *et al.* (2002) – during testing, we vary the parameter $\mathbf{a}$ along a direction $\boldsymbol{\beta}$. That is, for small $\lambda$

we set the parameter value to $\alpha + \lambda\beta$ and perform spectral clustering, selecting $\lambda$ such that the (weighted) distortion obtained after application of the spectral clustering algorithm of Figure 13.2 is minimal.

In our situation, there are two natural choices for the direction of search. The first is to use $\beta = \alpha/\|\alpha\|$, i.e. we hold fixed the direction of the parameter but allow the norm to vary. This is natural for diagonally-scaled Gaussian kernel matrices. The second solution, which is more generally applicable, is to use the gradient of the individual cost functions, i.e. let $\mathbf{g}_n = (dF(W_n(\alpha), E_n)/d\alpha) \in \mathbb{R}^F$. If we neglect the effect of the regularization, at optimality, $\sum_n \mathbf{g}_n = 0$. We take the unit-norm direction such that $\sum_n (\beta^\top \mathbf{g}_n)^2$ is maximum, which leads to choosing $\beta$ as the largest eigenvector of $\sum_n \mathbf{g}_n \mathbf{g}_n^\top$.

## 13.4.4 Handling very Large Similarity Matrices

In applications to vision and speech separation problems, the number of data points to cluster can be enormous: indeed, even a small $256 \times 256$ image leads to more than $P = 60\,000$ pixels while three seconds of speech sampled at $5\,\text{kHz}$ leads to more than $P = 15\,000$ spectrogram samples. Thus, in such applications, the full matrix $\mathbf{W}$, of size $P \times P$, cannot be stored in main memory. In this section, we present approximation schemes for which the storage requirements are linear in $P$, for which the time complexity is linear in $P$, and which enable matrix-vector products to be computed in linear time. See Section 13.6.3 for an application of each of these methods to speech separation.

For an approximation scheme to be valid, we require the approximate matrix $\widetilde{\mathbf{W}}$ to be symmetric, with non-negative elements, and with a strictly positive diagonal (to ensure in particular that $D$ has a strictly positive diagonal). The first two techniques can be applied generally, while the last method is specific to situations in which there is natural one-dimensional structure, such as in speech or motion segmentation.

### (a) Sparsity

In applications to vision and related problems, most of the similarities are local, and most of the elements of the matrix $\mathbf{W}$ are equal to zero. If $Q \leq P(P + 1)/2$ is the number of elements less than a given threshold (note that the matrix is symmetric so just the upper triangle needs to be stored), the storage requirement is linear in $Q$, as is the computational complexity of matrix-vector products. However, assessing which elements are equal to zero might take $O(P^2)$. Note that when the sparsity is low, i.e. when $Q$ is large, using a sparse representation is unhelpful; only when the sparsity is expected to be high is it useful to consider such an option.

Thus, before attempting to compute all the significant elements (i.e. all elements greater than the threshold) of the matrix, we attempt to ensure that the resulting number of elements $Q$ is small enough. We do so by selecting $S$ random elements of the matrix and estimating from those $S$ elements the proportion of significant elements, which immediately yields an estimate of $Q$.

If the estimated $Q$ is small enough, we need to compute those $Q$ numbers. However, although the total number of significant elements can be efficiently estimated, the indices of those significant elements cannot be obtained in less than $O(P^2)$ time without additional assumptions. A particular example is the case of diagonally-scaled Gaussian kernel matrices, for which the problem of computing all non-zero elements is equivalent to that of finding pairs of data points in a Euclidean space with distance smaller than a given threshold. We can exploit classical efficient algorithms to perform this task (Gray and Moore 2001).

If $\mathbf{W}$ is an element-wise product of similarity matrices, only a subset of which have a nice structure, we can still use these techniques, albeit with the possibility of requiring more than $Q$ elements of the similarity matrix to be computed.

### (b) Low-rank Non-negative Decomposition

If the matrix $\mathbf{W}$ is not sparse, we can approximate it with a low-rank matrix. Following Fowlkes *et al.* (2001), it is computationally efficient to approximate each column of $\mathbf{W}$ by a linear combination of a set of randomly chosen columns: if $I$ is the set of columns that are selected and $J$ is the set of remaining columns, we approximate each column $\mathbf{w}_j$, $j \in J$, as a combination $\sum_{i \in I} \mathbf{H}_{ij} \mathbf{w}_i$. In the Nyström method of Fowlkes *et al.* (2001), the coefficient matrix $\mathbf{H}$ is chosen so that the squared error on the rows indexed by $I$ is minimum, i.e. $\mathbf{H}$ is chosen so that $\sum_{k \in I} (\mathbf{w}_j(k) - \sum_{i \in I} \mathbf{H}_{ij} \mathbf{w}_i(k))^2$. Since $\mathbf{W}$ is symmetric, this only requires knowledge of the columns indexed by $I$. The solution of this convex quadratic optimization problem is simply $\mathbf{H} = \mathbf{W}(I, I)^{-1} \mathbf{W}(I, J)$, where for any sets $A$ and $B$ of distinct indices $\mathbf{W}(A, B)$ is the $(A, B)$ block of $\mathbf{W}$. The resulting approximating matrix is symmetric and has a rank equal to the size of $I$.

When the matrix $\mathbf{W}$ is positive-semidefinite, then the approximation remains positive-semidefinite. However, when the matrix $\mathbf{W}$ is element-wise non-negative, which is the main assumption in this chapter, then the approximation might not be and this may lead to numerical problems when applying the techniques presented in this chapter. In particular the approximated matrix $\mathbf{D}$ might not have a strictly positive diagonal. The following low-rank non-negative decomposition has the advantage of retaining a pointwise non-negative decomposition, while being only slightly slower. We use this decomposition in order to approximate the large similarity matrices, and the required rank is usually in the order of hundreds; this is to be contrasted with the approach of Ding *et al.* (2005), which consists in performing a non-negative decomposition with very few factors in order to potentially obtain directly cluster indicators.

We first find the best approximation of $\mathbf{A} = \mathbf{W}(I, J)$ as $\mathbf{VH}$, where $\mathbf{V} = \mathbf{W}(I, I)$ and $\mathbf{H}$ is element-wise non-negative. This can be done efficiently using algorithms for non-negative matrix factorization (Lee and Seung 2000). Indeed, starting from a random positive $\mathbf{H}$, we perform the following iteration until convergence:

$$\forall i, j, \mathbf{H}_{ij} \leftarrow \frac{\sum_k \mathbf{V}_{ki} \mathbf{A}_{kj} / (\mathbf{VH})_{kj}}{\sum_k \mathbf{V}_{ki}}. \tag{13.8}$$

The complexity of the iteration in Equation (13.8) is $O(M^2 P)$, and empirically we usually find that we require a small number of iterations before reaching a sufficiently good solution. Note that the iteration yields a monotonic decrease in the following divergence:

$$D(\mathbf{A} \| \mathbf{VH}) = \sum_{ij} \left( \mathbf{A}_{ij} \log \frac{\mathbf{A}_{ij}}{(\mathbf{VH})_{ij}} - \mathbf{A}_{ij} + (\mathbf{VH})_{ij} \right).$$

We approximate $\mathbf{W}(J, J)$ by symmetrization, i.e. $\mathbf{W}(J, I)\mathbf{H} + \mathbf{H}^\top \mathbf{W}(I, J)$. (For a direct low-rank symmetric non-negative decomposition algorithm, see Ding *et al.* (2005)). In order

Table 13.1 Performance on synthetic datasets: clustering errors (multiplied by 100) for method without learning (but with tuning) and for our learning method with and without tuning, with $N = 1$ or 10 training datasets; $D$ is the number of irrelevant features

| $D$ | No learning | Learning w/o tuning | | Learning with tuning | |
|---|---|---|---|---|---|
| | | $N = 1$ | $N = 10$ | $N = 1$ | $N = 10$ |
| 0 | 0 | 15.5 | 10.5 | 0 | 0 |
| 1 | 60.8 | 37.7 | 9.5 | 0 | 0 |
| 2 | 79.8 | 36.9 | 9.5 | 0 | 0 |
| 4 | 99.8 | 37.8 | 9.7 | 0.4 | 0 |
| 8 | 99.8 | 37 | 10.7 | 0 | 0 |
| 16 | 99.7 | 38.8 | 10.9 | 14 | 0 |
| 32 | 99.9 | 38.9 | 15.1 | 14.6 | 6.1 |

to obtain a better approximation, we ensure that the diagonal of $\mathbf{W}(J, J)$ is always used with its true (i.e. not approximated) value. Note that the matrices $\mathbf{H}$ found by non-negative matrix factorization are usually sparse.

The storage requirement is $O(MP)$, where $M$ is the number of selected columns. The complexity of the matrix-vector products is $O(MP)$. Empirically, the average overall complexity of obtaining the decomposition is $O(M^2 P)$.

### 13.4.5 Simulations on Toy Examples

We performed simulations on synthetic datasets involving two-dimensional datasets similar to that shown in Figure 13.3, where there are two rings whose relative distance is constant across samples (but whose relative orientation has a random direction). We add $D$ irrelevant dimensions of the same magnitude as the two relevant variables. The goal is thus to learn the diagonal scale $\mathbf{a} \in \mathbb{R}^{D+2}$ of a Gaussian kernel that leads to the best clustering on unseen data. We learn $\mathbf{a}$ from $N$ sample datasets ($N = 1$ or $N = 10$), and compute the clustering error of our algorithm with and without adaptive tuning of the norm of $\mathbf{a}$ during testing (cf. Section 13.4.3) on ten previously unseen datasets. We compare with an approach that does not use the training data: $\mathbf{a}$ is taken to be the vector of all 1s and we again search over the best possible norm during testing (we refer to this method as 'no learning'). We report results in Table 13.1. Without feature selection, the performance of spectral clustering degrades very rapidly when the number of irrelevant features increases, while our learning approach is very robust, even with only one training dataset.

## 13.5 Speech Separation as Spectrogram Segmentation

The problem of recovering signals from linear mixtures, with only partial knowledge of the mixing process and the signals – a problem often referred to as *blind source separation* – is a central problem in signal processing. It has applications in many fields, including speech processing, network tomography and biomedical imaging (Hyvärinen *et al.* 2001). When the problem is over-determined, i.e. when there are no more signals to estimate (the sources) than signals that are observed (the sensors), generic assumptions such as statistical independence of the sources can be used in order to demix successfully (Hyvärinen *et al.* 2001). Many interesting applications, however, involve under-determined problems (more sources than sensors), where more specific assumptions must be made in order to demix. In problems involving at least two sensors, progress has been made by appealing to sparsity assumptions (Jourjine *et al.* 2000; Zibulevsky *et al.* 2002).

However, the most extreme case, in which there is only one sensor and two or more sources, is a much harder and still-open problem for complex signals such as speech. In this setting, simple generic statistical assumptions do not suffice. One approach to the problem involves a return to the spirit of classical engineering methods such as matched filters, and estimating specific models for specific sources – e.g. specific speakers in the case of speech (Jang and Lee 2003; Roweis 2001). While such an approach is reasonable, it departs significantly from the desideratum of 'blindness.' In this section we present an algorithm that is a blind separation algorithm – our algorithm separates speech mixtures from a single microphone without requiring models of specific speakers.

Our approach involves a 'discriminative' approach to the problem of speech separation that is based on the spectral learning methodology presented in Section 13.4. That is, rather than building a complex model of speech, we instead focus directly on the task of separation and optimize parameters that determine separation performance. We work within a time–frequency representation (a spectrogram), and exploit the sparsity of speech signals in this representation. That is, although two speakers might speak simultaneously, there is relatively little overlap in the time–frequency plane if the speakers are different (Jourjine *et al.* 2000; Roweis 2001). We thus formulate speech separation as a problem in segmentation in the time–frequency plane. In principle, we could appeal to classical segmentation methods from vision (see, e.g., Shi and Malik 2000) to solve this two-dimensional segmentation problem. Speech segments are, however, very different from visual segments, reflecting very different underlying physics. Thus we must design features for segmenting speech from first principles.

## 13.5.1  Spectrogram

The spectrogram is a two-dimensional (time and frequency) redundant representation of a one-dimensional signal (Mallat 1998). Let $\mathbf{f}[t]$, $t = 0, \ldots, T - 1$ be a signal in $\mathbb{R}^T$. The spectrogram is defined via windowed Fourier transforms and is commonly referred to as a short-time Fourier transform or as Gabor analysis (Mallat 1998). The value $(\mathbf{U}f)_{mn}$ of the spectrogram at time window $n$ and frequency $m$ is defined as $(\mathbf{U}f)_{mn} = (1/\sqrt{M}) \sum_{t=0}^{T-1} \mathbf{f}[t]\mathbf{w}[t - na]e^{i2\pi mt/M}$, where $\mathbf{w}$ is a window of length $T$ with small support of length $c$, and $M \geqslant c$. We assume that the number of samples $T$ is an integer multiple of $a$ and $c$. There are then $N = T/a$ different windows of length $c$. The spectrogram is thus an $N \times M$ image which provides a redundant time-frequency representation of time signals[3] (see Figure 13.4).

---

[3] In our simulations, the sampling frequency is $f_0 = 5.5$ kHz and we use a Hanning window of length $c = 216$ (i.e. 43.2 ms). The spacing between windows is equal to $a = 54$ (i.e. 10.8 ms). We use a 512-point FFT ($M = 512$). For a speech sample of length 4 seconds, we have $T = 22\,000$ samples and then $N = 407$, which yields $\approx 2 \times 10^5$ spectrogram samples.

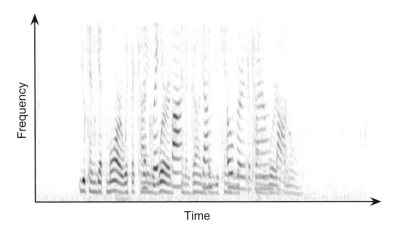

Figure 13.4 Spectrogram of speech (two simultaneous English speakers). The gray intensity is proportional to the amplitude of the spectrogram.

**Inversion** Our speech separation framework is based on the segmentation of the spectrogram of a signal $\mathbf{f}[t]$ in $R \geqslant 2$ disjoint subsets $A_i$, $i = 1, \ldots, R$ of $[0, N-1] \times [0, M-1]$. This leads to $R$ spectrograms $\mathbf{U}_i$ such that $(\mathbf{U}_i)_{mn} = \mathbf{U}_{mn}$ if $(m, n) \in A_i$ and zero otherwise. We now need to find $R$ speech signals $\mathbf{f}_i[t]$ such that each $\mathbf{U}_i$ is the spectrogram of $\mathbf{f}_i$. In general there are no exact solutions (because the representation is redundant), and a classical technique is to find the minimum $\ell_2$ norm approximation, i.e. find $\mathbf{f}_i$ such that $\|\mathbf{U}_i - \mathbf{U}\mathbf{f}_i\|^2$ is minimal (Mallat 1998). The solution of this minimization problem involves the pseudo-inverse of the linear operator $\mathbf{U}$ (Mallat 1998) and is equal to $\mathbf{f}_i = (\mathbf{U}^*\mathbf{U})^{-1}\mathbf{U}^*\mathbf{U}_i$, where $\mathbf{U}^*$ is the (complex) adjoint of the linear operator $\mathbf{U}$. By our choice of window (Hanning), $\mathbf{U}^*\mathbf{U}$ is proportional to the identity matrix, so that the solution to this problem can simply be obtained by applying the adjoint operator $\mathbf{U}^*$. Other techniques for spectrogram inversion could be used (Achan *et al.* 2003; Griffin and Lim 1984; Mallat 1998).

### 13.5.2 Normalization and Subsampling

There are several ways of normalizing a speech signal. In this chapter, we chose to rescale all speech signals as follows: for each time window $n$, we compute the total energy $e_n = \sum_m |\mathbf{U}\mathbf{f}_{mn}|^2$, and its 20-point moving average. The signals are normalized so that the 90th percentile of those values is equal to 1.

In order to reduce the number of spectrogram samples to consider, for a given pre-normalized speech signal, we threshold coefficients whose magnitudes are less than a value that was chosen so that the resulting distortion is inaudible.

### 13.5.3 Generating Training Samples

Our approach is based on the learning algorithm presented in Section 13.4. The training examples that we provide to this algorithm are obtained by mixing separately-normalized speech signals. That is, given two volume-normalized speech signals, $\mathbf{f}_1$ and $\mathbf{f}_2$, of the same

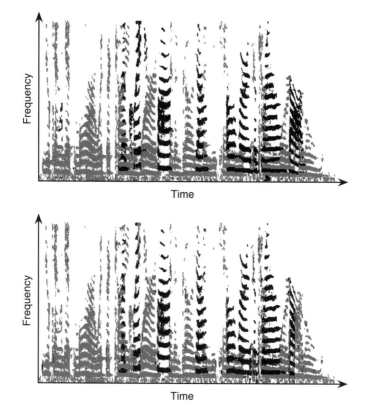

Figure 13.5    (Top) Optimal segmentation for the spectrogram of English speakers in Figure 13.4 (right), where the two speakers are 'black' and 'gray'; this segmentation is obtained from the known separated signals. (Bottom) The blind segmentation obtained with our algorithm.

duration, with spectrograms $\mathbf{U}_1$ and $\mathbf{U}_2$, we build a training sample as $\mathbf{U}^{\mathrm{train}} = \mathbf{U}_1 + \mathbf{U}_2$, with a segmentation given by $z = \arg\min\{\mathbf{U}_1, \mathbf{U}_2\}$. In order to obtain better training partitions (and in particular to be more robust to the choice of normalization), we also search over all $\alpha \in [0, 1]$ such that the $\ell_2$ reconstruction error obtained from segmenting/reconstructing using $z = \arg\min\{\alpha\mathbf{U}_1, (1 - \alpha)\mathbf{U}_2\}$ is minimized. An example of such a partition is shown in Figure 13.5 (top).

## 13.5.4    Features and Grouping Cues for Speech Separation

In this section we describe our approach to the design of features for the spectral segmentation. We base our design on classical cues suggested from studies of perceptual grouping (Cooke and Ellis 2001). Our basic representation is a 'feature map', a two-dimensional representation that has the same layout as the spectrogram. Each of these cues is associated with a specific time scale, which we refer to as 'small' (less than 5 frames), 'medium' (10 to 20 frames), and 'large' (across all frames). (These scales will be of particular relevance to the design of numerical approximation methods in Section 13.6.3.) Any given feature is not sufficient for separating by itself; rather, it is the combination of several features that makes our approach successful.

### (a) Non-harmonic Cues

The following non-harmonic cues have counterparts in visual scenes and for these cues we are able to borrow from feature design techniques used in image segmentation (Shi and Malik 2000).

- **Continuity.** Two time-frequency points are likely to belong to the same segment if they are close in time or frequency; we thus use time and frequency directly as features. This cue acts at a small time scale.

- **Common fate cues.** Elements that exhibit the same time variation are likely to belong to the same source. This takes several particular forms. The first is simply *common offset* and *common onset*. We thus build an offset map and an onset map, with elements that are zero when no variation occurs, and are large when there is a sharp decrease or increase (with respect to time) for that particular time–frequency point. The onset and offset maps are built using oriented energy filters as used in vision (with one vertical orientation). These are obtained by convolving the spectrogram with derivatives of Gaussian windows (Shi and Malik 2000).

  Another form of the common fate cue is *frequency co-modulation*, the situation in which frequency components of a single source tend to move in sync. To capture this cue we simply use oriented filter outputs for a set of orientation angles (eight in our simulations). Those features act mainly at a medium time scale.

### (b) Harmonic Cues

This is the major cue for voiced speech (Bregman 1990; Brown and Cooke 1994; Gold and Morgan 1999), and it acts at all time scales (small, medium and large): voiced speech is locally periodic and the local period is usually referred to as the pitch.

- **Pitch estimation.** In order to use harmonic information, we need to estimate potentially several pitches. We have developed a simple pattern matching framework for doing this that we present in Bach and Jordan (2006). If $S$ pitches are sought, the output that we obtain from the pitch extractor is, for each time frame $n$, the $S$ pitches $\omega_{n1}, \ldots, \omega_{nS}$, as well as the strength $y_{nms}$ of the $s$th pitch for each frequency $m$.

- **Timbre.** The pitch extraction algorithm presented in Appendix C (Gold and Morgan 1999) also outputs the spectral envelope of the signal. This can be used to design an additional feature related to timbre which helps integrate information regarding speaker identification across time. Timbre can be loosely defined as the set of properties of a voiced speech signal once the pitch has been factored out (Bregman 1990). We add the spectral envelope as a feature (reducing its dimensionality using principal component analysis).

**(c) Building Feature Maps from Pitch Information**

We build a set of features from the pitch information. Given a time–frequency point $(m, n)$, let

$$s(m, n) = \arg\max_s \frac{y_{nms}}{(\sum_{m'} y_{nm's})^{1/2}}$$

denote the highest energy pitch, and define the features

$$\omega_{ns(m,n)}, \; y_{nms(m,n)}, \; \sum_{m'} y_{nm's(m,n)}, \; \frac{y_{nms(m,n)}}{\sum_{m'} y_{nm's(m,n)}} \; \text{and} \; \frac{y_{nms(m,n)}}{(\sum_{m'} y_{nm's(m,n)})^{1/2}}.$$

We use a partial normalization with the square root to avoid including very low energy signals, while allowing a significant difference between the local amplitude of the speakers.

Those features all come with some form of energy level and all features involving pitch values $\omega$ should take this energy into account when the similarity matrix is built in Section 13.6. Indeed, this value has no meaning when no energy in that pitch is present.

# 13.6 Spectral Clustering for Speech Separation

Given the features described in the previous section, we now show how to build similarity matrices that can be used to define a spectral segmenter. In particular, our approach builds *parameterized* similarity matrices, and uses the learning algorithm presented in Section 13.4 to adjust these parameters.

## 13.6.1 Basis Similarity Matrices

We define a set of 'basis similarity' matrices for each set of cues and features defined in Section 13.5.4. Those basis matrices are then combined as described in Section 13.6.2 and the weights of this combination are learned as shown in Section 13.4.

For non-harmonic features, we use a radial basis function to define affinities. Thus, if $\mathbf{f}_a$ is the value of the feature for data point $a$, we use a basis similarity matrix defined as $\mathbf{W}_{ab} = \exp(-\|\mathbf{f}_a - \mathbf{f}_b\|^2)$. For a harmonic feature, on the other hand, we need to take into account the strength of the feature: if $\mathbf{f}_a$ is the value of the feature for data point $a$, with strength $y_a$, we use $\mathbf{W}_{ab} = \exp(-\min\{y_a, y_b\}\|\mathbf{f}_a - \mathbf{f}_b\|^2)$.

## 13.6.2 Combination of Similarity Matrices

Given $m$ basis matrices, we use the following parameterization of $\mathbf{W}$ : $\mathbf{W} = \sum_{k=1}^{K} \gamma_k \mathbf{W}_1^{\alpha_{j1}} \times \cdots \times \mathbf{W}_m^{\alpha_{jm}}$, where the products are taken pointwise. Intuitively, if we consider the values of similarity as soft Boolean variables, taking the product of two similarity matrices is equivalent to considering the conjunction of two matrices, while taking the sum can be seen as their disjunction. For our application to speech separation, we consider a sum of $K = 2$ matrices. This has the advantage of allowing different approximation schemes for each of the time scales, an issue we address in the following section.

### 13.6.3 Approximations of Similarity Matrices

The similarity matrices that we consider are huge, of size at least $50\,000 \times 50\,000$. Thus a significant part of our effort has involved finding computationally efficient approximations of similarity matrices.

Let us assume that the time–frequency plane is vectorized by stacking one time frame after the other. In this representation, the time scale of a basis similarity matrix $\mathbf{W}$ exerts an effect on the degree of 'bandedness' of $\mathbf{W}$. Recall that the matrix $\mathbf{W}$ is referred to as band-diagonal with bandwidth $B$, if for all $i$, $j$, $|i - j| \geqslant B \Rightarrow \mathbf{W}_{ij} = 0$. On a small time scale, $\mathbf{W}$ has a small bandwidth; for a medium time scale, the band is larger but still small compared with the total size of the matrix, while for large scale effects, the matrix $\mathbf{W}$ has no band structure. Note that the bandwidth $B$ can be controlled by the coefficient of the radial basis function involving the time feature $n$.

For each of these three cases, we have designed a particular way of approximating the matrix, while ensuring that in each case the time and space requirements are *linear* in the number of time frames, and thus linear in the duration of the signal to demix.

- **Small scale.** If the bandwidth $B$ is very small, we use a simple direct sparse approximation. The complexity of such an approximation grows linearly in the number of time frames.

- **Medium and large scale.** We use a low-rank approximation of the matrix $\mathbf{W}$, as presented in Section 13.4.4. For mid-range interactions, we need an approximation whose rank grows with time, but whose complexity does not grow quadratically with time (see Section 13.4.4), while for large scale interactions, the rank is held fixed.

### 13.6.4 Experiments

We have trained our segmenter using data from four different male and female speakers, with speech signals of duration 3 seconds. There were 15 parameters to estimate using our spectral learning algorithm. For testing, we use mixes from speakers which were different from those in the training set.

In Figure 13.5, for two English speakers from the testing set, we show an example of the segmentation that is obtained when the two speech signals are known in advance (top panel), a segmentation that would be used for training our spectral clustering algorithm, and in the bottom panel, the segmentation that is output by our algorithm.

Although some components of the 'black' speaker are missing, the segmentation performance is good enough to obtain audible signals of reasonable quality. The speech samples for these examples can be downloaded from http://www.di.ens.fr/ fbach/speech/. On this web site, there are several additional examples of speech separation, with various speakers, in French and in English. Similarly, we present segmentation results for French speakers in Figure 13.6. Note that the same parameters were used for both languages and that the two languages were present in the training set. An important point is that our method does not require knowing the speakers in advance in order to demix successfully; rather, it is only necessary that the two speakers have distinct pitches most of the time (another, but less crucial, condition is that one pitch is not too close to twice the other one).

A complete evaluation of the robustness of our approach is outside the scope of this chapter; however, for the two examples shown in Figure 13.5 and Figure 13.6, we can

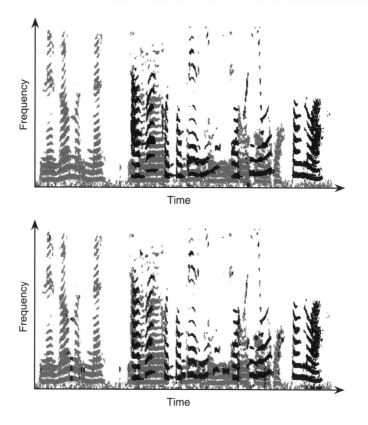

Figure 13.6   (Top) Optimal segmentation for the spectrogram of French speakers in Figure 13.4 (right), where the two speakers are 'black' and 'gray'; this segmentation is obtained from the known separated signals. (Bottom) The blind segmentation obtained with our algorithm.

compare signal-to-noise ratios for various competing approaches. Given the true signal $s$ (known in our simulation experiments) and an estimated signal $\hat{s}$, the signal-to-noise ratio (SNR) is defined as $SNR = (\|s - \hat{s}\|^2)/(\|s\|^2)$, and is often reported in decibels, as $SNR_{dB} = -10\log_{10}(\|s - \hat{s}\|^2)/(\|s\|^2)$. In order to characterize demixing performance, we use the maximum of the signal-to-noise ratios between the two true signals and the estimated signals (potentially after having permuted the estimated signals). In Table 13.2, we compare our approach ('Clust'), with the demixing solution obtained from the segmentation that would serve for training purposes ('Bound') (this can be seen as an upper bound on the performance of our approach). We also performed two baseline experiments: (1) in order to show that the combination of features is indeed crucial for performance, we performed $K$-means clustering on the estimated pitch to separate the two signals ('Pitch'). (2) in order to show that a full time-frequency approach is needed, and not simply frequency-based filtering, we used Wiener filters computed from the true signals ('Freq'). Note that to compute the four

Table 13.2 Comparison of signal-to-noise ratios

|  | Bound | Clust | Pitch | Freq |
|---|---|---|---|---|
| English (*SNR*) | 2.3% | 6.9% | 31.1% | 33.4% |
| English (*SNR$_{dB}$*) | 16.4 | 11.6 | 5.1 | 4.8 |
| French (*SNR*) | 3.3% | 15.8% | 35.4% | 40.7% |
| French (*SNR$_{dB}$*) | 14.8 | 8.0 | 4.5 | 3.9 |

SNRs, the 'Pitch' and 'Freq' methods need the true signals, while the two other methods ('Clust' and 'Bound') are pure separating approaches.

From the results in Table 13.2, we see that pitch alone is not sufficient for successful demixing (see the third column in the table). This is presumably due in part to the fact that pitch is not the only information available for grouping in the frequency domain, and in part to the fact that multi-pitch estimation is a hard problem and multi-pitch estimation procedures tend to lead to noisy estimates of pitch. We also see (the fourth column in the table) that a simple frequency-based approach is not competitive. This is not surprising because natural speech tends to occupy the whole spectrum (because of non-voiced portions and variations in pitch).

Finally, as mentioned earlier, there was a major computational challenge in applying spectral methods to single microphone speech separation. Using the techniques described in Section 13.6.3, the separation algorithm has linear running time complexity and memory requirement and, coded in Matlab and C, it takes 3 minutes to separate 4 seconds of speech on a 2 GHz processor with 1 GB of RAM.

# 13.7 Conclusions

In this chapter, we have presented two sets of algorithms – one for spectral clustering and one for learning the similarity matrix. These algorithms can be derived as the minimization of a single cost function with respect to its two arguments. This cost function depends directly on the eigenstructure of the similarity matrix. We have shown that it can be approximated efficiently using the power method, yielding a method for learning similarity matrices that can cluster effectively in cases in which non-adaptive approaches fail. Note in particular that our new approach yields a spectral clustering method that is significantly more robust to irrelevant features than current methods.

We applied our learning framework to the problem of one-microphone blind source separation of speech. To do so, we have combined knowledge of physical and psychophysical properties of speech with learning algorithms. The former provide parameterized similarity matrices for spectral clustering, and the latter make use of our ability to generate segmented training data. The result is an optimized segmenter for spectrograms of speech mixtures. We have successfully demixed speech signals from two speakers using this approach.

Our work thus far has been limited to the setting of ideal acoustics and equal-strength mixing of two speakers. There are several obvious extensions that warrant investigation. First, the mixing conditions should be weakened and should allow some form of delay or echo. Second, there are multiple applications where speech has to be separated from non-stationary noise; we believe that our method can be extended to this situation. Third, our framework is based on segmentation of the spectrogram and, as such, distortions are inevitable since this is a 'lossy' formulation (Jang and Lee 2003; Jourjine *et al.* 2000). We are currently working on post-processing methods that remove some of those distortions. Finally, while the running time and memory requirements of our algorithm are linear in the duration of the signal to be separated, the resource requirements remain a concern. We are currently working on further numerical techniques that we believe will bring our method significantly closer to real-time.

# References

Achan K, Roweis S and Frey B 2003 Probabilistic inference of speech signals from phaseless spectrograms. *Advances in Neural Information Processing Systems 16*. MIT Press.

Bach F and Harchaoui Z 2008 Diffrac: a discriminative and flexible framework for clustering. *Advances in Neural Information Processing Systems 20*. MIT Press.

Bach FR and Jordan MI 2006 Learning spectral clustering, with application to speech separation. *Journal of Machine Learning Research* **7**, 1963–2001.

Bar-Hillel A, Hertz T, Shental N and Weinshall D 2003 Learning distance functions using equivalence relations. *Proceedings of the International Conference on Machine Learning (ICML)*.

Bathe KJ and Wilson EL 1976 *Numerical Methods in Finite Element Analysis*. Prentice-Hall.

Bertsimas D and Tsitsiklis J 1997 *Introduction to Linear Optimization*. Athena Scientific.

Bie TD and Cristianini N 2006 Fast sdp relaxations of graph cut clustering, transduction, and other combinatorial problems. *Journal of Machine Learning Research* **7**, 1409–1436.

Blatt M, Wiesman M and Domany E 1997 Data clustering using a model granular magnet. *Neural Computation* **9**, 1805–1842.

Bregman AS 1990 *Auditory Scene Analysis: The Perceptual Organization of Sound*. MIT Press.

Brown GJ and Cooke MP 1994 Computational auditory scene analysis. *Computer Speech and Language* **8**, 297–333.

Chan PK, Schlag MDF and Zien JY 1994 Spectral K-way ratio-cut partitioning and clustering. *IEEE Transactions on Computer-Aided Design of Integrated Circuits* **13**(9), 1088–1096.

Chung FRK 1997 *Spectral Graph Theory*. American Mathematical Society.

Comon P and Golub GH 1990 Tracking a few extreme singular values and vectors in signal processing. *Proceedings of the IEEE* **78**(8), 1327–1343.

Cooke M and Ellis DPW 2001 The auditory organization of speech and other sources in listeners and computational models. *Speech Communication* **35**(3–4), 141–177.

Cour T, Gogin N and Shi J 2005 Learning spectral graph segmentation *Workshop on Artificial Intelligence and Statistics (AISTATS)*.

Cristianini N, Shawe-Taylor J and Kandola J 2002 Spectral kernel methods for clustering *Advances in Neural Information Processing Systems 14*. MIT Press.

Dhillon I, Guan Y and Kulis B 2004 A unified view of kernel k-means, spectral clustering and graph cuts. University of Texas, Computer Science. *Technical Report #TR-04-25*.

Ding C, He X and Simon. HD 2005 On the equivalence of nonnegative matrix factorization and spectral clustering. *Proceedings of the SIAM International Conference on Data Mining (SDM)*.

Edelman A, Arias TA and Smith ST 1999 The geometry of algorithms with orthogonality constraints. *SIAM Journal on Matrix Analysis and Applications* **20**(2), 303–353.

Filippone M, Camastra F, Masulli F and Rovetta S 2008 A survey of kernel and spectral methods for clustering. *Pattern Recognition* **41**(1), 176–190.

Fowlkes C, Belongie S and Malik J 2001 Efficient spatiotemporal grouping using the Nyström method. *Proceedings of the IEEE Conference on Computer Vision and Pattern Recognition (ECCV)*.

Gold B and Morgan N 1999 *Speech and Audio Signal Processing: Processing and Perception of Speech and Music*. John Wiley & Sons.

Golub GH and Loan CFV 1996 *Matrix Computations*. Johns Hopkins University Press.

Gray AG and Moore AW 2001 N-Body problems in statistical learning. *Advances in Neural Information Processing Systems 13*. MIT Press.

Griffin D and Lim J 1984 Signal estimation from modified short-time Fourier transform. *IEEE Transactions on Acoustics, Speech and Signal Processing* **32**(2), 236–243.

Gu M, Zha H, Ding C, He X and Simon H 2001 Spectral relaxation models and structure analysis for K-way graph clustering and bi-clustering. Pennsylvania State University, Computer Science and Engineering. *Technical Report*.

Hastie T, Tibshirani R and Friedman J 2001 *The Elements of Statistical Learning*. Springer-Verlag.

Higham D and Kibble M 2004 A unified view of spectral clustering. University of Strathclyde, Department of Mathematics. *Technical Report 02*.

Hubert LJ and Arabie P 1985 Comparing partitions. *Journal of Classification* **2**, 193–218.

Hyvärinen A, Karhunen J and Oja E 2001 *Independent Component Analysis*. John Wiley & Sons.

Jang GJ and Lee TW 2003 A maximum likelihood approach to single-channel source separation. *Journal of Machine Learning Research* **4**, 1365–1392.

Jourjine A, Rickard S and Yilmaz O 2000 Blind separation of disjoint orthogonal signals: Demixing N sources from 2 mixtures. *Proceedings of the IEEE International Conference on Acoustics, Speech, and Signal Processing (ICASSP)*.

Kamvar SD, Klein D and Manning CD 2003 Spectral learning. *Proceedings of the International Joint Conference on Artificial Intelligence (IJCAI)*.

Lee DD and Seung HS 2000 Algorithms for non-negative matrix factorization. *Advances in Neural Information Processing Systems 12*. MIT Press.

Magnus JR and Neudecker H 1999 *Matrix Differential Calculus with Applications in Statistics and Econometrics*. John Wiley & Sons.

Mallat S 1998 *A Wavelet Tour of Signal Processing*. Academic Press.

Martin D, Fowlkes C, Tal D and Malik J 2001 A database of human segmented natural images and its application to evaluating segmentation algorithms and measuring ecological statistics. *Proceedings of the International Conference on Computer Vision (ICCV)*.

Meila M and Heckerman D 2001 An experimental comparison of several clustering and initialization methods. *Machine Learning* **42**(1), 9–29.

Meila M and Shi J 2002 Learning segmentation by random walks. *Advances in Neural Information Processing Systems 14*. MIT Press.

Meila M and Xu L 2003 Multiway cuts and spectral clustering. University of Washington, Department of Statistics. *Technical Report*.

Mitchell TM 1997 *Machine Learning*. McGraw-Hill, New York.

Ng AY, Jordan MI and Weiss Y 2002 On spectral clustering: analysis and an algorithm. *Advances in Neural Information Processing Systems 14*. MIT Press.

Overton ML and Womersley RS 1993 Optimality conditions and duality theory for minimizing sums of the largest eigenvalues of symmetric matrics. *Mathematical Programming* **62**, 321–357.

Roweis ST 2001 One microphone source separation. *Advances in Neural Information Processing Systems 13*. MIT Press.

Scott G and Longuet-Higgins HC 1990 Feature grouping by relocalisation of eigenvectors of the proximity matrix. *Proceedings of the British Machine Vision Conference*.

Shental N, Zomet A, Hertz T and Weiss Y 2003 Learning and inferring image segmentations using the GBP typical cut algorithm. *Proceedings of the International Conference on Computer Vision (ICCV)*.

Shi J and Malik J 2000 Normalized cuts and image segmentation. *IEEE Transactions on Pattern Analysis and Machine Intelligence* **22**(8), 888–905.

Szummer M and Jaakkola T 2002 Partially labeled classification with Markov random walks. *Advances in Neural Information Processing Systems 14*. MIT Press.

Tishby N and Slonim N 2001 Data clustering by Markovian relaxation and the information bottleneck method. *Advances in Neural Information Processing Systems 13*. MIT Press.

von Luxburg U, Bousquet O and Belkin M 2005 Limits of spectral clustering. *Advances in Neural Information Processing Systems 17*. MIT Press.

Wagstaff K, Cardie C, Rogers S and Schrödl S 2001 Constrained K-means clustering with background knowledge. *Proceedings of the International Conference on Machine Learning (ICML)*.

Weiss Y 1999 Segmentation using eigenvectors: a unifying view. *Proceedings of the IEEE International Conference on Computer Vision (ICCV)*.

Xing EP and Jordan MI 2003 On semidefinite relaxation for normalized k-cut and connections to spectral clustering. EECS Department, University of California, Berkeley. *Technical Report UCB/CSD-03-1265*.

Xing EP, Ng AY, Jordan MI and Russell S 2003 Distance metric learning, with application to clustering with side-information. *Advances in Neural Information Processing Systems 15*. MIT Press.

Yu SX and Shi J 2002 Grouping with bias. *Advances in Neural Information Processing Systems 14*. MIT Press.

Yu SX and Shi J 2003 Multiclass spectral clustering. *Proceedings of the International Conference on Computer Vision (ICCV)*.

Zha H, Ding C, Gu M, He X and Simon H 2002 Spectral relaxation for K-means clustering. *Advances in Neural Information Processing Systems 14*. MIT Press.

Zibulevsky M, Kisilev P, Zeevi YY and Pearlmutter BA 2002 Blind source separation via multinode sparse representation. *Advances in Neural Information Processing Systems 14*. MIT Press.

# Index

a posteriori, 132, 195, 198, 202
acoustic modeling, 117, 127, 133, 163, 167
acyclic, 28, 31, 120, 121, 123
adaptation, 132, 133, 170, 178, 195, 198, 199,
    202, 203, 211, 212
area under curve (AUC), 176, 180, 182,
    184–187, 189–192
Augmented Statistical Models, 83, 86, 92, 97
authentication, 195, 196

Bayesian factor analysis, 214, 219
bigram, 119, 124, 125, 130, 131
blind source separation, 247

client, 195–197, 199, 201, 202, 205, 208, 210,
    212–214, 216
clustering, 221–223, 225, 227–230, 233–237,
    239, 244–247
Conditional Maximum Likelihood, 83, 96,
    102, 103, 108
Conditional Model, 127
Conditional Random Fields, 94, 143
confusion network, 94, 97

decoding graph, 125, 126, 130
detection error trade-off (DET), 202, 205, 209
determinization, 120, 121, 124, 126, 169
discriminant function, 107, 141, 142, 196, 200
discriminative models, 94, 134, 189
discriminative training, 6, 9, 67, 93, 95, 101,
    102, 105, 107, 112, 117, 118, 133,
    160, 180
dual coordinate ascent, 25
dynamic kernels, 9, 83, 84, 86–88, 93, 94, 97
dynamic programming, 54, 56, 63, 72, 108,
    142, 144, 149, 153, 169, 184

eigenvalue, 224, 225, 227, 231–233
eigenvector, 224, 225, 232, 233, 237
enrollment data, 202

equal error rate, 202
error decomposition, 17
expectation maximization (EM), 5, 6, 102,
    103, 108, 176, 178, 189, 198, 208
expected performance curve (EPC), 202

failure transition, 124
false acceptance, 197
false rejection, 178, 197
feature selection, 134, 144, 155, 222, 236
finite state, 120
finite state transducer (FST), 169
Fisher information matrix, 90, 206, 207, 209
Fisher kernel, 84, 88, 90, 205, 206, 209–212,
    217, 219
Fisher score, 216
forced alignment, 9, 53, 54, 57, 59, 64, 66,
    165, 167, 182

Gaussian, 5, 6, 9, 66, 69, 78, 79, 86, 87, 89,
    95, 96, 102, 103, 105, 106, 108,
    109, 126, 128, 143, 178, 184, 188,
    195, 197–199, 202, 203, 207, 218,
    222, 223, 234, 237, 239, 243
Gaussian Mixture Model (GMM), 5, 86, 89,
    103, 105, 106, 110, 178, 188, 191,
    195, 199, 205, 208, 210–213,
    216–218
generalized linear discriminant sequence
    kernel (GLDS), 204, 205, 213,
    218, 219
generative kernel, 87–91
generative methods, 93
generative model, 54, 83, 88, 89, 97, 130, 133,
    155, 178, 199, 205, 209
GMM supervector linear kernel (GSLK),
    210–215, 217–219

half total error rate (HTER), 202, 203, 218
Hamming distance, 107, 110–112, 150

---

*Automatic Speech and Speaker Recognition: Large Margin and Kernel Methods*    Joseph Keshet and Samy Bengio
© 2009 John Wiley & Sons, Ltd